Large Turbo-Generators

Malfunctions and Symptoms

LARGE TURBO-GENERATORS

Malfunctions and Symptoms

Isidor Kerszenbaum
Geoff Klempner

CRC Press
Taylor & Francis Group
Boca Raton London New York

CRC Press is an imprint of the
Taylor & Francis Group, an **informa** business

CRC Press
Taylor & Francis Group
6000 Broken Sound Parkway NW, Suite 300
Boca Raton, FL 33487-2742

First issued in paperback 2020

ISBN-13: 978-1-4987-0702-2 (hbk)
ISBN-13: 978-0-367-65590-7 (pbk)

Library of Congress Cataloging-in-Publication Data

Names: Kerszenbaum, Isidor, author. | Klempner, Geoff, author.
Title: Large turbo-generators : malfunctions and symptoms / authors, Isidor Kerszenbaum and Geoff Klempner.
Description: Boca Raton : Taylor & Francis Group, CRC Press, [2016] | Includes bibliographical references and index.
Identifiers: LCCN 2016006135 | ISBN 9781498707022 (alk. paper)
Subjects: LCSH: Turbogenerators--Maintenance and repair. | System failures (Engineering)
Classification: LCC TK2765 .K47 2016 | DDC 621.31/3--dc23
LC record available at https://lccn.loc.gov/2016006135

Visit the Taylor & Francis Web site at
http://www.taylorandfrancis.com

and the CRC Press Web site at
http://www.crcpress.com

To our families:

Jackie, Livi, and Yigal Kerszenbaum,

Susan Klempner,

and

to the operators, technicians, and engineers in the power stations around the world who keep the lights on, and the power flowing around the clock, and through adverse conditions for the benefit of everyone.

Contents

Preface

It is an indisputable fact that a large electric power system comprising thousands of generators feeding tens of thousands of miles of transmission and distribution lines, through countless transformers, breakers, and myriad associated devices, and controlled with the most sophisticated instruments and computer programs, represent the most complex engineering system in operation ever created.

Not only is the *electric power system* complex, but it is the organism most critical to the support of modern life, as we know it. One of the core components of any electric power system is generation, where the fuel, be it fossil, nuclear, hydraulic, wind, or any other, is converted into electricity. Of the many types of generators associated with these sources of generation, the most widely utilized are the combustion and steam turbine-driven generators, or simply, the turbogenerators.

Although all turbogenerators are distinguished by their cylindrical rotor, there are numerous variants, be it in their age, way of cooling its components, insulation characteristics, materials used in the construction of their key components, excitation arrangement and so on. Some are operated continuously (base load), and in others the load changes daily or several times during the day. The typical generator is designed to operate about 30 years; however, it is not uncommon to have its life extended for two or more decades. Given that over such long periods of time technologies and materials have changed immensely, it is common to see an abundance of diversity in design and construction of these machines. This book covers the following type of turbogenerators:

1. Air-cooled
2. Indirect hydrogen-cooled
3. Directly (inner) hydrogen-cooled (stator and rotor)
4. Directly (inner) water-cooled stators
5. Directly (inner) water-cooled rotors
6. Two poles
7. Four poles

Note: Directly (inner) oil-cooled stators are not specifically discussed herein, but the vast majority of the subjects included in this book also apply to those generators, with the exception of the stator winding oil cooling system.

Turbogenerators are found in generating stations, where operators are responsible for their daily operation, assisted by electricians, mechanics, technicians, and engineers. A vital challenge for the operator is to keep the machine under control for both normal and abnormal transient system conditions. In this, he/she is aided by monitoring and protection instrumentation; nevertheless, it is up to the operator to validate instrumentation behavior and maintain maximum availability of the unit to the grid. In doing so, the operator/technician/engineer is confronted with a large number of machine reactions to normal and abnormal operating conditions. Sometimes, the power plant personnel must troubleshoot a problem quickly to return a generator to service as soon as possible, and in most cases, significant sums of money depend on how fast a generator can be brought back online for the purpose of delivering power to the grid and its customers.

With deregulation of the electric power industry, many small nonutility (so called third party) power producers operate small fleets of generators without the benefit of in-house expertise. To make up for that, they depend heavily on OEMs and independent contractors and consultants. In many instances, large utilities have also seen their expertise dissipate to a large extent due to a refocus of management priorities. All these developments are occurring at the same time that these units are called to operate in more onerous environments such as *two-shifting* and load-following, as guided by *economic dispatch*. All these developments, together with an aging fleet of generators, are placing a significant burden on the operator.

This book is written with the operator and other power plant personnel responsible for the operation of the generators in mind. As mentioned earlier, not every company, much less power plants, retains an individual with the breadth and depth of expertise required for troubleshooting all of its units, which may be very different in their design and construction characteristics. With their expertise over many years, the authors' intent is to facilitate operators' understanding of developing adverse conditions with their generators, their causes, and mechanisms of defense. Hopefully, this knowledge can provide an edge in troubleshooting a problem more effectively and faster than otherwise would be the case while avoiding preventable damage to the unit and loss of production. We anticipate that this book will be not only useful to the operator in the power plant, but also to the design engineer and system operations engineers who may find its information useful from time to time. The authors are aware that although it is impossible to cover in any single book all possible conditions that may be encountered during the operation of a generator, they have strived to include a wealth of information that covers most common situations. And, hopefully, they may afford the reader enough skills to elucidate any other information not covered in this book. Although we have tried to cover each topic as comprehensively as possible, while at the same time keeping it simple and adequate for the purpose on hand, the book should not be seen as a guide to detail troubleshooting, but as an additional tool to facilitate finding the true causes of any given malfunction. In each case in which a real problem is approached, a whole number of very specific issues only relevant to that very unique machine come into light. These can never be anticipated or known and thus fully described in a book. Thus, we recommend using this book as a general reference source, or as a starting point in the interpretation of a developing problem, but the reader should always obtain adequate on-the-spot expertise when dealing with a particular malfunction.

The organization of the material in the book is covered in Chapter 1.

Isidor Kerszenbaum
Irvine, California

Geoff Klempner
Toronto, Ontario, Canada

Caveat Emptor

As with any similar book, caution *must* be exercised by the reader. The general nature of the contents of this book are not intended to serve as specific opinion or advice, similar to those proffered by a consultant well versed on a particular topic trying to solve a specific problem. No two machines are the same, and so no two outcomes may be the same, even if symptoms may appear to be pointing to the same problem. The purpose of this book is to aid in the identification of a malfunction, be it degradation or a failure, based on general knowledge accumulated over many years. It is up to the reader to build up on the book's material and other sources as well as personal experience, and finally on his/her independent judgment to reach an opinion as to the cause of a specific problem on hand, and the different pertinent responses.

Acknowledgments

The contents of a book such as this one are impossible to learn in the classroom. They are the result of personal experience accumulated over many years of working with large turbogenerators, but, for the most, they are the result of the invaluable long-term contribution of coworkers and associates. Each author was motivated and supported during his career by a number of outstanding individuals who had a major influence in their professional development. Attempting to mention all these people would lead to the unintended omission of some.

The authors are indebted to CRC Press for reviewing the proposal for the book and supporting its publication. They also express their sincere gratitude to the technical reviewers of the proposal, Dr. Mohamed El-Sharkawi and Tom Baker, and in particular the technical reviewers of the final draft, Relu Ilie and Nils Nilsson, who not only put long hours into the review of the book, but also provided very helpful insights and comments. The authors also thank the members of the editorial department at CRC Press, and all others involved in the publication of the book for their support in making its publication possible.

Finally, the authors thank their immediate families for the continuous support and encouragement.

List of Acronyms

AUT	auxiliary unit transformer
AVR	automatic voltage regulator
BCB	back of core burning
CEFS	core-end flux shield
CR	control room
CT	current transformer
CTG	combustion turbine generator
DCS	distributed control system
DFR	digital fault recorder
DO	dissolved oxygen
EL-CID	electromagnetic core imperfection detector
EMI	electromagnetic interference
FFC	field flashing circuit
GCM	generator condition/core monitor
GITV	grid-induced torsional vibration
GVPI	global vacuum pressure impregnation
HFG	high-frequency generator
Hipot	high potential test
HV	high voltage
IPB	isophase bus
IR	insulation resistance
LCM	life cycle management
MOT	main output transformer
MV	medium voltage
NDE	nondestructive examination
OEM	original equipment manufacturer
PDA	partial discharge activity
PI	polarization index
PMG	permanent magnet generator
PT	potential transformer
PTFE	polytetrafluoroethylene
RCW	rotor cooling water
RCWS	rotor cooling water system
RF	radio frequency
RFM	rotor flux monitor
RTD	resistance temperature detector
SCADA	supervisory control and data acquisition system
SCW	stator cooling water
SCWS	stator cooling water system
SUT	step-up transformer

TC thermocouple
TECW total enclosed water cooled
TST thermal sensitivity test
UAT unit auxiliaries transformer (also known as AUT)
UEL underexcitation limiter
UMP unbalanced magnetic pull
VDU visual display unit
VPI vacuum pressure impregnation
VT voltage transformer (same as PT)

1

Book Organization

Operators obtain information about the operation of the generator via a number of avenues:

- Panel annunciators
- Video display units fed from SCADA, DCS, or similar systems
- Computer printouts
- Gauges and meters at the control room (CR) or at the plant
- Alarms from monitoring or protection systems
- Protective device *flags* or *signals*
- Through the senses:
    - Sound (e.g., abnormal noise emanating from the machine)
    - Smell (e.g., smell of burning insulation, or ozone in open air-cooled generators)
    - Sight (e.g., sparks on collector rings)
    - Touch (e.g., excessive vibration of the turbine deck)
- Offline test results (e.g., IR,[1] PI,[2] Hipot,[3] etc.)
- Online test results (e.g., TST,[4] GITV,[5] etc.)

All of the aforementioned sources of information can be grouped as shown in Figure 1.1.

Plant engineers can obtain additional information, such as

- Machine history (equivalent operational hours; performed maintenance; previous faults; past operational events)
- Records related to sister units; known problems from OEM informational letters
- Pertinent information from technical literature, conference, and so forth

1 IR—Insulation resistance.
2 PI—Polarization index.
3 Hipot—High potential test.
4 TST—Thermal sensitivity test.
5 GITV—Grid-induced torsional vibration measurements.

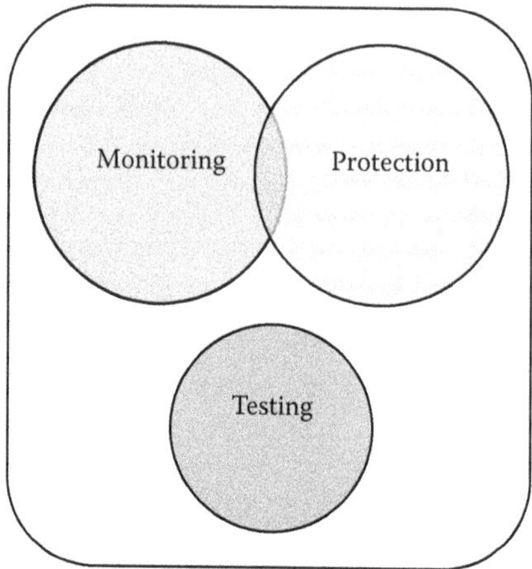

FIGURE 1.1
Operator sources of information about the condition of a given generator.

A succinct list of protective and monitoring functions will be presented in Chapters 2 and 3, respectively. Figure 1.1 shows that there is some gray area where protection and monitoring functions are shared. These will be described also in Chapters 2 and 3. Although tests are also advocated as being part of the troubleshooting tools, they are not covered in this book.

In addition to naming and briefly describing the typical generator protective functions, the purpose of Chapter 2 is mainly to illuminate the concept of *protective zone*. Understanding the *protective zone* concept, by which the *reach* of a given protective device is set, station personnel can optimize the effort of troubleshooting a given malfunction following a protection trip and/or alarm.

Chapter 3 incorporates a brief narrative describing available monitoring functions and techniques. The monitored parameters are a key input to Chapter 4, where the *symptoms* are described. While by definition *protection* is understood as a response to an abnormal condition that is too fast for operator intervention, *monitoring* is, on the other hand, the mechanism by which operators are appraised of a developing condition with their equipment, and can react properly with the purpose of resolving the issue with the generator online, or removing the unit from operation on a timely and deliberate basis. As it will be described in Chapter 3, a plethora of monitoring instrumentation translates into a more effective and opportune operator reaction to a developing anomaly. Obviously, there is a tradeoff between expenditure in monitoring capabilities and expected benefits.

Chapters 4 and 5 are the core of this book. Chapter 4 contains descriptions of all *symptoms* included in the book. By *symptom* it is meant *an objective evidence of a physical disturbance indicating an abnormal and/or unexpected condition*. Symptoms are primarily obtained from monitoring instrumentation or through the senses. One can also qualify flags, alarms, and trips from *protective devices* as symptoms; these are captured in Chapter 2 ("Protective Functions"). Chapter 5 contains descriptions of all *malfunctions* included in the book. By *malfunction* it is meant *a degradation or failure of a function, process or component, so that it fails to perform correctly its designed-for duty.* All *symptoms* and *malfunctions* are grouped in such a way that they fall under major generator components, such as stator, rotor, hydrogen systems, and so on.

Chapter 6 explains how symptoms and malfunctions may relate, and how these relationships can be taken advantage of, for the purpose of expediting troubleshooting.

Chapter 7 presents a number of ready-to-use forms that the operator and/or engineer can use to type in all the relevant variables that have alarm and trip values, so that they are handy during a troubleshooting activity.

2

Protective Functions

2.1 Difference between Protection and Monitoring Functions

As explained in Chapter 1, oversight functions for the main generator can be classified as *protection, monitoring,* and *testing. Testing* simply indicates carrying out tests on the machine, mostly offline. The difference between protection and monitoring can be described best by the statement found in the 2014 release of the revised IEEE Std. 1129 [2].

> *Protection*: The process of observing a system, and automatically initiating an action to mitigate the consequences of an operating condition that has deviated from established acceptable performance criteria, such as alarms, runbacks,[1] and trips.
>
> *Monitoring*: The process of observing a system by basic and advanced methods, with the purpose of verifying that its parameters lie within prescribed limits.

Some oversight functions are clearly identified as protective functions (e.g., sudden and large Volts/Hertz events and phase-to-phase short-circuits), because the event is too fast for the operator to react in time and execute a remedial action, and possible consequences too severe. Others, such as winding temperature and shaft voltage are clearly within the realm of monitoring, because they provide ample time for the operator to react and take defensive actions based on other information he/she gathers. Monitoring can also set alarms on, but automatic tripping is a distinctive feature of protection systems. There are a few that fall in the intersection of protection and monitoring. One such example is rotor vibration. If a trip value exists, it is almost always a high value, where operation of the machine is not recommended. Typically, the vibration increases at a pace that allows the operator to respond before the trip limit is reached; but more importantly, vibration is often trended, which is a clear distinction from variables that are not trended and that lead to the actuation of a protective relay. Another example is underexcitation monitoring, which uses the same device as loss-of-field protection; the former alarms and the latter trips. The difference in the two is the level of sensitivity of the settings.

[1] Runback—Action of automatically reducing the power output of a generator.

2.2 Basic Protection Philosophy

As shown in Chapter 1, Figure 1.1, *protection* is one of the three fundamental supervisory functions on the behavior of the generator; and the most vital of the three. Protection devices are designed to separate the generator from the grid and if necessary, shut it down (including its prime mover), following a number of abnormal operating conditions or equipment malfunction. The response by the protective relay(s) must be automatic, fast, coordinated with other protective devices, and most importantly, it must be very reliable; that is to say, it must trip when required to do so, and not trip unnecessarily. Protection systems will be seldom called to operate, but when they do, it is crucial they do it properly. The goals of a protection system are

- Protect against injury
- Minimize damage to equipment
- Minimize disruption of the supply of electric power

Protection does not constitute just a single relay, but a system, comprising the protective relays—either as independent (discrete) devices or in multifunction units, wiring from CTs and PTs (sometimes called VTs), communication to SCADA or similar systems, and wiring to circuit breakers, auxiliary relays. Protection also includes coordination with other similar systems, so that operation by one relay or protective function selectively removes one piece of faulty machinery or system from the grid and does not remove other equipment unnecessarily. Finally, protection also contains all the required *settings* that assure proper relay operation and coordination. Hence, for a protection system to operate correctly, both the settings of each relay (protective function) should be correct, as well as the coordination with other relays.

Unfortunately, given that they operate infrequently, protection systems sometimes tend to be overlooked. This means that maintenance is not pursued with the vigor it requires, resulting in potential for very deleterious consequences. One of the most common reasons for catastrophic failure of equipment in power systems is the failure of protection systems to operate properly, or to operate at all. The main generator and step-up transformer, as well as auxiliary transformers, are among the most critical and expensive electrical apparatuses in the power plant. In addition, the time to find a proper replacement after a sudden catastrophic failure may be counted in months, if not years, for the very large units. It is therefore of foremost importance that power plants retain a good predictive maintenance program for their protection systems and plant personnel are trained on the significance of each protective device and its intended function.

It should also be noted that electrical systems change with additions of circuits, modern protection equipment availability, and other alterations; hence, protection settings should be reviewed periodically to make sure that tripping will occur when intended, and not to occur when not needed.

2.3 Typical Protective Functions Applied to Large Turbogenerators

In *protection,* relays are designed to recognize very specific abnormalities, and actuate as required. For instance, some relays will recognize the existence of higher-than-normal currents and respond accordingly; others may be designed to recognize a large frequency deviation from nominal.[2] The capacity for detecting a very specific condition is called *protective function.*

Protective functions can be grouped according to the type of faults they protect against. These types of faults or abnormal conditions can be categorized as follows:

- Ground faults
- Phase-to-phase and three-phase faults
- Overvoltage
- Overexcitation
- Overfluxing
- Thermal faults (overload, lack of cooling)
- Loss of field
- Loss of synchronism (out-of-step/pole slip)
- Unbalanced currents
- Motoring
- Abnormal frequencies
- Inadvertent energization
- Circuit breaker failure
- Grid faults/abnormal conditions
- Loss of hydrogen pressure or purity in hydrogen-cooled machines
- Excessive vibration
- Loss of seal oil and/or lube oil

[2] Nominal or *fundamental* frequency indicates the frequency to which the generator is designed for, typically 50 or 60 Hz.

Protective functions are given a number in addition to a name. For instance, the overcurrent protective function is known as "51." These numbers are unique for each function, and with minor differences, harmonized across all standards, in all countries. Table 2.1 contains most of the protective functions that can be found in a generator protection scheme. Not all of these functions will be designed into the protection system of every generator; some may be left out, depending on the size and criticality of the unit. Although some other adaptations of the protection functions in Table 2.1 can be found, most units will be protected exclusively by some, if not most of the functions on the table.

2.4 Primary and Backup Protective Functions

The typical generating unit consists of very costly components. A generator or a turbine not only represents a substantial investment in capital, but its repair, not to mention replacement, may necessitate many months of production downtime. Hence, there is a commanding requirement that when needed, protection systems operate fast and exactly as designed for. Given the very low cost of protective devices when compared with the cost of the protected equipment and lost production, it is common, in particular in the case of the larger and more critical machines, to have installed a *primary* and a *backup* relay for each function.

Backup protection can be in the form of an additional identical relay. For example, installing two multifunction protection boxes[3] will duplicate all protective functions. If the boxes are from the same OEM, with the functions set to the same values, then they will back up each other. In that case the concept of *primary protection* or *backup protection* is irrelevant.

Another form of backup protection is when one protective function or combination thereof, provide backup to a single (primary) function. Table 2.2 shows a number of combinations of protective functions providing backup to some primary protective function. As it can be seen, duplication of protective functions of the same type, together with the natural backup provided by other functions when operating alone or in groups, offer a resilient protection system, as long as set correctly and maintained properly.

When a large fault occurs inside the generator, it is not uncommon for more than one relay function to operate, either by sending a trip signal or by *flagging*.[4] This multiplicity of information is very useful to the operator. When more than one relay (protective function) operates, the probability that one is confronted with a real event and not a degraded protection system is very high.

---

3 A multifunction protection *box* or relay is an instrument containing a number of discrete protective functions, packaged in a single *box*.

4 Protective relays have indicators called *flags*. They activate the flag when the monitored variables reached certain conditions, even if it does not result in a trip. They are also called *signals* by certain vendors, in particular in new electronic devices.

TABLE 2.1

Protective Function, Device Number, and Protected Zone

Device Number	Protective Function Description	Protected Zone or Equipment
15	Speed or frequency-matching device.	Provides some positive feedback during the critical activity of synchronizing to the grid, thus protecting the generator.
21	Distance protection—Backup for generator zone phase faults.	Protects generator, isolated phase bus (or lead cables) up to and including the HV winding of the step-up transformer and auxiliary transformers. Reach may be more or less than that, depending on the relay's settings.
24	Volts/Hertz protection—Senses an overfluxing condition.	Protects generator stator core and step-up and auxiliary transformers' cores from damage due to overfluxing.
25	Synchronizing or sync-check protection—Works together with the #15 function to facilitate smooth synchronization.	Protects generator and turbine components from malsynchronization.
27	Undervoltage protection—Senses a condition where the voltage of a given bus is below a normal range.	Protects generator from synchronizing on a *dead* bus. Mainly used to sense a loss of voltage in the station auxiliary systems. It may alarm and in some cases initiate an automatic bus transfer activity.
27H	Undervoltage protection at the generator neutral—Senses a condition where the third harmonic component of the neutral voltage is below certain level. Some schemes use the *59H* (a third harmonic overvoltage relay operating in reverse mode, accomplishing the same thing).	Protects the stator winding and stator core from grounds. It may also react to grounds in the lead cables or IPB, as well as the low voltage windings of the step-up and auxiliary transformers.
32	Reverse power protection—Reacts to a condition where the active power flow is from the grid to the generator.	Protects against *motoring* of the generator, either as a synchronous or induction machine. It can be part of a scheme to separate the unit from the grid in an orderly manner to avoid overspeed, as part of a *sequential tripping* scheme.

(Continued)

TABLE 2.1 (*Continued*)

Protective Function, Device Number, and Protected Zone

Device Number	Protective Function Description	Protected Zone or Equipment
40	Loss-of-field protection—Senses the condition when the excitation is too low to maintain synchronous operation with the grid.	Protects many generator components from a loss of synchronization. Fault may be located anywhere in the excitation circuit, including field winding, brushgear (if brushed), exciters, rectifiers, excitation cables or buses, AVRs and any other excitation component. Result of loss of field is the generator operating as an induction generator.
46	Stator unbalanced current protection—Senses condition when the stator negative-sequence current is higher than a given threshold.	The protected components are the rotor forging, wedges, retaining rings, and parts of the amortisseur.
49	Stator thermal protection—Senses when the temperature of various stator components (core, winding, etc.) is above the normal range.	Protects specific components where the TCs and/or RTDs are mounted.
50/51	Instantaneous/timed overcurrent protection—Reacts to sudden and/or large increases of stator current due to short circuits.	Protects the generator from large overcurrent events.
50GN/51GN	Instantaneous/timed overcurrent protection at secondary of the grounding transformer—Reacts to sudden and/or large increases in neutral current.	Protects the stator winding from faults in the generator and generator voltage system that result in sudden and/or large ground currents through the neutral of the generator.
50BF/51BF	Overcurrent function used as part of the *circuit breaker-failure* scheme.	Protects the main generator (and transformers if the main breaker is on the HV side of the SUT) for a case when the generator breaker fails to open during a fault-clearing operation.

(*Continued*)

TABLE 2.1 (*Continued*)

Protective Function, Device Number, and Protected Zone

Device Number	Protective Function Description	Protected Zone or Equipment
51TN	Time overcurrent protection—Reacts to currents above normal range flowing via the neutral of the wye-connected HV winding of the SUT.	Protects the SUT from ground faults in the SUT and HV systems. Protects the generator from large negative-sequence currents due to short circuits on the SUT or close by on the grid.
51V	Voltage-controlled (voltage-restrained) time overcurrent protection—Reacts to short circuits in the vicinity of the station and at the station.	Protects the generator from short circuits at the station or on the grid, in close vicinity to the station. The reach into the grid can be set, so not to trip the unit unnecessarily for *far-away* faults that otherwise would trip the 51 overcurrent relay.
59	Overvoltage protection—Reacts to output voltages above the rated maximum (typically: 105% of rated).	Protects the generator from overvoltage and overfluxing. When part of the protection scheme, it is normally used only to alarm.
59GN	Overvoltage protection—Reacts to an increase in voltage in the secondary of the grounding transformer.	Protects the stator winding from ground faults in the generator and generator voltage system that result in ground currents through the neutral of the generator (and thus voltage increase across the secondary's resistance). The fault may be located anywhere between the neutral of the generator, including the grounding transformer, to the delta-connected windings (generator side) of the SUT and auxiliary transformer(s).
59BN	Zero-sequence voltage protection—Senses a ground fault on an otherwise ungrounded bus. Also known as 59BG.	During backfeeding via the SUT with the generator breaker or generator links open, this function protects the equipment from a short circuit to ground. The protected zone includes all the equipment between the open links or breaker to the delta-connected windings of the SUT and auxiliary transformer(s).

(*Continued*)

TABLE 2.1 (*Continued*)

Protective Function, Device Number, and Protected Zone

Device Number	Protective Function Description	Protected Zone or Equipment
60	Voltage balance protection—Detects blown PT fuses or open secondary circuits.	Protects the generator against an unnecessary trip due to loss of a PT fuse or open secondary circuit. Most probably, the fault will be with the PT circuit (such as a fuse, short-circuit between secondary wires and/or to ground, or an open secondary circuit).
61	Time overcurrent protection—Detects turn-to-turn faults in generator windings.	Protects stator against shorted turns in the stator winding. Mainly in hydraulic-driven generators or small turbogenerators. Large turbogenerators (the subject of this book) have a single turn; thus this protection is not used.
62BF	Circuit breaker failure protection.	Protects the main generator (and transformers if the main circuit breaker is on the HV side of the SUT) for a case when the generator circuit breaker fails to open during a fault-clearing operation.
64F	Ground detector based on voltage sensing—Detects grounds on the excitation power circuit.	Protects against short circuits to ground on the excitation power circuit. The fault can be on the rotor winding, brushgear, dc cables/busses, or exciter.
64S/64R	Ground fault protection of stator/rotor based on voltage injection.	Protects 100% of stator/rotor against ground faults.
67	Phase directional overcurrent protection—Detects flow of current toward the generator. Used as part of the *inadvertent energization* scheme. Also implemented by a combination of 50/27 functions.	Protects the unit against inadvertent energization while standing still or on turning gear.

(*Continued*)

TABLE 2.1 *(Continued)*

Protective Function, Device Number, and Protected Zone

Device Number	Protective Function Description	Protected Zone or Equipment
78	Loss of synchronism protection.	Protects the generator against a loss of synchronism. Can be considered backup to the loss of field relay (#40).
81 O/U	Over/under frequency protection—Detects deviation from the nominal frequency.	Protects the unit components from operating at frequencies other than the rated frequency of the unit.
86	Hand-reset lockout auxiliary relay.	Used to lock the system following the operation of a protective relay. Must be reset manually.
87G	Differential protection of the generator.	Protects the generator from short-circuits between phases on the stator winding. Protected zone includes the stator winding and the portion of bus/bushings between the CTs connected to the 87G at both sides of the unit.
87T	Differential protection of the SUT, or AUTs.	Protects the transformer against a phase-to-phase fault inside the transformer. The wye-connected winding will also operate for a phase-to-ground fault.
87U	Unit differential protection. Also indicated as 87O in some protective schemes.	Protects the generator, SUT, AUTs, and HV breaker for a phase-to-phase short circuit. Protected zone where fault is located is between the CTs connected to the 87U.
Excitation protective functions	The exciter and AVR most often have built-in a number of protective functions, such as the maximum excitation limit, minimum excitation limit, and maximum V/Hz limit.	Protects the generator against overexcitation and overfluxing events, as well as operation outside the capability curve or below the steady-state stability limit. These protective functions must be set in such a manner that they coordinate with the other protective functions shown in this table.

AUT, auxiliary unit transformer.

TABLE 2.2

Primary and Backup Protective Functions

Type of Fault	Primary Protective Function	Backup Function(s)
Stator ground faults	27H, 50GN, 51GN, 59GN, 64S	27H, 50GN, 51GN, 59GN
Stator phase short-circuits	87G	87U, 21, 50, 51V, 46
Excitation ground faults	64F, 64R	40 and/or 78 (for certain double grounds)
Loss of excitation	40	51, 78
Overfluxing	24	59 (when synchronized to the grid), maximum V/Hz in excitation system
Overexcitation	Maximum excitation limit in excitation system.	24, 59
Circuit breaker failure	50BF/51BF/62BF	If the breaker failure scheme fails, clearing the fault will normally involve other nondedicated relays such as line protection, or other protective functions that will operate after the initial fault develops further into a bigger fault

2.5 Identifying the Various Protected Zones

In Table 2.1, the column titled *Protected Zone or Equipment* identifies the reach and coverage of a given protective function. *Equipment* pinpoints to the protected gear, which in this case always includes the generator. On the other hand, *zone* relates to all the equipment (apparatuses, lines, cables, etc.) that if experiencing a fault, will activate the protective function relay. The zone can be seen as also in a geographical context. For instance, a *distance* relay is called as such because its activation zone can be given in kilometers (of transmission line). In the case of a differential protection, the zone will be all that is between the CTs.

Figure 2.1 shows the zone protected by the generator differential protection (87G). A fault between phases on the busses, windings, generator circuit breaker, and IPB[5] (or lead cables) between the CTs, will cause the 87G to actuate. Faults on either CT or wiring between CTs and the relay (secondary circuit) may also cause actuation of the relay. Obviously a failure of the relay itself may cause an unintended actuation.

5 IPB—Isolated phase bus.

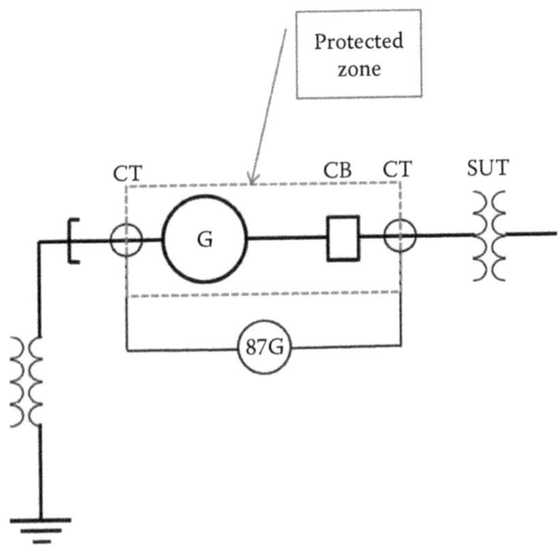

FIGURE 2.1
Generator differential *protected zone.* The MV generator circuit breaker (CB) is included in the protected zone. The neutral grounding transformer is shown between the generator's neutral point and earth, and is not included in the protected zone. Neither is the SUT.

Figure 2.2 shows the zone protected by the generator differential protection when the breaker is on the high-voltage (HV) side of the step-up transformer (SUT). As can be seen by inspection of the figure, in this scheme the generator breaker is not included in the protected zone; only the main generator is. Comments related to the CTs, secondary circuit, and relay integrity remain as explained above.

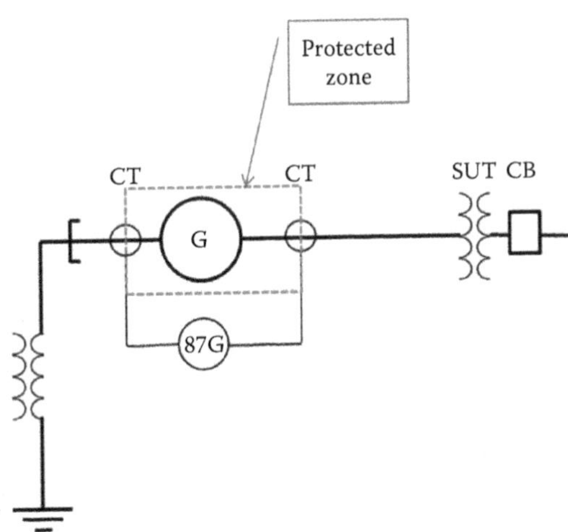

FIGURE 2.2
Generator differential *protected zone.* The generator circuit breaker (CB) is *not* included in the protected zone.

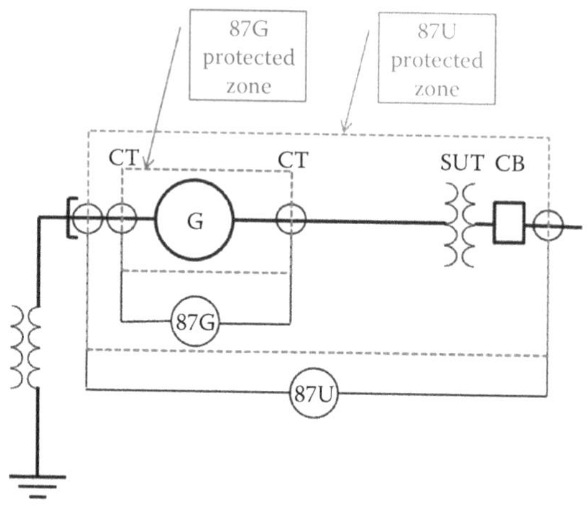

FIGURE 2.3
Generator differential and unit differential *protected zones*. The generator circuit breaker (CB) is *not* included in the protected zone of the generator differential (87G), but is included in the protected zone of the unit differential protection (87U).

Figure 2.3 shows the zones protected by the 87G and 87U differential protection relays (protective functions), respectively. It can be seen that the 87U serves as backup to the 87G for its protected zone, while providing additional coverage that includes the main transformer and main circuit breaker. In practical terms, the unit differential protection will, in most cases, also include the auxiliary transformers.

Figure 2.4 shows the zone protected by a distance relay, such as the one implementing protective function #21, or obtaining a similar goal with a

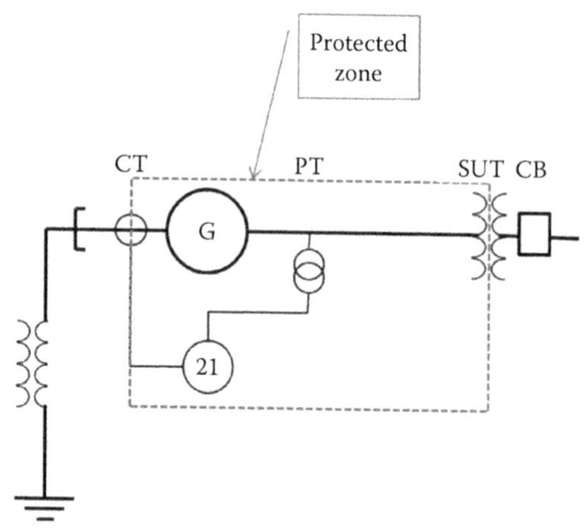

FIGURE 2.4
Zone protected by a backup *distance* protection relay.

voltage-controlled current relay (51V). Unlike the case with the differential protection where the protected zone is well defined by the location of the CTs feeding the 87 relay, the zone protected by the 21 or 51V can be expanded or shrank by the settings of the relay. Figure 2.4 shows the case where the reach includes the medium-voltage (MV) winding of the SUT.

Another example of *protected zone* is shown in Figure 2.5. In this case the turbine(s) is(are) included together with the generator, as well as any other not-shown rotating component in the unit's train, such as a rotary exciter. Moreover, even motors supplied via auxiliary unit transformers (not shown on the figure) are also protected for abnormal frequency when the circuit breaker is tripped. In this case, changing relay settings will not change the protected zone, but will change the threshold values that will trigger actuation of the 81 under/over frequency relay.

Figure 2.6 shows the protected zone by the stator ground protection. Typically, the stator ground protection is implemented at the secondary of the grounding transformer by an overcurrent relay and/or an overvoltage relay. Also, a third harmonic relay is commonly applied. In later years, stator ground protection has been implemented by the third harmonic relays monitoring the third harmonic voltages at both the neutral and lead side of the machine. Finally, low-frequency/low-voltage signal injection is also used as an active circuit implementation of a stator ground fault detection technique covering 100% of the stator winding. Figure 2.7 shows the case, typical of some small peaking units, or some CTGs,[6] where two or more generators are connected to a single bus before the SUT. In this case, the ground protection relays may not distinguish a ground fault on one generator from a ground fault on the other generator.

Some protection devices, such as the V/Hz (overfluxing) protection (24), protect very specific components of the generator and/or other unit

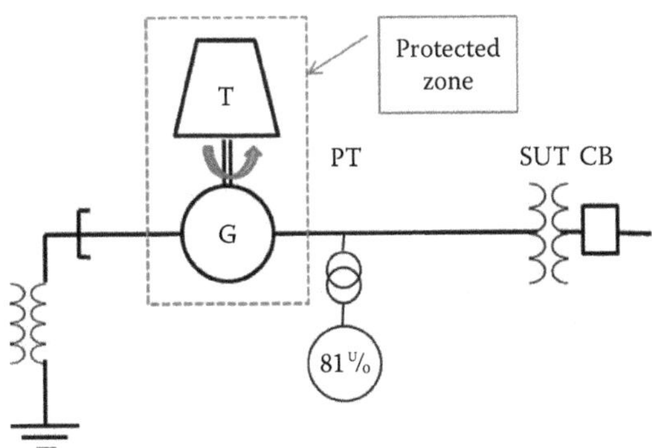

FIGURE 2.5
Zone protected by an over/under frequency relay.

[6] CTG—Combustion turbine generator.

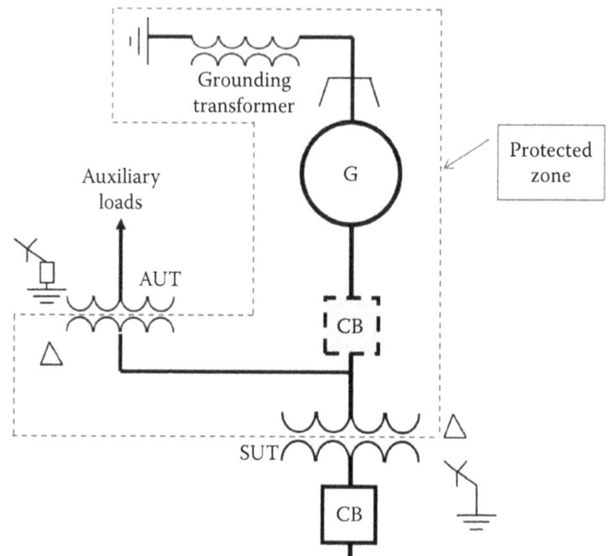

FIGURE 2.6
Zone protected by a stator ground protection relay.

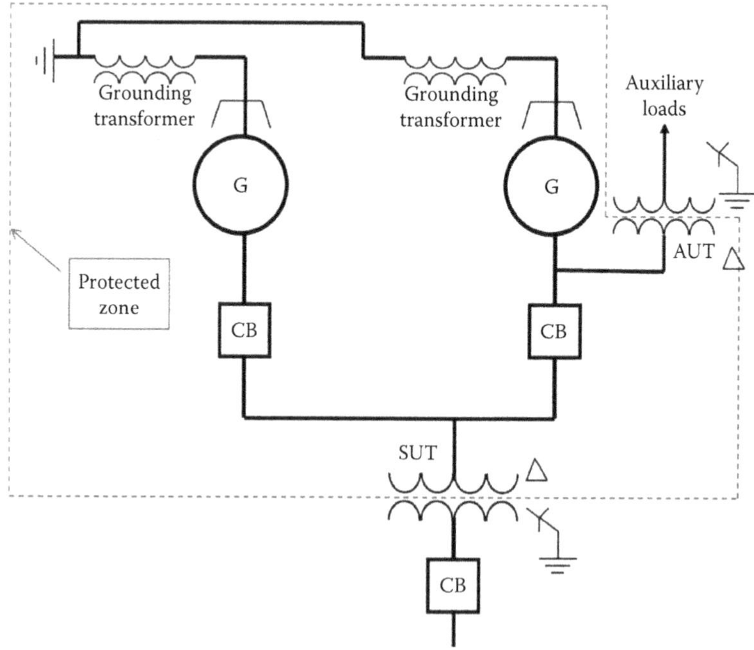

FIGURE 2.7
Zone protected by stator ground protection. A ground fault anywhere in the protected zone will result in activation of the ground relay. Positive identification of a stator ground as belonging to one or the other alternator may be impossible with most protection schemes.

equipment. In the case of the V/Hz protective function, the protected zone includes the generator core, as well as the core of all auxiliary transformers and the step-up transformer.

Negative-sequence current protection (46) protects some rotor components that are at, or close to, the surface of the rotor. It can also protect against metal fatigue of long turbine blades, such as those encountered in large low-pressure turbines of nuclear-powered generators. Another example of generator/ turbine zone protection is tripping for pole slipping, which can also damage the long blades that are mounted on the last-stage of the turbine.

Temperature-based relays protect specific generator components, such as the stator and rotor windings, core, and so forth, from overheating damage. Same applies to the excitation system.

Synchronizing (25) and sync-check (15) relays protect the entire train from damage due to malsynchronization. Windings, couplings, bearings, forging, turbine blades, and other components are prone to damage or degradation due to a bad case of synchronizing out of step.

The examples above are not exhaustive; there are a number of protective functions that are not covered. Nevertheless, the reader can obtain their protective reach (zone) in the form of equipment covered, by following the same method of analysis employed above.

2.6 How to Identify Probable Causes from the Actuation of Protection Relays

Unfortunately, it is not uncommon for plant personnel to misinterpret the causes for a generator trip initiated by the protection system. Sometimes, this leads operators to close the circuit breaker on a remaining fault, causing additional damage that could have been avoided. This results in needless additional delays in returning the unit to operation because of late discovery of the faulty component. For example, *meggering*[7] the stator after a V/Hz event and subsequent trip will not provide any insight onto the condition of the stator core, which is the component most prone to failure due to overfluxing. Similarly, looking for a short circuit in the switchyard immediately after the stator ground protection relay has actuated, probably is not the most effective way of addressing the problem.

Sometimes it is quite obvious after a protection trip, which is the failed component. For example, a flashover in the brushgear compartment will be readily recognizable by smell and sight. In many other cases, such as short circuits inside totally enclosed machines, the only practical way for ascertaining the location where a fault occurred is by carrying out electric testing

[7] *Meggering* of *megger test* is common jargon for an insulation resistance test.

and/or inspection of the machine internals after some level of disassembly. Where to start looking, or what to start testing, is a key factor in setting the duration of the forced outage. For instance, carrying out an insulation resistance test of the stator winding requires separating the generator winding from the neutral transformer, and removing links to the IPB in generators without a generator breaker. These activities add time to the outage. If a high potential test is desired, the amount of disassembly required to perform the test might be greater. It would be a waste of time to do all of that if careful analysis of the protective relay(s) actuation indicates that the fault is somewhere else other than on the stator. The rational approach is to first look inside the protected zone(s) of the actuated protective relay(s) that tripped and/or flagged.

It is true that on occasion, during a large fault, protection relays that have protective zones outside the faulted area may unselectively trip and/or flag (for instance, differential protection operation due to large through-fault currents, because of uneven saturation of the CTs, or the relay setting being too sensitive). Nevertheless, in most cases careful inspection of the protected zones of the actuated relays, in addition to the sequence of events (hopefully recorded by the unit's DFR[8] or similar instrumentation), serves as a compass in the search for the faulty component and fault location. Figure 2.8 shows how

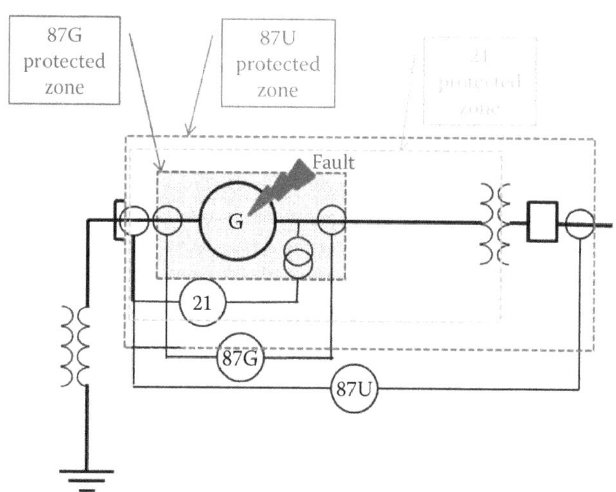

FIGURE 2.8
Shown on the figure are three protected zones: by the 87G generator differential protection relay; by the 87U unit differential protection relay and by the 21 distance protection relay. If all three relays trip, or if some trip and the others flag, for a phase-to-phase short circuit inside the generator, then the red dotted area represents the intersection of all three zones, and it the most probable to contain the fault.

[8] DFR—Digital fault recorder.

the protected zones activated during a fault determine the most logical place to initiate the search.

As indicated in Figure 2.8 by the red dotted area, which is the intersection of all three protected zones, that is the area most probable to contain the fault, if all three relays actuate, either tripping or a combination of tripping and alarming (flagging). Hence, it makes little sense to start undoing the connections of the main step-up transformer with the purpose of performing electric tests searching for the fault. Obviously after a severe fault, a conservative approach may include some testing/inspection of equipment outside the protected zone (e.g., that may be affected by large *through* short-circuit currents).

Some protective functions provide coverage to very specific components of the generator. For instance, Figure 2.9 shows the protected zone (rather the protected components) by the overfluxing (V/Hz) relay (24). In this particular and unfortunate case, although the components that may be directly impacted are easily identified, the level of degradation is much more difficult to determine.

Figure 2.10 shows a conceptualization of the process by which the location of a faulty component may be found. Table 2.1 provides the reader with hints as to what might be the most probable location of the fault of damaged components. For additional information on all protective functions, see References [1–5] at the end of this chapter.

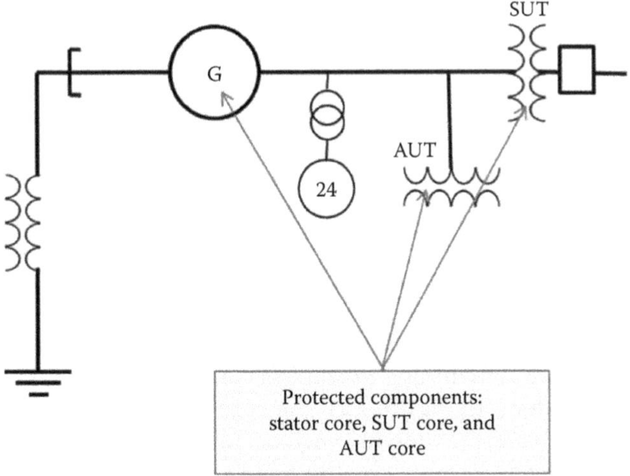

FIGURE 2.9
The equipment protected by the volts-per-hertz relay. The relay also protects the cores of transformers, because in this case the main breaker is on the HV side of the SUT.

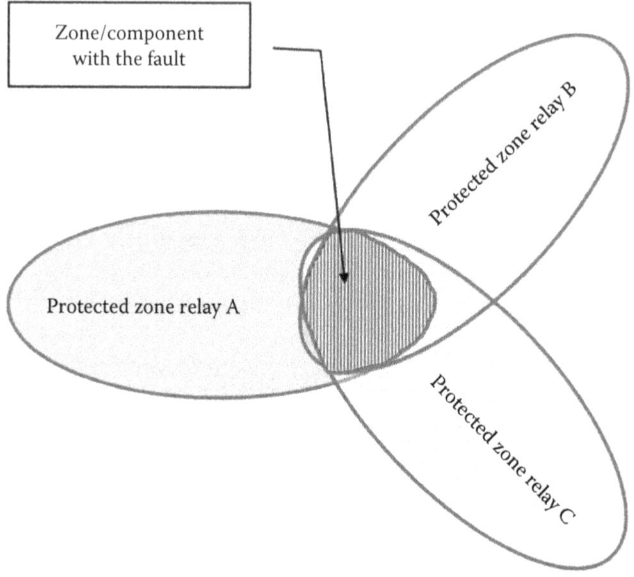

FIGURE 2.10
The intersection of the protected zones provides a clue as to where the fault might have happened.

2.7 Essential Checks before Returning a Unit to Service after a Protection Trip

As mentioned in the previous sections, it is not unheard about a unit being restarted after a protection-initiated trip, without an exhaustive investigation of the causes that led to the actuation of the protective relay(s). The *spurious trip* explanation is often overused, resulting, in some cases, in very serious damage to the equipment, not to mention compromising the safety of the personnel. Having hand-reset lockout relays for internal malfunctions, and self-reset relays for grid-related faults goes some distance in addressing this concern. It behooves each generating station to have clear written procedures outlining the process operators and technical personnel must follow before returning a unit to service, following a trip initiated by its protection system. In Figure 2.11, the reader can find an algorithm that may provide the basis for charting such a procedure, to assuring that a unit will not be returned to service before a meticulous effort is carried out for the purpose of verifying that all systems and components are good-to-go. The algorithm presented therein should be adjusted to the particular protective scheme found at each plant.

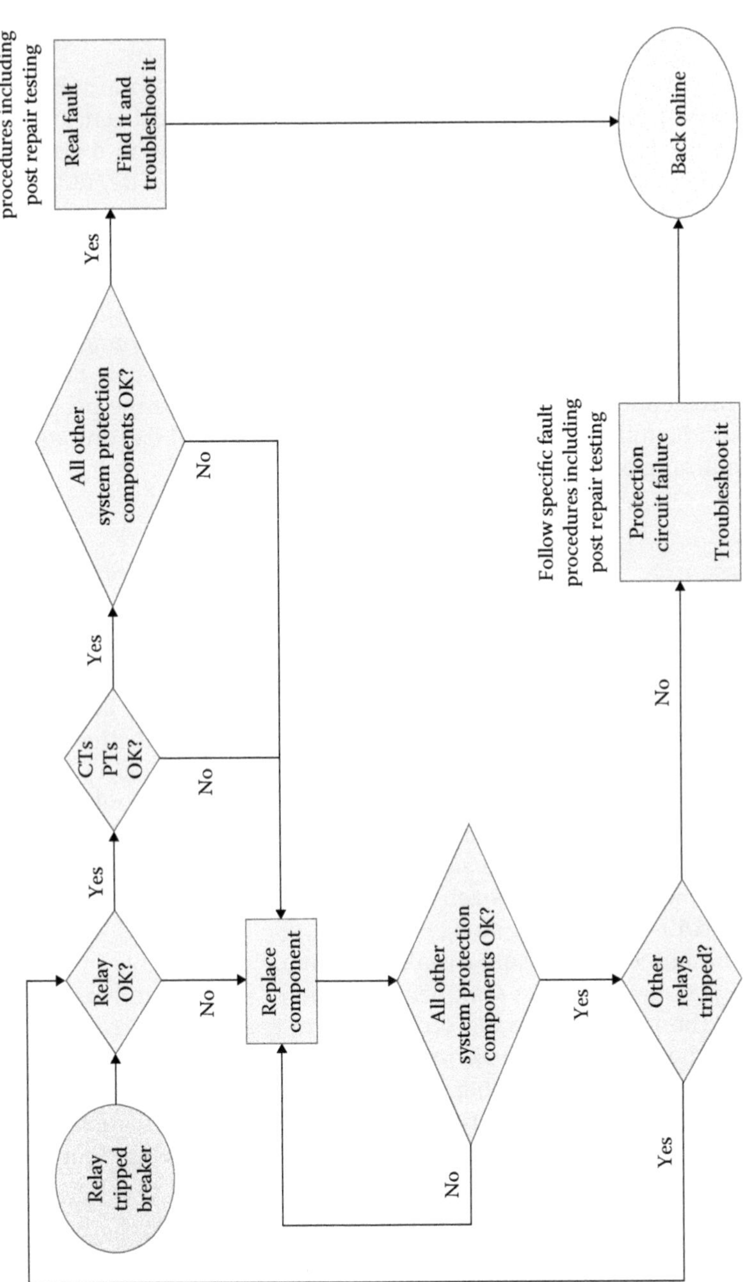

FIGURE 2.11
A possible algorithm representing steps to be taken after a unit is tripped by its protective relays, to assure that the machines are returned to operation without having first investigated and the cause for the trip removed.

2.8 Essential Checks before Returning a Unit to Service after Receiving an Alarm

Figure 2.12 presents an algorithm that may be part of every alarm-response procedure in a power plant. The algorithm offers a due diligence method that prevents misdiagnosing an alarm as *spurious* or due to a *bad sensor,* or some other nonchalant reason, and therefore averting a degraded component condition from developing into a major fault. Unfortunately, it is not uncommon for alarms generated by the protection system to be dismissed as not being the result of a real malfunction within the equipment. Obviously, grid-initiated alarms may indicate a transient abnormal condition with the grid that does not require further investigation (e.g., a fault close to the plant on a transmission line that was immediately cleared by the line-protection relays). Nevertheless, even in this case the proper process should be followed to vet the protection system and the unit as being fit for remaining online.

2.9 Alarm Response Procedures

During the many years of work with power plants of all types, the authors have come across instances in which there was a glaring lack of written procedures addressing how operators should react in the presence of certain alarms. In some cases the response is *by the seat of the pants.* Unfortunately, this approach sometimes leads to undesirable results. Written alarm-response procedures are helpful in averting the inconsistency in the responses to the same alarm by different operators, and to allow for a thought-out approach leading to minimum disruption to production and/or equipment damage.

There are numerous sensors installed on many components in the plant that can send alarms to the annunciator panel or display units in the control room; the notion of having written alarm-response procedures apply to many of them. When it comes to the main generator, which is the subject of this book, there are a number of alarms whose disposition may benefit from the existence of a written procedure. The next few subsections discuss just a few of those alarms that are often found without having clear written instructions on how to dispose of them.

2.9.1 Rotor Grounds

There are numerous and very diverse practices among power plants for responding to rotor winding ground faults. Some units are tripped automatically for a field ground that reaches a given value (e.g., 1000 Ω), others

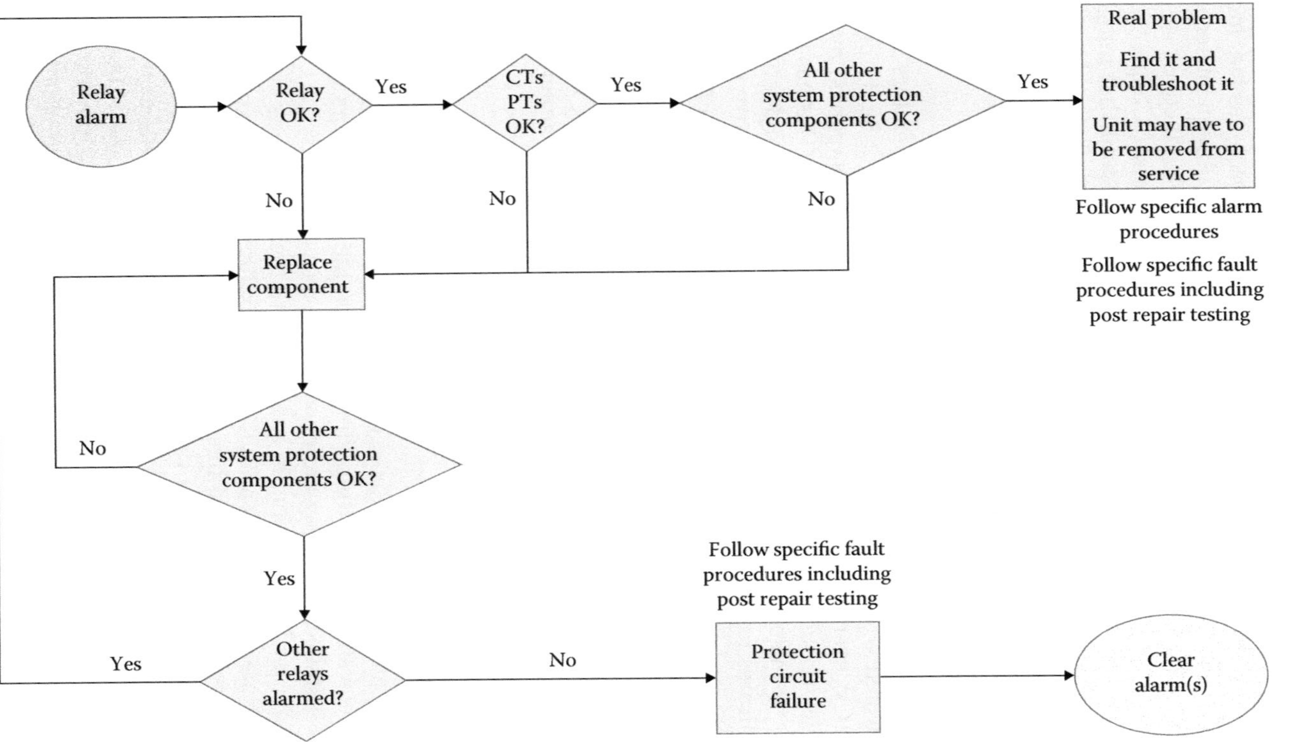

FIGURE 2.12
A possible algorithm representing steps to be taken after an alarm is generated by its protective relays, to assure that the machine and protective system are reliable for continuing operation online.

immediately start troubleshooting the problem, and the unit is manually removed from service if the fault cannot be fixed or fixed within certain time limits, and some others are left to run for long periods of time (even months) before the problem is addressed. This last approach has led, more than once, to serious damage to the field (rotor) from a subsequent second ground, given the likelihood for a second ground developing increases after a first short circuit to ground establishes a ground reference. Chapter 4 in Reference [3] includes a typical written operating procedure that may be gleaned for cues on how to write an effective written procedure addressing a number of alarm responses, including a field ground.

2.9.2 Winding Temperature

One unfortunate recurring avoidable situation is when operators misinterpret elevated winding temperatures as being the result of faulty probes (RTDs or TCs), even in cases where these readings are accompanied by that of other sensors, such as generator condition monitor alarms. This has led, in some cases, to major damage requiring months of downtime and large capital expenditures for repairs. On the other hand, one can identify himself/herself with the hesitation to removing a large unit from operation without clear evidence that an elevated temperature reading is valid, that is, that it denotes a real developing malfunction. This is one of those situations where a well-written procedure can prevent undesirable consequences. The procedure should be written well in advance, with solid engineering input and managerial acquiescence, and without the time pressure encountered during a developing problem.

2.9.3 End-Winding Vibration

Another area where practices deviate significantly from each other is when addressing elevated end-winding vibrations. On most generators there are no sensors installed providing online vibration measurements; thus, operators have no knowledge of what is going on with their machine's endwindings. By the time it becomes clear that the endwindings may be experiencing high vibrations (through visual inspection and/or testing during outages), serious degradation might have occurred. In those fortunate cases where units have installed end-winding vibration sensors that allow measuring of the end-winding vibration online, many do not take advantage of the existence of the probes to read and trend the vibration, and those that do trend them, in certain occasions do not have clear written procedures about limits and adequate remedial actions. Reference [2] may be a source of useful information on this subject.

2.9.4 Rotor Vibration

This is also an area where there is a very diverse approach to how to address the problem of increasing vibrations. Most, if not all, generators that fall under the scope of this book have sensors installed that provide online real-time measurement of rotor vibrations; most even follow the same international standards. However, there are differing approaches about how they respond to a significant increase in vibrations. This is also a good example where having written procedures in place addressing the problem of increasing vibration levels can be very helpful in a time of need.

References

1. IEEE Std C37.102-2012. *IEEE Guide for AC Generator Protection.*
2. IEEE Std. 1129-2014. *IEEE Guide for Online Monitoring of Large Synchronous Generators (10 MVA and Above).*
3. G. Klempner and I. Kerszenbaum. *Handbook of Large Turbo-Generator Operation and Practice*, 2nd edition, Chapter 6, Wiley, Hoboken, NJ, 2008.
4. D. Reimert. *Protective Relaying for Power Generating Systems*, CRC Press, Boca Raton, FL, 2006.
5. T. Baker. *Electrical Calculations and Guidelines for Generating Stations and Industrial Plants*, CRC Press, Boca Raton, FL, 2012.

3

Monitored Parameters

3.1 Typical Monitored Variables in Large Turbogenerators

There are a multitude of parameters that are monitored on any given generating unit within the range of ratings covered in this book (typically synchronous generators with cylindrical rotors rated 10 MVA and larger). In a nuclear or large fossil power plant, the monitored points will reach into the hundreds. In this book, only those variables related to the generator, excitation, and generator auxiliaries are covered. In the sections below, the monitored variables are segregated based on to which internal component they belong to, as well as those monitoring the entire generator performance. A brief description is added to each one. The list is by no means complete, but it covers the most important parameters. It is incumbent to the operators, plant technicians, and engineers to possess a good understanding of how the readings from the monitored equipment fit into the big picture, that is, how the excursion of any monitored parameter outside the typical or specified range reflects a problem with the equipment. And in addition, what response is desirable for bringing operation and/or machine conditions back to normal, or in some cases, reducing load or removing the unit from operation.

Most of the parameters monitored online in real-time will also trigger an alarm when they leave their regular or specified operating range. A very few variables, such as bearing vibration, may be also wired to trip the generator and the entire unit if the readings divert too far from their normal range. Other points that may be wired in such a way that online monitoring is readily achievable, are, in practice, monitored (and maybe also trended) on a periodic basis (for instance, stator end-winding vibration or partial discharge activity [PDA]). Some parameters are never monitored, though it is always a good practice to monitor everything that is available for monitoring, if the proper probes are installed in the machine.

Given the aforementioned, it is important when specifying a new generator, or a major rework, such as a new stator winding, or new core and winding, that the customer also specifies which sensors must be installed. For example, if one wants RTDs embedded in the core for measuring core temperatures at various locations, they must be installed during the assembly of the core, when the laminations are being stacked. Once stacking of the core is finished and it has been compressed, there is no possibility of inserting the temperature probes, and less desirable walk-arounds must be used, without much benefit. A similar situation arises when requiring RTDs or TCs installed in the slots, between top and bottom stator bars. Once the top conductor bars have been installed, it is not possible to mount the temperature sensors; they must be installed before the top bars are placed in the slots.

Reference [1] is the IEEE standard that covers the topic of monitoring large synchronous generators. Much and good information can be found therein.

3.2 General Machine Variables

3.2.1 Stator Terminal Voltage

A key parameter defining the rating of the generator, the terminal voltage has a permissible range of ±5% of nominal, per most national and international standards. This may be changing around the world, based on new grid requirements. In any case, if the machine is specified by the owner to have a wider range of operating output voltage than provided by the applicable standard(s), this can be accommodated—for a price—by the OEM. Operators must not allow the stator terminal voltage, under any circumstance, to remain above or below its specified rating. Doing otherwise risks serious damage to the machine, such as overfluxing of the stator core, and/or overheating of the stator or rotor windings. Most typical monitoring is done by a voltage indication in the control room and/or an alarm for deviation from the permissive operating region. Very seldom a voltage deviation with the unit online leads to an automatic trip. Operators typically control the terminal voltage by adjusting the excitation current.

3.2.2 Stator Terminal Current

This is also a key parameter defining the rating of the generator. The stator current, also called *armature current*, is derived from the following equation:

$$\textbf{Rated stator current (amperes)} = \frac{\textbf{Rated MVA}}{\sqrt{3} \times \textbf{Rated terminal line voltage (kV)}}$$

The stators windings are designed to carry the rated stator current. Any current higher than that most probably will cause the temperature of the stator bars to rise, exceeding the insulation temperature class limit, with a

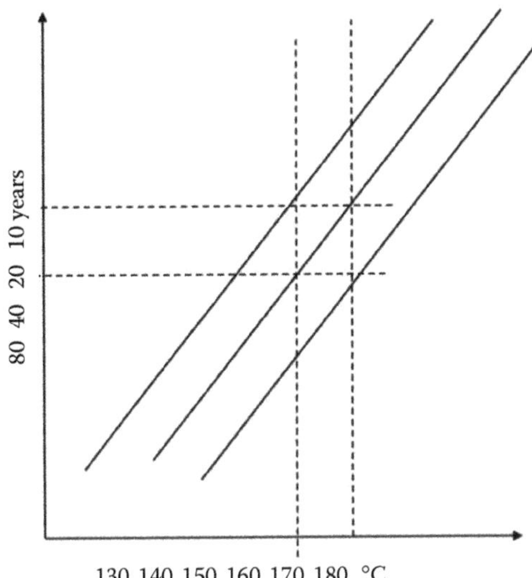

FIGURE 3.1
Arrhenius graph showing the dependency of the expected life of the insulation (vertical axis) on the temperature of the conductor.

consequential loss of expected life of the insulation, per Arrhenius formula, as shown in Figure 3.1.

As it can be seen on the graph, the expected life of the insulation is approximately halved for each 8°C–10°C increase in temperature over System Class temperature. Hence, operators must never allow operation above rated stator terminal current for any length of time. Occasionally, the stator current may exceed its nominal value due to grid transients and other abnormal conditions (such as during a short circuit outside the generator); however, the windings are designed to withstand such short overreach, as long as they remain within the withstand capability of the insulation; these capabilities are spelled out in the relevant standards, such as Reference [2].

Stator current is always monitored in machines within the rating covered herein, by gauges in the control room panels or VDUs.[1] Operators typically control stator current by changing the reactive output of the generator, by reducing the rotor field current when the machine is operating in the over-excited region (lagging power factor), or by increasing rotor field current when the machine is in the leading power factor region. A less desirable but sometimes needed action is to reduce the active (real) power output of the machine (by lowering the output of the prime mover).

Figure 3.2 shows the stator and rotor current withstand capabilities according to the IEEE C50.13 standard [2]. From the curves, the withstand capability discrete values of Table 3.1 are obtained. See Reference [3] for additional discussion.

[1] VDU—Visual display unit.

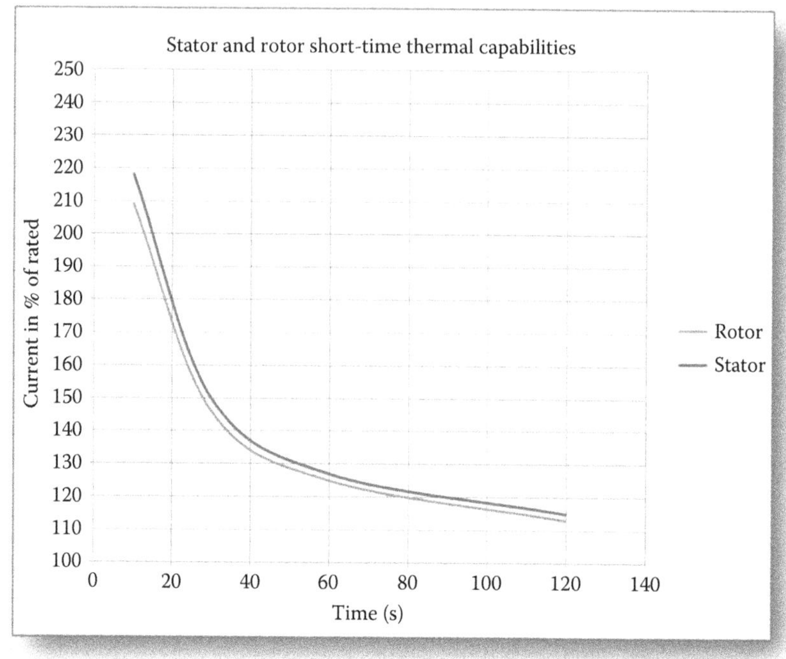

FIGURE 3.2
Stator and rotor winding withstand capability for short-time thermal overcurrent. (From IEEE Std C50-13-2014, *IEEE Standard for Cylindrical-Rotor 50 Hz and 60 Hz Synchronous Generators Rated 10 MVA and Above.*)

TABLE 3.1
Discrete Values for the Stator and Rotor Windings Short-Time Thermal Withstand Capabilities

Stator				
Time (s)	10	30	60	120
Stator current (% or rated)	218	150	127	115
Rotor Field				
Time (s)	10	30	60	120
Rotor field voltage (% of rated)	209	146	125	113

3.2.3 Generator Real and Reactive Power Output

Rated real and reactive power output are defined at a single point on the capability curve, where the rotor and stator limiting curves meet in the lagging region of the capability curve. However, while the real and reactive power outputs can assume any value within the capability curve, there are strict limits on the maximum output that can be achieved. The limits are defined by the envelope of the capability curve of the generator. In theory, the entire region of the capability curve is good for operation. Nevertheless, with respect to active (real) power output, there are additional limits imposed by the prime mover, and/or boiler of fossil fueled plants, and/or steam generator in nuclear plants, for the case of steam-driven turbines, or by the combustion turbine. With respect to reactive power, there are limits imposed by the steady-state stability limit (SSSL) and additional stability margins in the underexcited region, and, in older or compromised machines, by concerns about core-end degradation, also known as *end-iron heating*.

Figure 3.3 shows a typical capability curve that includes limits to the active and reactive power.

3.2.4 Generator Frequency

All large synchronous generators are designed to operate at a single frequency reflected by a unique rotational speed. The relation between frequency and speed is:

$$\textbf{Rated speed} = \frac{\textbf{Rated frequency} \times \textbf{120}}{\textbf{Number of poles}}$$

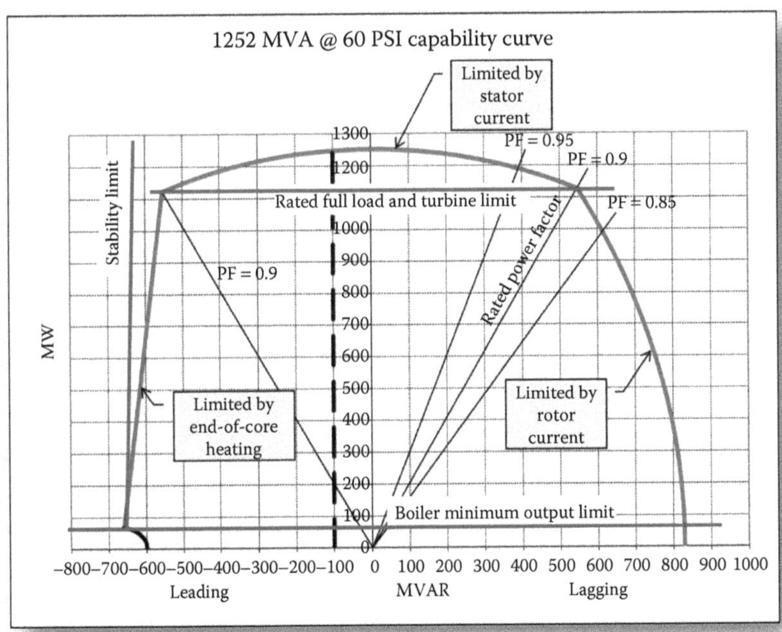

FIGURE 3.3
This capability curve for a 4-pole hydrogen-cooled generator is shown in the *English View*, that is, with the active power on the vertical axis and reactive power on the horizontal axis, and the lagging reactive power appearing on the right side.

There is a very small margin around the nominal frequency where operation is allowed. For larger excursions, the time where operation is allowed is very short. In fact, although the generator could withstand some excursion from the nominal frequency, the turbine is in general much more limiting; in particular this is so with large low-pressure steam turbines with long blades.

Monitoring of the frequency is by gauges and/or digital display on control room VDUs. In the case of large grids, often generators do not alarm or trip for frequency excursions, because once the unit is synchronized to the grid, the frequency of the generator is governed by the frequency of the grid. Alarms may exist when the generator is a large source of power in a small grid. In general (and in particular for generators feeding large grids), the only time frequency (speed) must be closely monitored and controlled is when the unit is accelerating toward synchronization. See Reference [3], Chapter 6 for more details.

3.2.5 Rotor Field Voltage and Current

In machines with sliprings (*brushed*), both rotor field voltage and current are easily monitored and often displayed in the CR.[2] In machines without sliprings (*brushless*), the rotor field current is observed indirectly by monitoring the exciter's output current. In the case of rotor field voltage from brushless generators, it can be monitored via auxiliary sliprings, or indirectly by monitoring the exciter's output voltage. Sometimes, only the rotor field voltage or the rotor field current alone are monitored at the CR. In many units the rotor field current and voltage are automatically used to calculate the temperature of the rotor winding (Reference [3]). In other cases, only exciter's output voltage and not the current are available for monitoring. Thus, there are a number of variations to how rotor field voltage and current can or cannot be monitored.

As with other variables, rotor field current and voltages have nominal values that must not be exceeded, other than during transient conditions, when the graph in Figure 3.2 applies. In generators without too many rotor field shorted-turns, keeping the operating point inside the capability curve will automatically keep the rotor field current and voltage in their permissive range. However, in cases where there are a significant number of shorted turns in the rotor field winding, a situation can be reached that, to attain high levels of lagging reactive power, the rotor requires field currents higher than rated. The AVR and/or exciter should have protection (usually limit functions) that is designed to avoid, and generate alarms, for this undesirable situation. Figure 5.15 in Reference [4] shows in a clear manner how the capability curve in the lagging region is reduced by the presence of shorted turns in the rotor field.

3.2.6 Volts per Hertz

As shown in Chapter 2, V/Hz (24) is one of the most critical protective functions safeguarding the generator from extremely damaging overfluxing

[2] CR—Control room.

events. However, in many stations, V/Hz is also monitored at the CR, which oftentimes includes an alarm. Operators must, upon receiving an alarm, return to the permissive operating region, most often than not by reducing the field current which lowers the output voltage. In the majority of cases the alarm, if not cleared soon by operator remedial action, is followed by an automatic trip of the generator. Keeping the generator from going above the maximum terminal voltage will eliminate the possibility of drifting into a condition of excessive volts-per-hertz; unless the frequency is substantially lower than nominal (a highly improbable occurrence when synchronized to a large grid). In this rare case, the V/Hz relay is the only device that can prevent the generator from being damaged due to overfluxing. The IEEE standard C50-13-2014 [2] includes a graph ("Operation over ranges of voltage and frequency") that shows the relationship between terminal voltage and operating frequency. Higher frequency deviation from rated frequency is only permitted for a smaller deviation of terminal voltage from nominal voltage; and vice versa. This tends to be a bigger problem during startup, when the excitation voltage is too high for the generator when it is operating at much lower speeds than rated speed, for such things as a turbine heat-soak, or rotor field pre-heating requirements.

3.2.7 Negative Sequence Stator Current

Similar to the V/Hz variable, negative sequence armature current is basically a parameter that is sometimes also monitored at the CR. The negative sequence current is, when monitored, compared with the maximum allowable according to the standards [3], and the rating of the machine. The monitoring loop/instrument may in most cases also alarm when the negative sequence current reaches its permissive continuous limit. In the vast majority of cases, continuous negative sequence currents are a product of some grid problem or intrinsic unbalance, and thus, do not lend themselves to be controlled by the operator of the generator. If that is the case, the only avenue for the operator to decrease the stator negative sequence current is by reducing load, either reactive or active power, or both. One must keep in mind that excessive negative sequence currents have a detrimental effect on the rotor of the generator. Figure 3.4 shows typical damage to the rotor due to large negative sequence currents.

3.2.8 Byproducts of Insulation Pyrolysis

There are many components in the generator that have *electrical insulation* as a key subcomponent. For example: insulation between the stator laminations, insulation of the stator conductors and insulation of the rotor field conductors. Insulation has a maximum safe operating temperature, and when this is exceeded, accelerated aging occurs following the Arrhenius law (Figure 3.1). If the temperature reaches levels well above their maximum permissive operating limits, pyrolysis may occur. This happens during conditions that may

FIGURE 3.4
The photo shows damage to a rotor wedge due to large negative sequence currents. The weld was between the wedge and the retaining ring (RR).

lead to short circuits or core damage. Pyrolysis releases particles that can be captured by the generator condition monitor (GCM), setting off an alarm indicating a developing abnormal thermal condition. Many machines have GCM devices installed, but others do not. The subject is covered in Chapters 4 and 5. Figure 3.5 shows such an instrument installed on a hydrogen-cooled generator, and Figure 3.6 shows one designed for air-cooled machines.

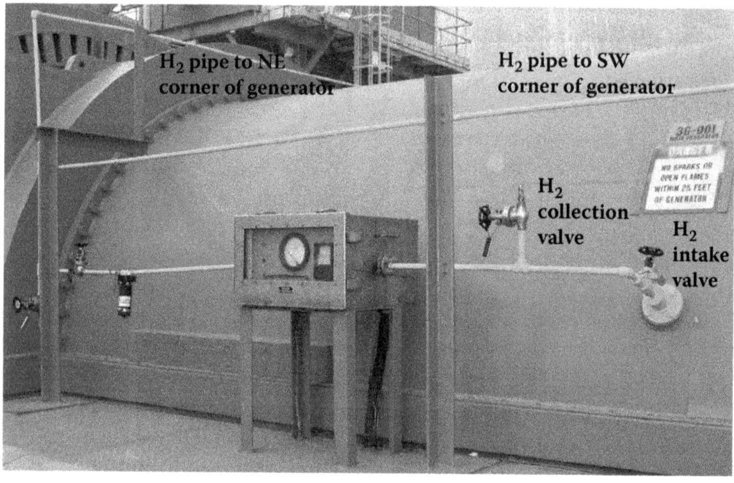

FIGURE 3.5
The photo shows an older generation of a generator condition monitor (GCM).

FIGURE 3.6
Generator condition monitor for an air-cooled machine. (Courtesy of the Environment One Corporation, Niskayuna, NY.)

3.2.9 Liquids inside the Generator

All large turbogenerators have a collection vessel under the casing that collects any liquid that may ingress the machine. Obviously, during operation only oil and/or water may leak from coolers, seals, windings, and become free inside the machine. The collection device in many cases also has a sight glass for visual confirmation that there is a liquid present, and in most cases there is also a mechanism (e.g., with a floater) that will trigger and alarm. Problems associated with the presence of free oil and/or water inside the machine, and operator response, are discussed further down, in Chapters 4 and 5 and in Reference [3]. The photograph in Figure 3.7 shows a liquid detector without a sight glass, wired for alarming in the presence of liquids. Figure 3.8 shows one with a sight glass.

3.2.10 Ozone

In air-cooled generators, PDA in the stator slots (where the conductor bars are located) has the potential to break down the air molecules, which is conducive to the generation of ozone. Ozone is a highly oxidizing agent that has the potential for damaging the outer layers of the stator insulation, leading to its accelerated degradation and eventual ground faults. Ozone is monitored with instruments that are designed for that particular function.

FIGURE 3.7
Generator liquid detector equipped with a sight glass.

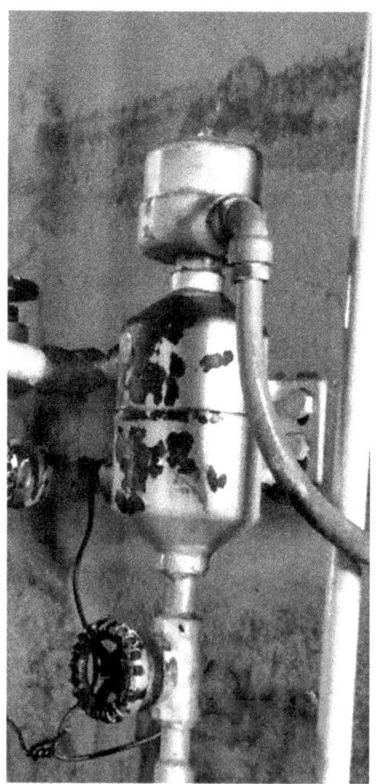

FIGURE 3.8
Generator liquid detector equipped with an alarming circuit.

The surge of ozone in a machine may be an indicator for increased PDA between slot and conducting bar, as well as corona discharge from the end-windings. Without a major scope of work, remedial action may be limited. Such a scope could include one or more of the following activities: removing the stator winding and recoating it with semiconducting paint or tape before reinstalling it, or rewinding with new bars, and rewedging.

3.3 Stator (Frame, Core, and Winding) Variables

3.3.1 Frame Movement and/or Vibration

It is unusual for the frame of a generator to be directly monitored for movement such as displacement or vibration. Yet, there are some cases where the machine is prone to such movement and in those cases, vibration sensors may be installed on the frame. Excessive frame vibration may over time lead to cracks, as shown in Figure 3.9. Long-term frame vibration needs to be addressed to avoid premature degradation of key components, such as frame, core, and stator windings. This topic is covered in Chapters 4 and 5 wand in Reference [3].

3.3.2 Core Vibration

This issue is also covered in Chapters 4 and 5, including operator response, and Reference [3]. As with the frame vibration (previous section), excessive

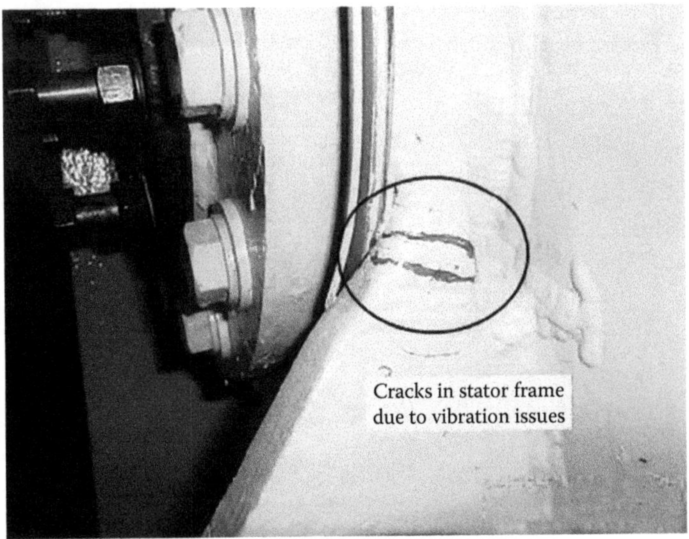

FIGURE 3.9
Vibration-induced frame crack.

core vibration is an undesirable condition. While most generators do not have vibration sensors installed on the core or frame, machines showing propensity to develop such condition may benefit from their installation, as shown in Figure 3.10.

3.3.3 Core Temperature

Most large generators have temperature probes (thermocouples or RTDs) embedded in their cores (Figure 3.11). The purpose is to monitor for core temperature at different spots of the core; in particular, the area of the teeth is susceptible to the development of hot spots and consequential damage to the interlaminar insulation. Some large machines have numerous temperature probes installed along the core, at different angles from the vertical, and at various depths. Monitoring all the available core temperature sensors is a good practice, because it can assist the operators in preventing serious long-term degradation of core and stator windings. Further discussion including operator response can be found in Chapters 4 and 5 and in Reference [3].

3.3.4 Flux Shield and Compression-Plate Temperatures

It is not uncommon to monitor the temperature of the flux shield in some large turbogenerators, if such a shield is installed, and/or to monitor the temperature of the compression or clamping ring. Figure 3.12 shows such a case.

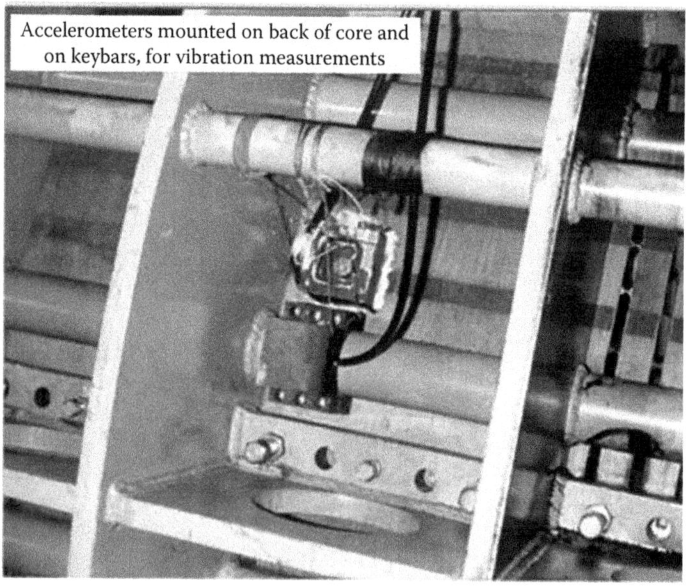

FIGURE 3.10
Vibration sensors installed on core and keybars for monitoring core vibration.

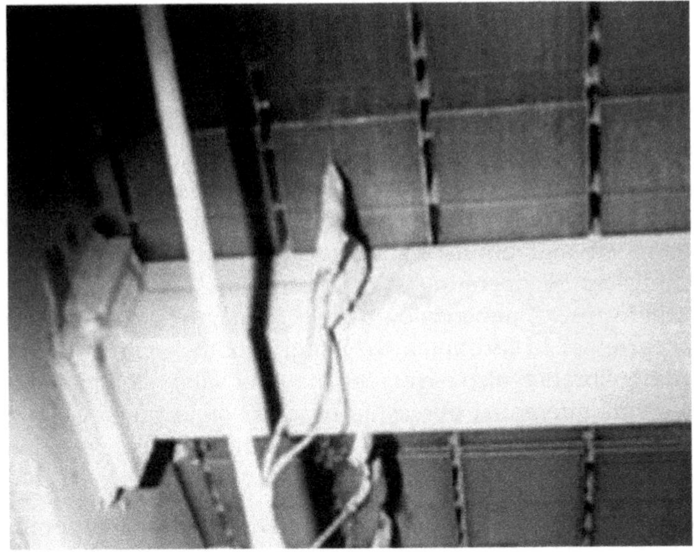

FIGURE 3.11
Temperature probes (thermocouples) monitoring the back of the core.

FIGURE 3.12
Temperature probe (thermocouples) monitoring the clamping plate.

Monitoring these components may be helpful during operation at leading power factors (underexcited condition), when the core ends tend to operate at elevated temperatures. If monitored, these temperatures should be set to trigger an alarm in case their maximum safe operating conditions are reached. Returning the machine to its safe operating region may require load reduction, or most probable, operating at a lesser underexcited mode.

3.3.5 End-Winding Vibration

End-winding vibration is a critical factor impacting the integrity of the stator winding. Many stations that did not have vibration sensors installed on their generator endwindings are now taking advantage of the modern fiber-optic technology to install them during planned machine overhaul outages. Reference [1] includes a weighty discussion on the subject of recommended limits of vibration. Excessive end-winding vibration can only be mitigated by major repair and offline testing activities. Temporary mitigation can be accomplished by operating at a reduced load (end-winding vibration, among other sources, depends on the electromagnetic forces developed by the stator currents). In addition to the integrity of the conductors, monitoring end-winding vibration also covers the manifolds in water-cooled windings, as well as circumferential busses and radial connections. Chapters 4 and 5 discuss this topic further. Figure 3.13 shows end-winding vibration sensors.

3.3.6 Stator Conductor Temperature and Differential Temperature

In generators with indirect stator conductor cooling (i.e., the main source of bar cooling is the core itself), temperature monitoring of the stator bars is accomplished by installing RTDs or TCs between the top and bottom bar in the slot area. Monitoring bar temperature is very important for the reasons

FIGURE 3.13
The photo shows stator endwinding vibration monitoring sensors. Shown is an earlier generation of sensors. Modern fiber-optical sensors are small and very light.

given above in Section 3.2.2. Generators with inner-cooled stator windings by hydrogen, water, or oil may not have, in some cases, RTDs embedded between top and bottom bars; however if they have them, they should also be monitored because they may provide important data about impending failures of the insulation. Unfortunately, many generating stations do not wire these temperature probes to a display unit, and thus, they are in the position that they may miss critical information. More about all this in Chapters 4 and 5 and Reference [3]. See Figure 3.14 for an example of a TC embedded between top and bottom bars.

Machines that are well monitored also have monitored the differential temperatures between all bars and/or group of bars. This is of particular importance in water-cooled stator windings, where the flow of cooling water may diminish in certain parts of the winding, for instance due to copper-corrosion buildup, hydrogen locking or a major failure of the plumbing.

3.3.7 Partial Discharge Activity

All large turbogenerators operate with voltages well above the inception voltage for partial discharge, that is, such machines undergo some level of PDA in the slot area and corona on the endwindings. Engineers spend considerable effort in designing windings that include materials and proper geometry to control this otherwise damaging electric activity, both inside the insulation and between the bars and the core. Nevertheless, PDA may occur due to a number of factors, most common aging of the insulation system, in

FIGURE 3.14
The arrow points to the wiring that connects to a TC installed between top and bottom bars in the slot.

particular, loose windings leading to the deterioration of the semiconducting outer layer of the bars, or drying out of the bulk insulation. Considering the propensity for these types of occurrences, early detection goes a long way in enabling plant personnel to plan and carry out remedial actions. Because of the strong oxidizing properties of ozone (a byproduct of PDA in air-cooled generators), these machines are more susceptible to this type of problem than hydrogen-cooled machines (see Figure 3.15).

PDA develops over months or years, and there is nothing that an operator can do to mitigate it, because it is mostly voltage-dependent. Any remedial action requires major rework, such as rewedging, application of new semiconducting material, reinsulating, or altogether replacement of the stator windings. The earlier the problem is addressed, the smaller the scope of the repair and the length of the downtime. Due to its slow development, PDA is in many cases not monitored online on a real-time basis, but by taking readings every few months.

Typically, PDA is monitored by either capacitive-type couplers connected directly to the terminals of the machine, which is the main method employed with air-cooled generators, or the slot-type coupler, mainly used on hydrogen-cooled machines. An alternative method for detecting PDA is EMI[3] technology, by mounting a wideband current transformer (RF-monitor) around the neutral of the generator, between the generator and the grounding transformer. See Figure 3.16 for a RF monitor installed on a hydrogen-cooled generator. Further discussion can be found in References [3,5], as well as in Chapters 4 and 5.

FIGURE 3.15
Stator conductor bars showing major erosion of the semiconductor layer and portions of the graded insulation due to severe PDA in an air-cooled generator rated at about 80 MW.

[3] EMI—Electromagnetic interference.

FIGURE 3.16
The photo shows a RF monitor for the detection of partial discharge activity installed around the cable leading from the neutral of a 636 MVA H_2-cooled generator to the grounding transformer. The photo was taken during a planned major overhaul outage.

3.4 Rotor Variables

3.4.1 Rotor Vibration

This is the one of the most basic monitored parameters in all large turbogenerators. Although in many cases the probes are wired to trip the unit for very high level of vibrations, it can be considered a monitoring system and not a protection one. In almost all cases, operators have the time to react to an increase in the level of vibrations by either reducing load (MVARs or MWs, depending on the source of vibrations), changing temperature of the lubrication and/or hydrogen seal oil, or removing the machine from service. See Chapters 4 and 5.

3.4.2 Rotor Winding Temperature

For those generators with sliprings, it is customary to calculate and display the rotor temperature as a function of the division of the rotor field voltage by the rotor field current. In generators that do not have sliprings, either there is no such parameter being monitored, or it is based on the existence of auxiliary sliprings in the exciter, or readings of the field voltage and current in solid-state exciters. It is unusual to wire this function as a trip function; however, when monitored, it allows the operators receiving an alarm to take corrective actions, depending on the cause for the increase in rotor field temperature. See Chapters 4 and 5 for operator response. As a side note, it is worthwhile mentioning that there are manufacturers of rotor temperature monitors that do not depend on the measurement of rotor field voltage and current, measuring directly the temperature of the winding in a given spot(s), and sending the information by telemetry. Although applied to a number of large hydro-generators, this technology has not penetrated the rotor field of turbogenerators to any noticeable degree.

3.4.3 Rotor Field Shorted Turns

According to some statistics, no less than 50% of all operating turbogenerators have at least one shorted turn in the rotor field winding. However, the percentage of machines requiring remedial action is much smaller. In fact, many machines will operate for decades with at least one shorted turn and show no sign of distress, while a few will require immediate attention. The reasons for that disparity can be found in the location of the short (which coil), number of shorts, number of poles, and type of cooling. However, when the machine starts showing what is called *var-sensitivity* or the inability to reach high levels of lagging reactive power because of excessive vibration or lack of sufficient excitation current, the presence of a sensor monitoring the rotor field for shorted turns can be critical in facilitating that search for the root of the problem, and planning for a solution. It is thus advantageous for operators of large turbogenerators to monitor the rotor winding for shorted turns, if not online-real-time, at least periodically. This is the reason why more and more stations take advantage of their rotor-out outages to install magnetic flux sensors in the airgap. Chapters 4 and 5 discuss possible operator responses to this situation. Figure 3.17 shows a typical RFM (rotor flux monitor).

3.4.4 Rotor Ground Faults

Theoretically, a generator can operate for a long time with a single ground fault on the rotor field winding. However, if a second ground fault becomes present anywhere in the excitation power circuit, the consequences can be disastrous (see Chapters 4 and 5 and Reference [3]). Although some generators include an automatic trip when the field winding develops a ground fault, many do not; most are wired only to alarm. For those that only alarm, monitoring the rotor ground resistance offers operators the opportunity to respond timely, preventing larger damage to the rotor. There is no difficulty in installing monitoring and protection instruments for rotor ground faults in machines with collector rings (sliprings). However, there is a challenge to do so in machines without those rings. In this case, access to the winding can be done via auxiliary rings, if available. Some manufacturers have developed instrumentation for monitoring rotors without directly wiring the rotor to the instrument, but by using telemetry or other form of wireless communication. Figure 3.18 shows such a transmitter.

3.4.5 Shaft Voltage and Current

The existence of voltages building up on the shafts of large electric machines has been a known phenomenon for as long as large rotating

FIGURE 3.17
Rotor flux monitor (RFM) installed over a stator wedge. The output wire is also visible.

FIGURE 3.18
Ground fault monitoring collar installed on the shaft of a brushless generator. (Image Courtesy of Accumetrics, Inc., Schenectady, NY.)

electric machines have been in operation. Also, the existence of ground currents as result of shaft voltages has been known for a long time, including their damaging effects on bearings. Voltages can be electrostatic in nature, or due to electromagnetic induction. References [3,6] have good discussions on the subject. What is interesting about shaft voltages and currents is that there are many possible sources for their existence. Some manufacturers have designed instruments that can record traces of the voltage and provide some direction toward identification of the source. Other instruments measure both shaft current and voltage with the same purpose. Controlling shaft voltages and currents is one very important design and operational task, because if not properly controlled, they can impair the operation of hydrogen seals and bearings to the extent they can lead to their failure. Thus, monitoring shaft voltages and currents provides the operators with an opportunity to respond proactively to any detected degradation. See References [7–9] for some interesting articles on this subject.

3.4.6 Torsional Oscillation

All rotating components in a large turbine-generating unit are susceptible to suffering mechanical damage from resonance induced by the grid, or by other units located in the same power plant. In particular, long blades on low-pressure steam turbines of high ratings are most susceptible to failure due to resonance. Fortunately, a relatively small number of power plants have units exposed to this type of phenomenon to a degree that requires real-time online monitoring for grid-induced torsional oscillations. In those cases where this risk is known, it is advantageous to install dedicated instrumentation for monitoring torsional vibration with the purpose of calculating the remaining fatigue life of the affected components (e.g., turbine blades). Currently, consideration is given to upgrading existing instrumentation used for monitoring other parameters (for instance, stator current), with the aim of also monitoring torsional oscillations. The ubiquitous PSS (power system stabilizer) was invented years ago for the purpose of counteracting one particular type of grid-induced oscillation: subsynchronous resonance (SSR). See Reference [10] for a comprehensive article on SSR.

3.5 Excitation Variables

Brushing is one of the most, if not the most, subtle processes in the generator, and where most of tear and wear occur on a daily basis. Good brushing depends on the environment (contamination), temperature, grade of brushes, current density in brushes, and so on. See the following list:

Conditions affecting brush performance:

- Humidity
- Contamination
- Ambient temperature (too low or too high)—Proper cooling

- Changing/mixing/wrong grade of brushes
- Altitude (barometric pressure)
- Unevenness of collector surface
- Brush pressure
- Brush snugness inside brush-holder
- Insulation condition at bottom of slip-rings
- Poor shunt connections
- Vibration (higher vibration → higher wear)
- Type of excitation (solid-state wears brushes somewhat faster than rotary excitation systems, such as amplidynes)
- The higher the harmonic content of the DC, the faster the wear of the brushes
- Axial misalignment of brushes with collectors

Deterioration of the brushing process can promptly lead to failure of the brushgear, sometimes catastrophic, with consequential loss of field and loss of synchronism. Figure 3.19 shows an example of a brushgear failure.

The condition of the brushes and collectors is only partially monitored by instrumentation; therefore, there is as yet no substitute for frequent visual inspections by electricians and operators.

FIGURE 3.19
Catastrophic damage to brushgear and rotor shaft of a large turbogenerator due to inadequate brushing (brushes not replaced and brushgear not cleaned on a timely basis).

3.5.1 Brushgear Temperature and Condition of Air Filter

Most generators have temperature sensors in the brushgear compartment that are wired to an alarm in the control room. Operator response may necessitate some load reduction, in particular the lowering of lagging reactive power output, but also may require removing the unit from service until the problem is resolved. Temperatures that go higher than normal may be the result of blocked filters. Hence, it is important that the daily electrician or operator rounds include inspection of the filters at the inlet of the cooling air-path to the brushgear. Some large units have pressure sensors that can monitor the condition of the air-filter by sensing the pressure inside the brushgear compartment.

3.5.2 Hydrogen in Brushgear Housing/Compartment

There have been a few recorded instances where hydrogen from the generator leaked into the brushgear compartment via the bore of the shaft exploded injuring workers.

Some stations have installed hydrogen detectors in their brushgear compartment to detect any accumulation of the hydrogen. Actually, the risk for hydrogen buildup in the brushgear housing is higher when the unit is not spinning or is on turning gear, but fully pressurized with hydrogen. This is so because there is no cooling airflow through the brushgear; hence, any hydrogen leak into the brushgear compartment may accumulate thereat. Therefore, it is very important that the H_2 monitors be operational under these conditions.

3.5.3 Water Leaks in the Exciter/Rectifier Cubicle

Large rotating exciters are often cooled by air, itself cooled in heat exchangers inside the exciter and/or rectifiers. In some instances, the conductors themselves are inner-cooled by water (see Figure 3.20). As in the main generator, a water leak inside the exciter or the rectifier holds the potential for causing an internal short-circuit, leading to a loss-of-field event. It is not common for water-cooled rectifiers to have a water leak detection system. The alternative is for operators to monitor the area daily through a transparent-plastic window on their covers. The main rotating exciter, on the other hand, may have a water detection scheme in place, which should be alarmed and monitored as it is done with the main generator. Operator remedial activity depends on the type, location and severity of the leak.

3.5.4 Exciter Temperature

Rotating exciters of large turbogenerators can be, by themselves, relatively large machines. For instance, the exciter of a 1500 MVA generator can be rated between 5 and 6 MW. Normally, totally enclosed and air-cooled, these machines have RTDs and/or TCs installed on their stator windings. Monitoring these temperature probes is as important as monitoring the

FIGURE 3.20
Water-cooled AC–DC busses belonging to a water-cooler rectifier mounted on a large rotating exciter. The photo taken during an outage shows the busses disconnected from the water hoses and the openings protected against foreign material intrusion.

temperature sensors in the main generator, given the lengthy downtime and loss of production in the event of an internal short-circuit. Note: temperature monitoring of steady-state excitation cabinets is no less essential than monitoring rotating exciters, though repair time (replacing a card or solid-state device) necessitates very little time, when compared to the length of a forced outage required for repairing a rotary exciter.

Some older generators can be found excited by an AC motor/DC generator set (MG Set) (Figure 3.21). These machines tend to be much smaller in size than the rotary exciter of a large unit; yet, a winding fault and subsequent repair-time can be costly in lost production. Thus, these units must also be monitored (temperature and vibration), so that operators can take preventive action in case of a deleterious trend of those parameters.

3.5.5 PMG/HFG Voltage Output

Many turbogenerators have a permanent magnet generator (pilot exciter) or high frequency generator feeding the exciter, in lieu of a battery to *flash* the system. The permanent magnets may lose remanence over time, or during an abnormal event, yielding less output voltage. Some operators monitor the output voltage online, while others do it infrequently with hand-held voltmeters. In either case, it is important to keep track of this parameter. Remanence degradation requires planning for re-magnetizing of the permanent magnets, which is a complex activity.

FIGURE 3.21
The photography shows a MG-set providing the excitation current to a turbogenerator. The drive-motor is in this case a squirrel-cage induction machine.

3.6 Hydrogen System Variables

There are too many variations in the implementation of the hydrogen system by different OEMs and even by the same OEM, for a detailed list of those parameters monitored in each one of them to be presented. Hence, the following list only includes those variables most likely to be monitored in a typical design:

- *Hydrogen purity*: The purpose is to reduce windage losses, mitigate the possibility of creating an explosive H_2–air mixture, and avoid the ingression of moisture into the machine. According to some sources, a reduction of purity from 95% to 90% can result in an increase in windage losses of about 33%; therefore, controlling hydrogen purity really *pays off*. It is desirable to keep H_2 purity about 98% or higher, if possible.
- *Dew point*: The purpose of having a low dew point is to keep moisture from condensing onto the internal components of the machine, with a potential for the rusting of *ferritic* retaining rings, and stress-corrosion cracking of 18-5 Mn-Cr *austenitic* retaining rings.
- *Hydrogen pressure*: In hydrogen-cooled generators, maintaining the pressure of this gas at this nominal value is critical in proper cooling of the stator and rotor windings. Therefore, monitoring the hydrogen pressure is a very important activity. A loss of pressure will typically trigger an alarm. Initially, operator response will be trying to raise the

pressure, and if that is not possible, then partially removing load or removing the unit from operation. Chapters 4 and 5 cover this issue.

- *Hot gas and cold gas temperatures of the hydrogen as it enters and leaves the cooler, and their difference*: These three variables are critical parameters to be monitored. They provide information about the cooling of the generator, as well as the condition of the hydrogen heat exchanger, including raw water flow/temperature. Operators can react in a number of ways to an increase in those temperatures, by adjusting the flow of raw water, and/or reducing load.
- *Hydrogen depletion*: Loss of hydrogen from the machine is monitored by measuring its consumption. This provides a bulk measure of the total leak. Other monitored parameters may offer some idea as to the source of the leak; for instance, monitoring H_2 leaking into the stator cooling water system, or into the primary side of the H_2 heat exchangers, or into the brushgear compartment.
- *Hydrogen leaking into primary side of the H_2 coolers*: This type of leak, in addition to a loss of hydrogen, can lead to gas locking of the primary side, reducing the flow of raw water to the cooler and lowering the cooler effectiveness. This condition is difficult to identify as the source of a global temperature increase of *cold gas* and thus, the windings. Gas locking may occur in the form of random transients, which makes the control of the situation most difficult.
- *Hydrogen leak into the SCWS*[4]: There are two key reasons for monitoring hydrogen leaks into the SCWS in machines with stators inner-cooled by water. First, major ingression of H_2 into the SCW side of the coolers can lead to gas locking of parts of the winding with dear consequences to the integrity of the insulation. Second, regardless of being at a lower pressure, water can ooze out into and along the insulation of the leaky bar due to osmosis, also with the potential for serious degradation to its ability to withstand the operational voltage. A high leak rate of H_2 into the SCW can impact both high and low oxygenated systems. In addition to combining with the O_2 and changing the chemistry of the water in unintended ways, it can lead to gas locking. Also, high rates of hydrogen leaks into the SCWS indicates that water may be wetting the insulation of the leaky bar, if the leak is on a conductor bar, or entering the machine if the leak is somewhere else (e.g., a cooler), increasing the dew point and lowering the hydrogen's purity.
- *Hydrogen loss to the seal oil*: It is normal for some amount of hydrogen to dissolve into the seal oil and then be removed (detrained) in the vacuum tank. However, too much of it may indicate that there is a problem with the hydrogen seals and/or seal-oil system.

[4] SCWS—Stator cooling water system.

- *Hydrogen heat-exchanger raw-water flow and pressure*: Controlling the flow of raw (primary) water through the hydrogen coolers is a key factor in maintaining the cold-gas temperature within its recommended range. Pressure and flow go together, thus, normally both are monitored. Operators have a range of options for responding to a degradation of those two parameters, such as changing conditions in the primary water system and/or reducing load.

3.7 Stator Cooling Water System Variables

As with the hydrogen system, there are too many variations in the implementation of the SCW system by different OEMs and even by the same OEM, for a detailed list to be presented of those parameters monitored in each one of them. Hence, the following list only includes the most likely variables to be found in a typical design:

- *Stator cooling water cold and hot (inlet and outlet of the cooler) temperatures*: The purpose of monitoring these temperature points is to ascertain the SCWS cooler(s) is (are) working according to specification. Measuring and displaying the difference between those two readings is also customary in certain stations and certainly very helpful. This monitoring provides an opportunity for the operators to not only respond promptly to deteriorating conditions of the SCWS, but also to identify the source of global temperature increase of the stator winding. Chapters 4 and 5 cover this topic.
- *Stator cooling water (inlet and outlet of the winding) temperatures*: The measurement arrangements differ widely among different designs. In every case, it is important to monitor all TCs and/or RTDs measuring the temperature of the bulk SCW at both ends and of every conducting bar at the outlet end. Close monitoring of these temperatures is key in detecting a deteriorating condition, and in taking adequate steps for mitigating any damage and loss of production. Also, it is a good practice to monitor the difference between temperatures at the inlet and outlet side of the coil, and the difference between diverse bars and/or groups of bars. All these differentials can provide insight into developing issues.
- *Stator cooling water flow rate*: Of similar importance as the SCWS temperatures, changes in the flow rates of the SCW provides clear indication of problems related to the SCWS pumps and motors, valves, major leaks and the possibility of gas-locking and/or other obstructions to the free flow of the water. Continuous monitoring of the SCW flow rate affords operators the opportunity to take timely defensive actions to avoid undue temperature increase of the stator winding.

- *Stator cooling water system pressures*: There is a long list of points along the SCWS where pressure readings convey a picture about the condition of the system. The following list presents the most typical points that can be monitored:
  - At the inlet and outlet manifolds, and then calculating and displaying their difference.
  - Inlet and outlet of coolers and then calculating and displaying their difference.
  - Differential pressure across filters/strainers.
  - Outlet of SCWS pump(s).
  - Pressure differential between the SCW at the *outlet* side of the stator winding and the hydrogen inside the machine.
- *Stator cooling water conductivity*: This parameter is of great importance. Too high conductivity can lead to electric tracking across the insulating hoses at both sides of the winding, with a possible ground trip. It is very important to monitor conductivity of the SCW in machines with alkalizers installed for controlling copper-corrosion buildup. Close monitoring and alarming allows the operators to correct a conductivity excursion by adjusting the operation of the alkalizer if one is present, replacing the resin bed, purging hydrogen, or whatever else is necessary.
- *Stator cooling water pH*: As explained above, the acidity level of the SCW has a direct impact on conductivity and copper corrosion buildup rates; thus, its monitoring provides the operator with the tool to control SCW chemistry. See Reference [3] for further discussion about the specifics of SCW chemistry.
- *Concentration of metals in the SCW, in particular copper and iron*: Monitoring the concentration of copper and iron in dissolved and/or solid form is important for the early detection of degradation of the SCWS and buildup of copper corrosion. In most stations, this is part of the SCW chemistry program for the monitoring of the SCW.
- *Dissolved oxygen content in the SCW*: There are two main designs for SCW systems: High oxygen content and low oxygen content. Anything in between has the potential for creating large amounts of copper corrosion leading to the plugging of the conductor bars, and even the need for a partial or total winding replacement. It is then of great importance to monitor the level of DO[5] in the SCW, and take immediate remedial action if it drifts beyond the limits recommended by the OEM. For further discussion, see [3,11–13].
- *Level of water in the SCWS makeup tank*: Loss of SCW must be automatically compensated by flow of water from the makeup tank to the SCWS. Hence, maintaining proper levels of treated water in the

[5] DO—Dissolved oxygen.

makeup reservoir, and monitoring that this is the case, is another important function of the SCWS-monitoring program, allowing operator timely response in case of a developing problem.

3.8 Rotor Cooling Water System Variables

There are very few generators with water inner-cooled rotors in North America at the time of writing this book, and no additional ones are expected anytime in the future. However there are some others in the rest of the world. Most of these machines have cooling water systems (RCWS) that are separated from the SCWS. Most of the parameters monitored in the SCWS have similar parameters that ought to be monitored in the RCWS, so they will not be repeated herein.

3.9 Seal-Oil System Variables

As with the hydrogen system, there are too many variations in the implementation of the seal-oil system by different OEMs and even by the same OEM, for a detailed list of those parameters monitored in each one of them to be presented. Hence, the following list only includes the most likely to be found in a typical design:

- *Hydrogen seal temperature*: Most important monitored parameter; it gives a direct measure of the temperature of the active part of the hydrogen seals. Maintaining this temperature within its designed-for range is critical for keeping hydrogen from leaking out and air from ingressing the generator, as well as avoiding seal rubs with the shaft and seal degradation.
- *Hydrogen seal-oil inlet and outlet temperatures*: By monitoring the seal-oil temperatures at both the inlet and outlet of the hydrogen seals, operators acquire information about the operation of the seal-oil heat exchanger(s) and the hydrogen seal itself.
- *Seal-oil flowrate*: A key monitored parameter. It gives information about the condition of the seal-oil pump, filter, and the presence of any obstruction in the system.
- *Seal-oil pressures*: There are a number of points along the seal-oil system where monitoring the pressure at those points yields important data about the integrity of the system. Those points are pump outlet pressure, inlet and outlet of the filter (and their difference), difference between the pressure of the hydrogen inside the generator and the seal oil at the seal outlet, seal-oil pressure at the air-side of the seal, and pressure in the vacuum tank.

- *Level of oil in the seal-oil tank*: Monitoring this level provides information about possible loss of sea-oil through leaks, and about having on hand enough reserves of seal oil to accommodate any such loss.
- *Seal-oil cooler primary side*: The hot seal-oil emerging from the hydrogen seals is cooled by flowing through a heat exchanger. Monitoring pressure, flow, and temperature of the primary side (raw water) temperature is important for maintaining proper operating condition of the hydrogen seals.

3.10 Lubrication System Variables

As with the hydrogen, SCW, and seal-oil systems, there are too many variations in the implementation of the lube-oil system by different OEMs and even by the same OEM, for a detailed list of those parameters monitored in each one of them to be presented. Hence, the following list only includes the most likely to be found in a typical design:

- *Bearing temperature*: The purpose of the lube-oil system is to maintain proper flow of cold oil to the bearings, so that they operate as designed for. The key indicator that *all is well* with the operation of the bearings is the temperature of the bearing white metal (Babbitt). Hence, measuring this temperature on both generator bearings provide the operators with a critical indication as to the condition of the bearings. Figure 3.22 shows a typical bearing temperature probe.
- *Lubricating-oil flow rate and temperature*: The Babbitt temperature is maintained within its specified operating range by the lubricating-oil flow rate and temperature; hence, monitoring those two variables allows operators to maintain proper lubrication and ascertain the integrity of the lube-oil system.
- *Lubricating-oil pressure*: Maintaining proper flow rate of the oil requires suitable pressure in the system. There are a number of key points along the system where pressure can be monitored, such as pump outlet, inlet to the bearings and inlet and outlet to the lube-oil filter, and if applicable, vacuum tank.
- *Level of the oil in the lubrication oil tank*: Monitoring this level provides information about possible loss of lube oil through leaks, and about having on hand enough reserves of lube-oil to accommodate any such loss.
- *Lubricating-oil cooler primary side*: The hot lube oil emerging from the bearings is cooled by flowing through a heat exchanger. Monitoring pressure, flow, and temperature of the primary side (raw water) temperature is important for maintaining proper operating condition of the bearings.

FIGURE 3.22
The photograph shows two temperature thermocouples on a sleeve bearing of a turbogenerator.

3.11 Identifying the Common Sources of Monitoring and Protection Variables

In Chapter 2, a discussion was included about the benefit of looking at the intersection between the various protective zones of relays that may alarm and/or trip during a particular event, for the purpose of identifying the source(s) of the problem. In practice, there is great benefit in looking at the intersection of all relevant monitored parameters, among themselves, and with any and all protective-functions zones that were triggered during a given occurrence. Any deviation of a monitored parameter from its normal range is a *symptom* of an abnormal condition or a *malfunction*. Success in troubleshooting a fault or a developing problem is largely facilitated by understanding the links between possible causes of these problems (the *malfunctions*) and their *symptoms*. Each *symptom* listed in Chapter 4 includes a list of *malfunctions* that can express themselves by that and perhaps by other symptoms. The reader should also be aware of similar relationships between protection responses in the form of alarms and/or trips, with the same malfunctions. A protection alarm and/or trip are also a symptom!

For clarification of the concept of intersection between protection zones with monitoring parameters, a few examples are included hereafter.

Case 1

Background: The machine in question is an air-cooled generator. During startup the unit tripped. Post-trip inspection revealed that the AVR malfunctioned, increasing field current well above open-circuit-field-current value, and that the V/Hz relay (24) actuated sending a trip signal to the breaker. The machine tripped at about 70% of nominal speed. Engineers and operators are concerned about any damage to the generator; however no such damage is immediately identified. The unit is returned to service. After a period of several weeks the shaft voltage, which is monitored periodically, is trending upward.

Discussion: Operators and plant engineers are interested in finding the source for the increasing shaft voltage. The shaft voltage increase is the *symptom*. Next, the relationships between the symptom and possible causes are established as shown on Figure 3.23.

Following inspection of Figure 3.23, plant personnel verify the integrity of the shaft grounding components/system, and the lack of shorted turns in the field by monitoring the RFM or by other means, such as checking var-sensitivity of the rotor. The rotor field winding is also checked for the presence of ground faults by inspecting the field ground resistance-measuring instrument, and found it is not grounded. There is also no apparent reason that the electrostatic potential generated on the turbine rotor has increased. Following the tests, the graph in Figure 3.24 is obtained.

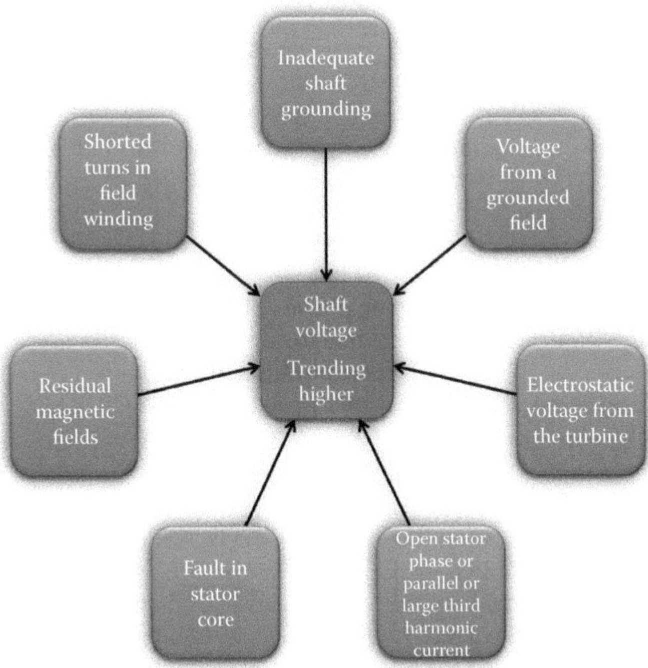

FIGURE 3.23
Graph showing the possible causes (in blue) for an increase in the shaft voltage.

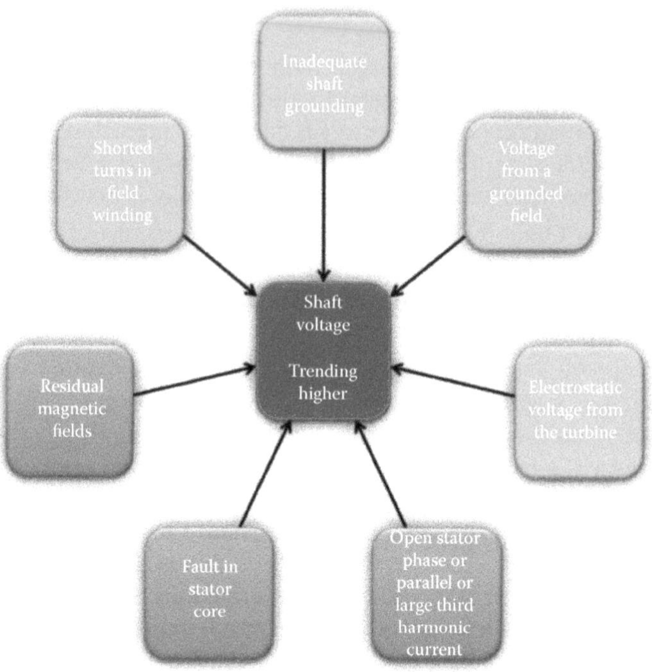

FIGURE 3.24
Graph showing the remaining possible causes for an increase in the shaft voltage (in blue). Causes that can be discounted based on the tests performed post-trip are shown in green.

Third harmonic currents of the magnitude that may cause a shaft voltage increase will most probably have a source that can be readily identified in a large system. It is a very improbable cause. On the other hand, open parallels of phases can, and do happen, but they also result in large negative sequence currents; and in the case of an open phase with the unit loaded, major voltage transients inside the machine always result in large internal faults. In the case of open parallel circuits, differentials in the temperature rise of various groups of conductors will become obvious and negative sequence current content may increase. So the graph in Figure 3.25 can be assumed to be correct.

To analyze the two remaining possible sources for the noted increase in shaft voltage, the relationships for each one are established and shown on Figures 3.26 and 3.27. Plant personnel can go through the list of secondary causation items for both remnant magnetic fields and a faulty stator core, and try to intersect each of the items on the list with the information obtained from the protection tripping the unit. It is relatively easy to visualize the connection between the protected zone/component of the V/Hz relay,

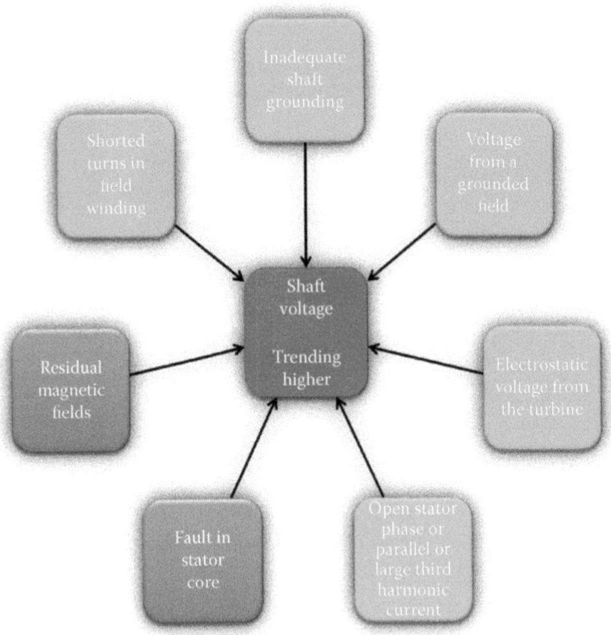

FIGURE 3.25
Graph showing the remaining possible causes for an increase in the shaft voltage (in blue). Causes that can be discounted based on the tests performed post-trip are shown in green.

which is the stator core, and the *overfluxing* item shown as one of the secondary causes for a fault in the stator core in Figure 3.27.

So in this case, station personnel, after a step-by-step cause–effect analysis including mapping the intersection of the available symptoms (the actuation of the #24 relay and the increasing trend in shaft voltage), are able to properly identify the source of the problem: an overfluxing event. This is shown schematically on Figure 3.28.

Case 2

Background: The large machine in question has a water inner-cooled stator winding. The stator winding has a thermocouple installed on each conductor bar measuring the outlet water temperature of the bar. During operation close to full load, one outlet TC started to trend upward. While operators were contemplating the reasons for this behavior, the machine's GCM went into alarm mode. This old GCM has been known to trigger false alarms, so its reliability is suspicious. Finally, hydrogen purity shows a small diminishing trend which is stronger and longer than typical for this unit. Operators and plant personnel must decide what to do about the situation.

Discussion: One possible troubleshooting avenue is to establish a cause–effect relationship between the available symptoms

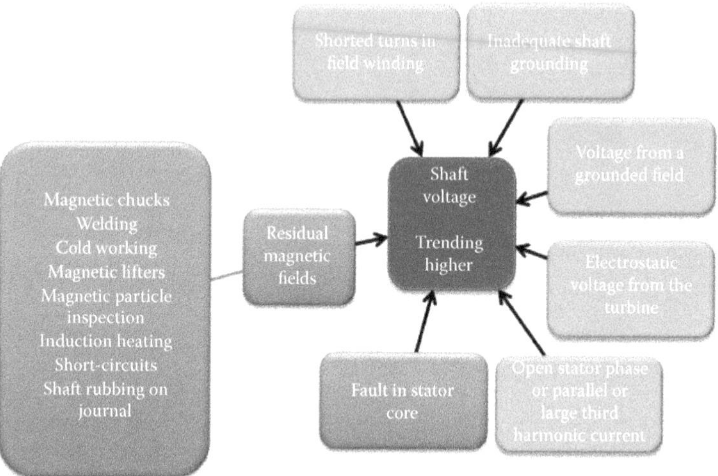

FIGURE 3.26
Graph showing secondary layer of possible causes for the existence of residual magnetic fields. Sources in green were already discounted.

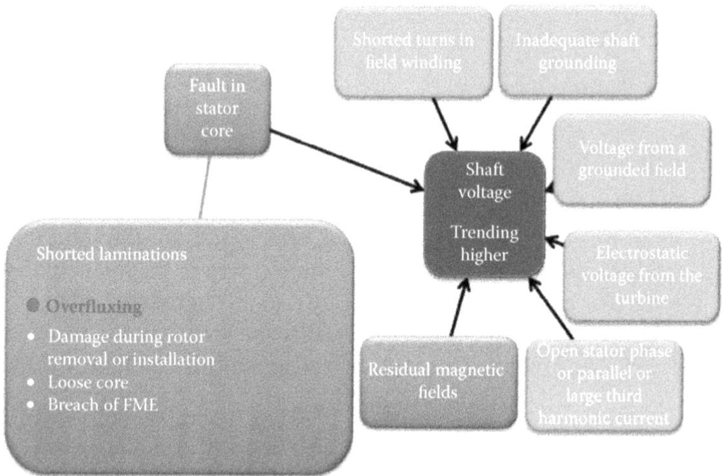

FIGURE 3.27
Graph showing secondary layer of possible causes for the existence of a faulty core.

(GCM alarming and the bar outlet-water temperature trending up) and possible causes. Figure 3.29 shows such a chart for an increase in outlet bar temperature, while Figure 3.30 shows a relationships chart for the GCM going into an alarm state, and Figure 3.31 for trending-down hydrogen purity.

Additional monitored points quickly discount several of the possible malfunctions in those charts. For example, global

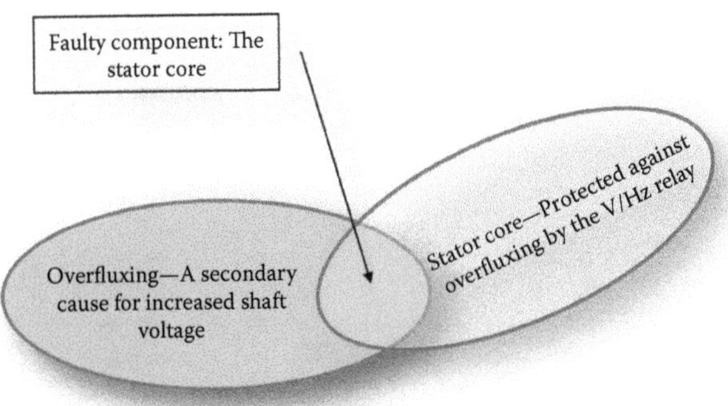

FIGURE 3.28
The intersection between the protection and monitoring symptoms leading to the root-cause of the problem: The overfluxing event.

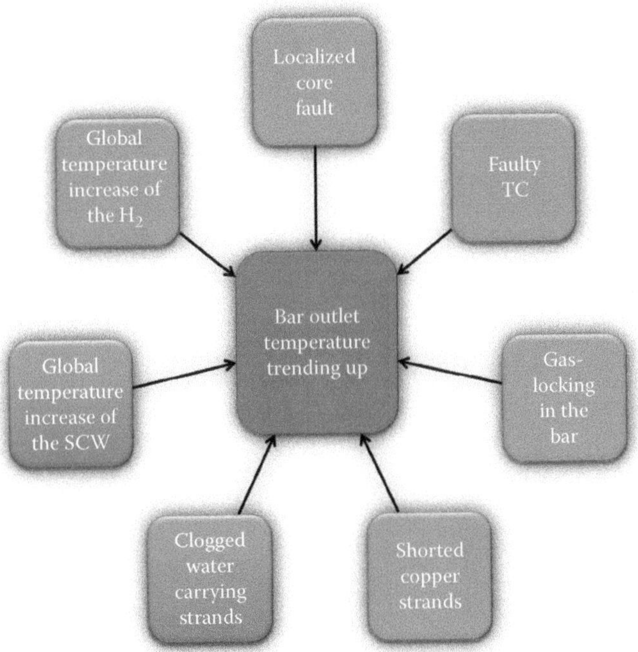

FIGURE 3.29
The chart for the outlet TC shows the relationship between possible malfunctions (in blue) and a common outcome or symptom (red). Each malfunction may lead to additional symptoms.

temperature rise of the hydrogen or SCW can easily be measured in several other monitored points and discarded as not relevant. The condition of the GCM can also be tested to verify it is working as it should. Unfortunately, verifying the integrity of the TC is not feasible, because it is unreachable, being inside the machine. Water in the generator can be checked at the collection point and

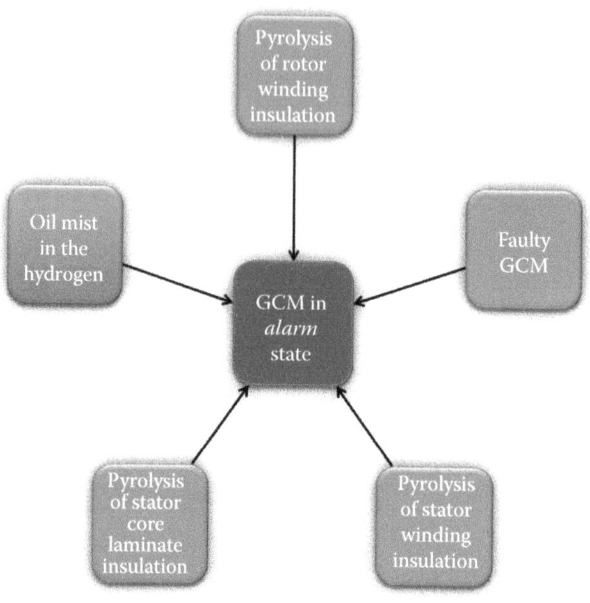

FIGURE 3.30
The chart for the GCM shows the relationship between possible malfunctions (in blue) and a common outcome or symptom (red). Each malfunction may lead to additional symptoms.

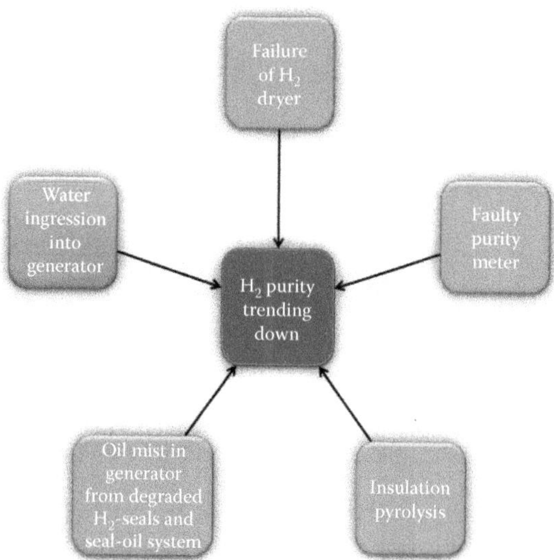

FIGURE 3.31
The chart for the hydrogen-purity trending down, shows the relationship between possible malfunctions (in blue) and a common outcome or symptom (red). Each malfunction may lead to additional symptoms.

discounted as a cause. Failure of the H_2 dryer can also be checked and discounted as a cause. Same for the H_2 purity meter.

So how may one look at this all? First of all, the probability that three different sensors (or the communication loops and annunciators) are all malfunctioning at the same time can be considered low; hence common sense may indicate the presence of a developing problem. How severe the problem and what is the required response, depends on what the problem is. Figure 3.32 shows

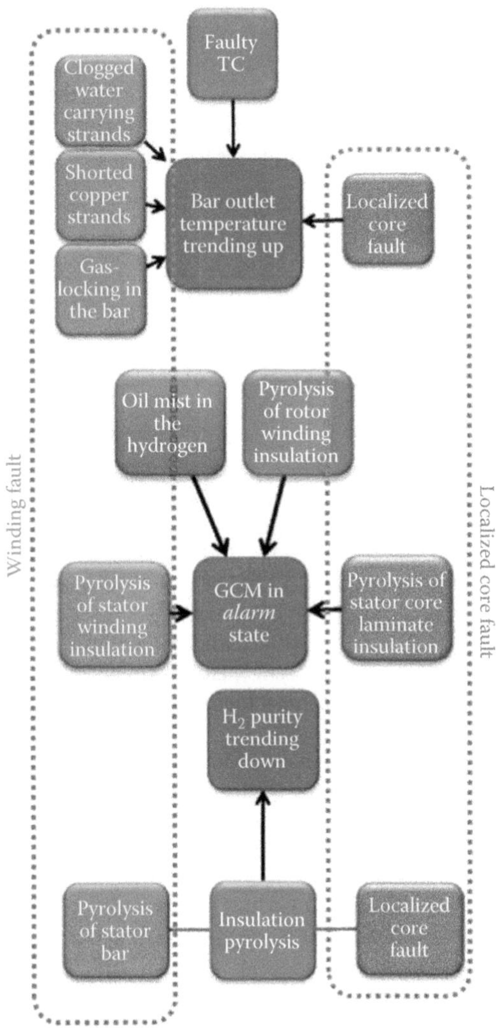

FIGURE 3.32
The figure shows two malfunctions common to all three previous charts. One is related to a conductor-bar fault, and the other to a localized core fault.

what is common to the graphs in Figures 3.29 through 3.31, after those malfunctions that could be discarded are removed. The figure shows that there are two different malfunctions that can be causes for the three symptoms: a fault in a conductor bar and a localized core fault. The common denominator is that both types of faults include pyrolysis of insulation, leading to the same type of symptoms. However, core faults that burn the insulation to this extreme are not load dependent; actually they are mainly voltage dependent; thus, the temperature of the fault will not change as much with load, as a winding fault. Hence, one way operators may try to discriminate between the core and a conductor bar as being the source of the fault, is by changing load. In any case, either core or winding fault, it is obvious that this is a serious condition that requires early removal of the unit from operation, once the presence of a real fault becomes evident.

References

1. IEEE Std 1129-2014. *IEEE Guide for Online Monitoring of Large Synchronous Generators (10 MVA and Above).*
2. IEEE Std C50-13-2014. *IEEE Standard for Cylindrical-Rotor 50 Hz and 60 Hz Synchronous Generators Rated 10 MVA and Above.*
3. G. Klempner and I. Kerszenbaum. *Handbook of Large Turbo-Generator Operation and Maintenance*, 2nd edition, Wiley, Hoboken, NJ, 2008.
4. A.J. Gonzalez, M.S. Baldwin, J. Stein, and N.E. Nilsson. *Monitoring and Diagnosis of Turbine-Driven Generators*, EPRI M&D Center, Prentice Hall, Englewood Cliffs, NJ, 1995.
5. G.S. Stone, E.A. Boulter, I. Culbert and H. Dhirani. *Electrical Insulation for Rotating Machines*, 2nd edition, Wiley, Hoboken, NJ, 2004.
6. EPRI-3002001758. *Turbine-Generator Topics for Plant Engineers: Residual Magnetism*, 2013.
7. P. Nippes, D. David, and A. Peniazev. Monitoring of shaft voltages and grounding current brushes, Presented at the *EPRI Motor and Generator Predictive Maintenance and Refurbishment Conference*, Orlando, FL, November 29, 1995.
8. C. Ammann, K. Reichert, R. Joho, and Z. Posedel. Shaft voltages in generators with static excitation system-problems and solutions, *IEEE Transactions on Energy Conversion*, 3(2), 409–419, June 1998.
9. P.I. Nippes. Magnetism and stray currents in rotating machinery, *Journal of Engineering for Gas Turbines and Power, Transactions of ASME*, 118, 225–228, January 1996.
10. D.H. Baker, G.E. Boukarim, R. D'Aquila, and R.J. Piwko. *Subsynchronous resonance studies and mitigation methods for series capacitor applications, Power Engineering Society Inaugural Conference and Exposition in Africa*, Durban, South Africa, IEEE, 2005.

11. EPRI TR-107137s. *Main Generator On-Line Monitoring and Diagnostics*, December 1996.
12. EPRI TR-1015669. *Turbine Generator Auxiliary System Maintenance Guides—Volume 4: Generator Stator Cooling System*, December 2008.
13. EPRI-TR-1014813v1. *Electrochemical Corrosion Potential (ECP) of Hollow Copper Strands in Water Cooled Generators—Volume 1: Measurements at Craig Station*, March 2007.

4

List of Symptoms

4.1 Definition of Symptom

By *symptom*, it means those signals and signs that are the direct result of a particular component degradation or failure. It may also be the result of an abnormal condition that not necessarily indicates a component failure. An example of such a case is when the grid exhibits unbalanced load conditions. These abnormal conditions may or may not result in a component failure or degradation, depending on its severity, measured in intensity, duration, and recurrence.

This chapter, with 97 symptoms (Table 4.1), contains a comprehensive but not exhaustive list of symptoms that may indicate the presence of one or another developing problem within the generator, defined in this book as *malfunctions*, which are listed in Chapter 5. The list of symptoms is grouped under different major components or systems of the generator, and with a short numerical identifier to help identify the component or system, as follows:

Component of System	Numerical Identifier Key
Core and frame	S-C&F-xx
Stator	S-S-xx
Rotor	S-R-xx
Excitation	S-E-xx
Lube-oil system	S-LOS-xx
Seal-oil system	S-SOS-xx
Hydrogen	S-H-xx
Stator cooling water system	S-SCW-xx

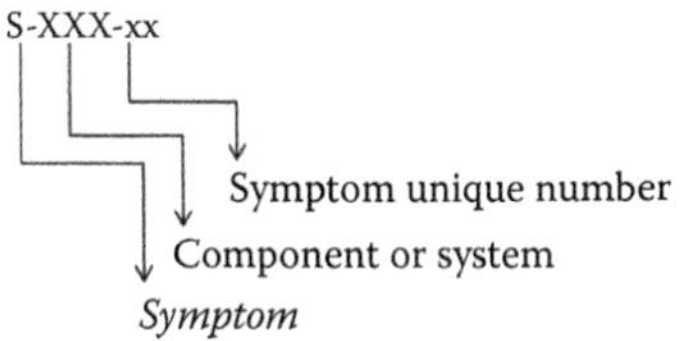

Section 4.2 lists all symptoms along with their corresponding explanation and related malfunctions in a tabulated form. It goes without saying that symptoms may adopt somewhat altered characteristics for different types of machines, and according to available monitors, protection systems, and the like. The knowledge and experience accumulated by the engineers, technicians, and, in particular, the operators of generating units is a major factor in their aptitude for identifying and differentiating between the many symptoms.

Note: Along the lists of symptoms and malfunctions, and in the body of the book, the term *field* is often found alone. *Field* is understood by many as a synonym for *rotor*. So, *rotor winding* may be found in this book as *field winding*. Same for rotor voltage (field voltage) and rotor current (field current).

4.2 List of Symptoms

TABLE 4.1

Global List of Symptoms

Symptom Identifier	Symptoms by System or Component
	CORE AND FRAME
S-C&F-01	GENERATOR CORE-CONDITION MONITOR—ALARM
S-C&F-02	GLOBAL TEMPERATURE OF CORE ABOVE NORMAL
S-C&F-03	HIGH VIBRATION OF CORE AND/OR FRAME
S-C&F-04	LIQUID IN GENERATOR ALARM
S-C&F-05	LOCAL CORE TEMPERATURE ABOVE NORMAL
S-C&F-06	TEMPERATURE OF CORE END ABOVE NORMAL
S-C&F-07	TEMPERATURE OF FLUX SHIELD ABOVE NORMAL
	STATOR
S-S-01	BYPRODUCTS FROM STATOR INSULATION PYROLYSIS
S-S-02	CURRENT IN ONE OR TWO PHASES ABOVE NORMAL
S-S-03	ELECTROMAGNETIC INTERFERENCE (EMI) ACTIVITY HIGH
S-S-04	HYDROGEN IN STATOR COOLING WATER (SCW) HIGH
S-S-05	LARGE DECREASE IN PARTIAL DISCHARGE ACTIVITY (PDA)
S-S-06	NEGATIVE-SEQUENCE CURRENTS ALARM AND/OR TRIP
S-S-07	OUTLET TEMPERATURE OF STATOR CONDUCTOR-BAR COOLANT ABOVE NORMAL (CASE 1)
S-S-08	OUTLET TEMPERATURE OF STATOR CONDUCTOR-BAR COOLANT ABOVE NORMAL (CASE 2)
S-S-09	OUTLET TEMPERATURE OF STATOR CONDUCTOR-BAR COOLANT BELOW NORMAL
S-S-10	OZONE SMELL NEAR AIR-COOLED GENERATORS

(Continued)

TABLE 4.1 (*Continued*)

Global List of Symptoms

Symptom Identifier	Symptoms by System or Component
S-S-11	PARTIAL DISCHARGE ACTIVITY (PDA) HIGH
S-S-12	PHASE CURRENT UNBALANCE
S-S-13	RADIO-FREQUENCY (RF) MONITOR HIGH
S-S-14	SLOT TEMPERATURE OF STATOR CONDUCTOR BAR—ABOVE NORMAL (CASE 1)
S-S-15	SLOT TEMPERATURE OF STATOR CONDUCTOR BAR—ABOVE NORMAL (CASE 2)
S-S-16	STATOR CURRENT ABOVE NOMINAL
S-S-17	STATOR GROUND ALARM AND/OR TRIP
S-S-18	TEMPERATURE OF PHASE CONNECTOR ABOVE NORMAL
S-S-19	TEMPERATURE OF STATOR TERMINAL ABOVE NORMAL
S-S-20	TERMINAL VOLTAGES UNBALANCED
S-S-21	VIBRATION OF STATOR ENDWINDING HIGH
	ROTOR
S-R-01	GROUNDED FIELD-WINDING ALARM AND/OR TRIP
S-R-02	INSULATION RESISTANCE LOW IN WATER-COOLED ROTORS
S-R-03	RESISTANCE OF SHAFT GROUNDING BRUSH—HIGH
S-R-04	SHAFT CURRENT ABOVE NORMAL
S-R-05	SHAFT VOLTAGE ABOVE NORMAL
S-R-06	SHAFT VOLTAGE BELOW NORMAL
S-R-07	SHORTED TURNS IN FIELD WINDING
S-R-08	TEMPERATURE OF BEARING ABOVE NORMAL
S-R-09	TEMPERATURE OF FIELD WINDING ABOVE NORMAL
S-R-10	VIBRATION OF ROTOR—HIGH
	EXCITATION
S-E-01	ANOMALOUS GENERATOR OUTPUT VOLTAGE
S-E-02	ERRATIC BEHAVIOUR OF GENERATOR FIELD TEMPERATURE
S-E-03	FIELD CURRENT TOO HIGH OFFLINE
S-E-04	FIELD CURRENT TOO HIGH ONLINE
S-E-05	FIELD CURRENT TOO LOW ONLINE
S-E-06	HIGH RATE OF BRUSH WEAR
S-E-07	INSUFFICIENT COOLING OF SOLID-STATE EXCITATION CABINETS
S-E-08	MVAR TRANSIENT
S-E-09	MW TRANSIENT
S-E-10	OPERATION WITH LEADING POWER FACTOR BEYOND ALLOWABLE REGION
S-E-11	OVERFLUXING (VOLTS PER HERTZ) TOO HIGH OFFLINE

(*Continued*)

TABLE 4.1 (*Continued*)

Global List of Symptoms

Symptom Identifier	Symptoms by System or Component
S-E-12	PRESSURE DROP ACROSS BRUSHGEAR FILTER—HIGH
S-E-13	ROTOR SPEED BELOW NOMINAL—ONLINE
S-E-14	SPARKING ON SLIPRINGS
S-E-15	TEMPERATURE IN SOLID-STATE EXCITATION CUBICLE—HIGH
S-E-16	TEMPERATURE OF SLIPRINGS/BRUSHGEAR ABOVE NORMAL
	LUBE-OIL SYSTEM (LOS)
S-LOS-01	DIFFERENTIAL PRESSURE ACROSS LUBE-OIL COOLER ABOVE NORMAL
S-LOS-02	FLOW OF COOLING WATER TO LUBE-OIL COOLERS BELOW NORMAL
S-LOS-03	FLOW OF LUBE OIL BELOW NORMAL
S-LOS-04	INLET TEMPERATURE OF LUBE-OIL COOLING WATER HIGH
S-LOS-05	PRESSURE OF LUBE OIL BELOW NORMAL
S-LOS-06	TEMPERATURE OF LUBE OIL AT BEARING OUTLET—HIGH
S-LOS-07	TEMPERATURE OF LUBE-OIL COOLER OUTLET ABOVE NORMAL
	SEAL-OIL SYSTEM (SOS)
S-SOS-01	COOLING-WATER PRESSURE AT INLET TO SEAL-OIL HEAT EXCHANGER—LOW
S-SOS-02	DIFFERENTIAL PRESSURE ACROSS SEAL-OIL FILTER ABOVE NORMAL
S-SOS-03	HEAT-EXCHANGER OUTLET TEMPERATURE OF SEAL OIL ABOVE NORMAL
S-SOS-04	INLET PRESSURE OF SEAL OIL BELOW NORMAL
S-SOS-05	PRESSURE IN SEAL-OIL DETRAINING TANK ABOVE NORMAL
S-SOS-06	PRESSURE OF SEAL OIL ABOVE NORMAL
S-SOS-07	SEAL-OIL COOLING-WATER INLET TEMPERATURE—HIGH
S-SOS-08	TEMPERATURE OF HYDROGEN SEAL ABOVE NORMAL
S-SOS-09	TEMPERATURE OF RAW WATER AT SEAL-OIL COOLER INLET ABOVE NORMAL
S-SOS-10	TEMPERATURE OF SEAL OIL ABOVE NORMAL
	HYDROGEN SYSTEM
S-H-01	COOLING-GAS HEAT-EXCHANGER DIFFERENTIAL TEMPERATURE—HIGH
S-H-02	DEGRADED HYDROGEN PURITY
S-H-03	ELEVATED HYDROGEN DEW POINT
S-H-04	HYDROGEN LEAK INTO THE ISOPHASE BUS (IPB)—ALARM

(*Continued*)

TABLE 4.1 (*Continued*)

Global List of Symptoms

Symptom Identifier	Symptoms by System or Component
S-H-05	HYDROGEN MAKE-UP RATE ABOVE NORMAL
S-H-06	HYDROGEN PRESSURE INSIDE GENERATOR BELOW NORMAL
S-H-07	INADEQUATE FLOW OF COOLING WATER TO AIR or H_2-COOLERS
S-H-08	INLET TEMPERATURE OF AIR OR HYDROGEN COOLING WATER—HIGH
S-H-09	PRESSURE OF HYDROGEN SUPPLY—LOW
S-H-10	TEMPERATURE OF COLD GAS ABOVE NORMAL
	STATOR COOLING WATER (SCW) SYSTEM
S-SCW-01	BLOCKAGE IN THE FLOW OF STATOR COOLING WATER (SCW)
S-SCW-02	BLOCKAGE OF STATOR COOLING WATER (SCW) HEAT EXCHANGER
S-SCW-03	CONDUCTIVITY OF STATOR COOLING WATER ABOVE NORMAL
S-SCW-04	COOLER OUTLET TEMPERATURE OF SCW ABOVE NORMAL
S-SCW-05	DIFFERENTIAL PRESSURE ACROSS SCW HEAT EXCHANGER ABOVE NORMAL
S-SCW-06	DIFFERENTIAL PRESSURE OF SCW ACROSS STATOR WINDING ABOVE NORMAL
S-SCW-07	DIFFERENTIAL PRESSURE OF SCW ACROSS STATOR WINDING BELOW NORMAL
S-SCW-08	INLET PRESSURE TO PRIMARY SIDE OF SCW HEAT EXCHANGER BELOW NORMAL
S-SCW-09	LOW LEVEL OF SCW IN THE H_2-DETRAINING TANK
S-SCW-10	LOW LEVEL OF SCW IN THE MAKE-UP TANK
S-SCW-11	PRIMARY WATER FLOW SCW COOLERS BELOW NORMAL
S-SCW-12	STATOR COOLING WATER FLOW BELOW NORMAL
S-SCW-13	STATOR COOLING WATER OUTLET PRESSURE BELOW NORMAL
S-SCW-14	STATOR COOLING WATER PRESSURE DROP ACROSS SCW FILTER/STRAINER ABOVE NORMAL
S-SCW-15	TEMPERATURE OF GLOBAL OUTLET STATOR COOLING WATER ABOVE NORMAL
S-SCW-16	TEMPERATURE OF PRIMARY (RAW) WATER AT INLET OF COOLERS ABOVE NORMAL

Table 4.2 includes a list of symptoms not specifically addressed in this book that operators may encounter. In general, the malfunctions in Chapter 5 include a few that also cover some of these additional symptoms. In other cases, it will require the operator to follow a similar type of analysis as carried out for symptoms described in Table 4.1.

TABLE 4.2

List of Symptoms Not Specifically Covered in the Book

LUBE-OIL SYSTEM (LOS)
Differential pressure across lube-oil filter high
Oil level in main lube-oil tank too low or too high
Lube-oil pump activated
SEAL-OIL SYSTEM (SOS)
Differential pressure between seal-oil hydrogen too high or too low (regulator problem?)
Oil level in main seal-oil tank too low or too high
Seal-oil pump activated
IF LOS AND SOS SHARE A COMMON OIL TANK
Oil level in the common tank too low or too high
HYDROGEN SYSTEM
Hydrogen purge alarm (usually following a SOS problem, fire, etc.)
Hydrogen dryer malfunction alarms
STATOR COOLING WATER SYSTEM (SCWS)
Chemistry issues such as deviation in the concentration of dissolved oxygen (DO), presence of copper ions and other metals, and electrochemical levels.

4.2.1 Core and Frame Symptoms

S-C&F-01 GENERATOR CORE/ CONDITION MONITOR—ALARM

APPLICABILITY:

- Air-cooled generators
- Hydrogen-cooled generators

DESCRIPTION:

Many large hydrogen-cooled generators and some air-cooled ones have core monitors also known as generator condition monitors or GCMs. These instruments continuously sample the air or hydrogen in the machine looking for particles (pyrolysis byproducts) that have become free as a result of the overheating (burning) of insulation, or electrical arcing in the presence of insulation. If such a condition is detected with levels above a given threshold, the GCM will be triggered into an alarm state. However, it is not uncommon for GCMs to generate, on occasion, spurious alarms. Thus, operators are required to validate the alarm to ensure that it is not the result of a malfunction of the instrument (Figure 4.1).

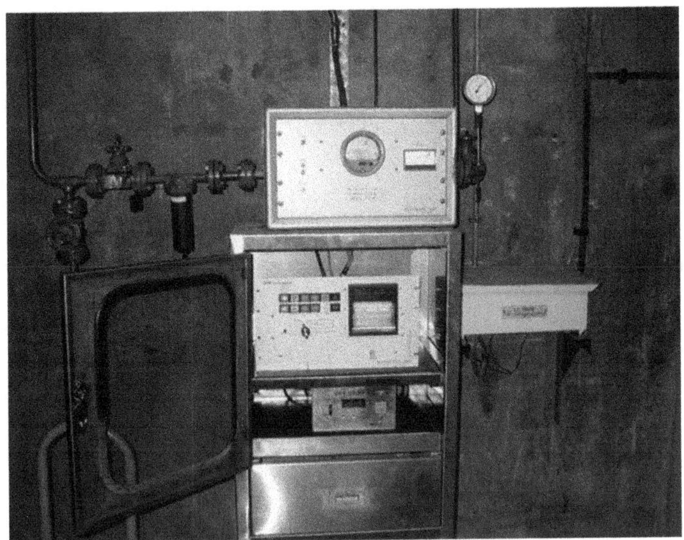

FIGURE 4.1
Generator core (condition) monitor.

Modern GCMs have a manual switch that allows the air or hydrogen to bypass the test chamber, so that the instrument's integrity can be ascertained following an alarm. Additionally, GCMs are equipped with a sampling device so that the particles collected can be analyzed. However, identification of the source of pyrolysis byproducts is an overwhelming challenge because there are so many different organic materials inside a generator. To help with this challenge and assist in making a positive identification of the source, some machines that have GCMs also are equipped with *tagging* compounds. Tagging compounds are made of known organic material compounds mixed with particular color epoxy paints, and applied to specific areas of the generator. Therefore, if an alarm is triggered, a sample of the air or hydrogen can be taken and analyzed. Depending on the tagging compound detected, identification of the area of burning may be achieved. The following areas are typically those inside the generator on which tagging compounds are applied (Figure 4.2):

- Stator core
- Turbine end—stator endwinding
- Collector end—stator endwinding
- Stator winding phase connectors
- Stator terminals
- Rotor body

RELATED MALFUNCTIONS	NUMBER
ARCING BETWEEN CORE AND KEYBARS	M-C&F-01
BLOCKAGE OF PARALLEL-PHASE CONNECTOR COOLING PATH	M-S-03
BLOCKAGE OF PHASE-CONNECTOR COOLING PATH	M-S-05
BLOCKAGE OF PHASE-COOLANT FLOW	M-S-04
BLOCKED COOLING VENT IN STATOR CORE	M-C&F-02
BURNING/PYROLYSIS OF CORE INSULATION	M-C&F-05
BURNING OF FLUX-SHIELD INSULATION	M-C&F-04
BURNING/PYROLYSIS OF ROTOR INSULATION	M-R-04
BURNING/PYROLYSIS OF STATOR WINDING INSULATION	M-S-08
BYPRODUCTS FROM BURNING INSULATION	M-C&F-06
GAS LOCKING OF PHASE COOLANT	M-S-22
INSUFFICIENT FLOW OF PHASE-PARALLEL COOLANT	M-S-34
INSUFFICIENT FLOW OF TERMINAL COOLANT	M-S-35
LOOSE CORE LAMINATIONS	M-C&F-16
TEMPERATURE OF A PARALLEL CIRCUIT ABOVE NORMAL	M-S-51

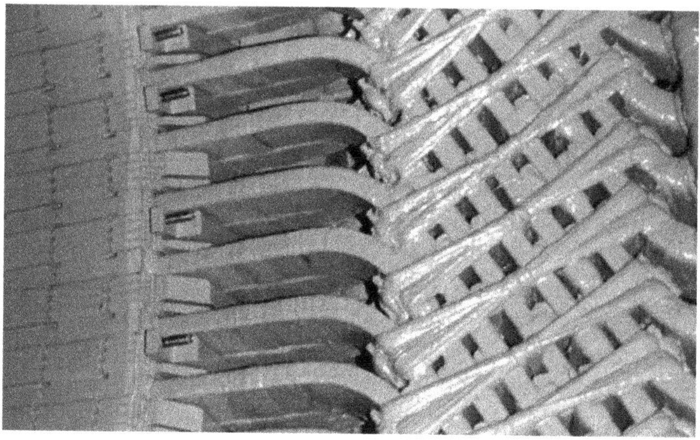

FIGURE 4.2
Generator stator core and endwinding coated with tagging compounds.

S-C&F-02 GLOBAL TEMPERATURE OF CORE ABOVE NORMAL

APPLICABILITY:

All generators

DESCRIPTION:

This symptom applies when the entire stator core is exhibiting temperatures above *normal*. By *normal*, it means temperatures that have previously been measured under similar terminal voltage and load conditions, or as specified by the OEM.

The most common reasons a stator core starts to overheat are listed below.

RELATED MALFUNCTIONS	NUMBER
BLOCKAGE IN THE FLOW OF STATOR COOLING WATER (SCW)	M-SCW-01
BLOCKED COOLING VENT IN STATOR CORE	M-C&F-02
GLOBAL TEMPERATURE OF CORE ABOVE NORMAL	M-C&F-09
HYDROGEN PRESSURE INSIDE THE GENERATOR BELOW NORMAL	M-H-09
INSUFFICIENT PERFORMANCE OF AIR OR HYDROGEN HEAT EXCHANGERS	M-H-12
OVERFLUXING OFFLINE	M-E-14
OVERFLUXING ONLINE	M-E-15

S-C&F-03 HIGH VIBRATION OF CORE AND/OR FRAME

APPLICABILITY:

All generators

DESCRIPTION:

This symptom applies when the vibration of the core and/or the frame of the machine are high, or the core and frame are vibrating out of phase with each other. Figure 4.3 shows a number of accelerometers installed on a core and its frame.

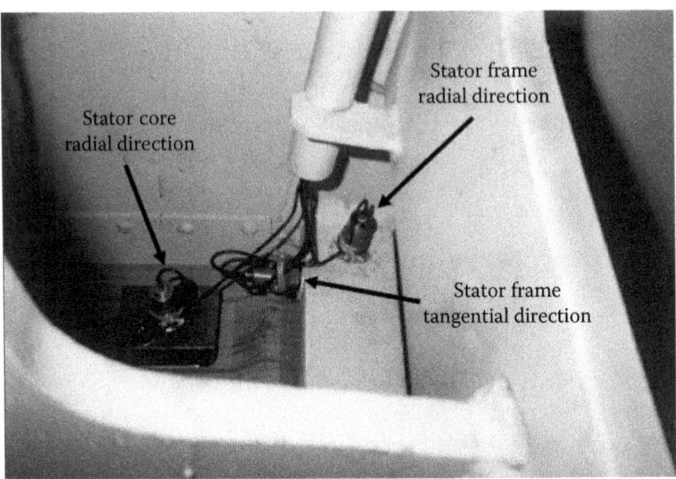

FIGURE 4.3
Core and frame accelerometers installation on a large turbogenerator. These are older types of sensors. Modern sensors are based on fiber-optic technology, making them lighter and with nonmetallic wiring.

It is not common practice to install vibration probes on the core or frame of a generator, unless there is some particular known vibration issue with the machine which makes it prudent to monitor vibration. However, when monitoring is carried out, both core and frame measurements should be taken at the same location(s), close to each other. The most common method is to install accelerometers on the back of the core, and accelerometers on the frame, next to each other. As well, the most common practice is to install both radial and tangential probes at each location.

A *high* vibration would generally be of a magnitude higher than values provided by the OEM, or Industry-accepted criteria. *High* can also be taken as higher than vibration normally measured in past occasions at the same position, under similar load conditions.

RELATED MALFUNCTIONS	NUMBER
HIGH VIBRATION OF CORE AND FRAME	M-C&F-10
LOOSE CORE LAMINATIONS	M-C&F-16
LOOSE GENERATOR FOOTING	M-C&F-17
SHORTED TURNS IN FIELD WINDING	M-R-24
THERMAL SENSITIVITY TEST	M-R-29
VIBRATION OF ROTOR—HIGH	M-R-30
VIBRATION OF STATOR ENDWINDING—HIGH	M-S-56
In machines with more than one hydrogen cooler, unbalance between the temperature of the cold gas in each cooler	N/A

S-C&F-04 LIQUID IN GENERATOR ALARM

APPLICABILITY:

All generators

DESCRIPTION:

This symptom applies when a *liquid in generator* alarm is received. See Figure 4.4 for a typical alarmed visual liquid detector.

Liquids inside the generator can be

- Raw service water from hydrogen or air coolers
- Oil from bearings
- Seal oil from hydrogen seals (in hydrogen-cooled machines)
- Stator cooling water (SCW; in machines with water inner-cooled stators)
- Stator cooling oil (in machines with oil inner-cooled stators)
- Rotor cooling water (in machines with water-cooled rotors)

FIGURE 4.4
Generator liquid level detector shown mounted against the wall, one deck below the turbine deck.

It is almost a universal feature for generators to have a system for collecting liquids spilled inside the machine, and to alarm if certain level is reached. The detection systems do not provide information about the nature of the liquid in the machine, only that there is liquid inside the generator. A sample must be taken and analyzed to determine what liquid it is.

RELATED MALFUNCTIONS	NUMBER
FAILURE OF OIL WIPER	M-R-12
INGRESSION OF OIL INTO GENERATOR	M-R-16
LEAK IN THE FLOW PATH OF STATOR COOLING WATER (SCW)	M-SCW-09
LEAK OF CONDUCTOR-BAR COOLING WATER INTO GENERATOR	M-S-36
LEAK OF HYDROGEN INTO H_2-COOLING WATER SYSTEM	M-H-13
LEAK OF STATOR COOLING WATER (SCW) FROM A FLUX SHIELD	M-C&F-14
LEAK OF STATOR COOLING WATER (SCW) INTO GENERATOR	M-S-37
LEAK OF STATOR COOLING WATER (SCW) INTO GENERATOR FROM TERMINALS	M-S-38
MISALIGNMENT OF HYDROGEN SEAL	M-R-19
MOISTURE IN HYDROGEN—HIGH	M-H-14
OIL & WATER LEAKS INSIDE THE GENERATOR	M-C&F-18
PHYSICAL BREACH OF HYDROGEN OR AIR HEAT EXCHANGERS	M-H-15
WATER LEAK INTO GENERATOR FROM AIR OR HYDROGEN COOLER	M-H-18

S-C&F-05 LOCAL CORE TEMPERATURE ABOVE NORMAL

APPLICABILITY:

All generators

Generator condition:

- Generator synchronized to the grid, or
- At rated speed, not synchronized, excitation ON, and voltage present at the terminals of the generator (i.e., open circuit)

This symptom applies when the temperature measured in a given spot of the stator core is above *normal.*

DESCRIPTION:

The stator core of a generator only generates heat when it is magnetized (secondary windage losses in core passages are neglected). Therefore, only when there is voltage at the machine's terminals, the core temperature may be higher than the temperature of the cooling air or hydrogen in the casing. By *higher than normal,* it means any temperatures higher than the maximum recommended values provided by the OEM, or on the rest of the core in identical locations (axial and radial), or as measured in the same core location in previous occasions under similar load conditions.

RELATED MALFUNCTIONS	NUMBER
LOCAL CORE FAULT	M-C&F-15
OVERFLUXING OFFLINE	M-E-14
OVERFLUXING ONLINE	M-E-15
TEMPERATURE OF A LOCAL CORE AREA ABOVE NORMAL	M-C&F-20

S-C&F-06 TEMPERATURE OF CORE END ABOVE NORMAL

APPLICABILITY:

All generators

DESCRIPTION:

In large units there is a number of temperature sensors (mainly resistance temperature detectors [RTDs]) embedded in the core, at various locations along the length of the stator and around its circumference. Furthermore, these sensors may be located in the stator core teeth, an inch or so below slot bottom, and further back in the yoke of the stator iron.

The temperatures measured will vary with load and power factor. Core-end regions tend, under leading power factors, to be hotter than the center part of the core. Therefore, comparisons must be made with readings taken under similar load conditions. The probes located the core end may also read somewhat different temperature values because all of these are not exactly at the same distance from the core end. However, a trend can easily be recognized by looking at them as a whole or by looking at each one individually.

This symptom applies whenever any of the core-end probes, or all together, show they are measuring a higher than *normal* temperature. By *normal*, it means a range of allowable temperatures (per OEM), or temperatures that were historically experienced in the same generator in previous occasions, under similar load conditions.

RELATED MALFUNCTIONS	NUMBER
FIELD CURRENT TOO LOW ONLINE	M-E-06
LOCAL CORE FAULT	M-C&F-15
OPERATION OUTSIDE CAPABILITY CURVE—ONLINE	M-E-11
OVERFLUXING OFFLINE	M-E-14
OVERFLUXING ONLINE	M-E-15
TEMPERATURE OF A LOCAL CORE AREA ABOVE NORMAL	M-C&F-20
TEMPERATURE OF CORE END ABOVE NORMAL	M-C&F-21

C-C&F-07 TEMPERATURE OF FLUX SHIELD ABOVE NORMAL

APPLICABILITY:

Larger generators in the 400 MW and above range (basically machines with water-cooled or hydrogen inner-cooled stator windings as a rough guide)

DESCRIPTION:

Large cores may have flux screens (Figure 4.5) protecting their stator core ends from axial stray flux—mainly during leading power factor operation. These screens (shields) may or may not be directly cooled by SCW. The large majority however, are usually hydrogen cooled only. Water-cooled flux screens are vendor specific and apparently not made anymore.

Some manufacturers use a flux shunt (Figure 4.6) rather than a copper shield on their larger machine designs. A shunt is basically an arrangement of additional stepped core iron at the end of the stator core. Smaller machines of the air-cooled or indirectly cooled H_2 type do not have enough stray flux to warrant using flux shielding.

In most cases, temperature sensors (usually embedded thermocouples) are installed on the shields transmitting information about their temperature. The location of the TCs is also generally on the *nose* of the flux shield, nearest to the inner diameter of the core end. This is the hottest

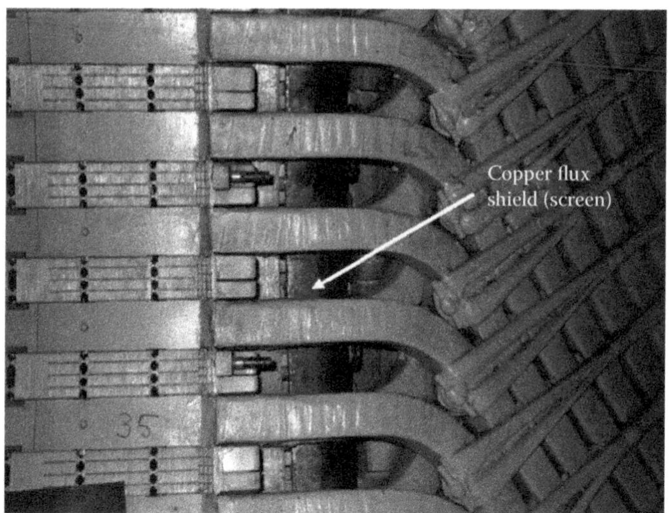

FIGURE 4.5
Copper flux shield shown at the end of the stator core.

FIGURE 4.6
Laminated flux shunt installed at the end of the stator core.

area of the flux shield due to the axial stray flux components from both the stator endwinding and the rotor endwinding.

By *normal*, it means temperatures typical for this machine and previously measured under similar load, terminal voltage, and ambient conditions. The temperature limit is usually selected as the same temperature as the stator core temperature limit and this is usually Class B, 130°C.

RELATED MALFUNCTIONS	NUMBER
BREACH OF THE FLUX-SHIELD COOLING PATH	M-C&F-03
FIELD CURRENT TOO LOW ONLINE	M-E-06
GAS LOCKING IN FLUX-SHIELD COOLING PATH	M-C&F-08
HYDROGEN PRESSURE INSIDE GENERATOR BELOW NORMAL	M-H-09
INSUFFICENT PERFORMANCE WITH AIR OR HYDROGEN HEAT EXCHANGERS	M-H-12
OPERATION OUTSIDE CAPABILITY CURVE—ONLINE	M-E-11
TEMPERATURE OF FLUX SHIELD ABOVE NORMAL	M-C&F-22

4.2.2 Stator Symptoms

S-S-01 BYPRODUCTS FROM STATOR INSULATION PYROLYSIS

APPLICABILITY:

All generators

DESCRIPTION:

The byproducts from stator insulation pyrolysis are basically the stuff released from burned organic materials inside the generator, that is, the insulation and packing blocks, and so on. These may be from the core insulation, stator or rotor windings, or even thermocouple or RTD insulation. The byproducts can be collected via a sampling gadget inside the GCM. Automatic identification of these materials is not a realistic option and so, the GCM only gives an alarm indicating that some organic material is/has burned inside the machine. To make the GCM more useful, *tagging* compounds were developed that can basically be painted on the organic surfaces of interest. When burning occurs and a GCM alarm comes in, a sample is taken and analyzed for these specific tagging compounds. That way, the area of burning can be identified before the machine is opened and a more focused inspection can be carried out.

RELATED MALFUNCTIONS	NUMBER
BURNING/PYROLYSIS OF STATOR WINDING INSULATION	M-S-08
BURNING/PYROLYSIS OF ROTOR INSULATION	M-R-04

S-S-02 CURRENT IN ONE OR TWO PHASES ABOVE NOMINAL

APPLICABILITY:

All generators when energized and synchronized to the system (grid)

DESCRIPTION:

Every generator is designed for a maximum (*nominal* or *rated*) stator current value which should not be exceeded during operation. A machine may have more than one rated value; for instance, one for each H_2 pressure shown on the capability curve(s). However, once the hydrogen pressure is set, the applicable nominal stator current value must not be exceeded.

A generator may undergo sudden and/or transient current overloads during grid disturbances. There is little or nothing the operators can do during a transient, given its short duration in the vast majority of cases. Depending on the severity of the overload, a relatively short occurrence can have significant negative lasting effects on a number of major components.

The situation covered under this symptom applies when one or two phases exhibit stator current values above the maximum allowable. Depending on the difference between the phase currents, this may trigger a *negative-sequence current* event that affects the rotor. Depending on the length of time and the magnitude of the unbalanced stator currents, significant damage can be sustained by the rotor. If all of the stator currents were to exceed rated values, this would be a simple overload condition.

Negative-sequence currents may also reach limiting values for a particular generator even if the currents in all three phases are below their maximum ratings. In any case, negative-sequence current events that encroach the withstand capability limits of a generator will trigger a protection alarm and if persisting, a trip (protective function #46). Negative-sequence current limits are set based on the maximum rated stator current (the withstand capability for continuous and transient negative-sequence currents based on the maximum rating of the machine).

RELATED MALFUNCTIONS	NUMBER
CURRENT UNBALANCE BETWEEN PHASE PARALLELS	M-S-13
CURRENT UNBALANCE BETWEEN PHASES	M-S-12
NEGATIVE-SEQUENCE CURRENTS	M-S-42
PHASE CURRENT HIGH	M-S-45
SUDDEN LARGE CHANGE IN GRID IMPEDANCE	M-E-18
TEMPERATURE OF A PARALLEL CIRCUIT ABOVE NORMAL	M-S-51
VOLTAGE OF GRID—UNBALANCED	M-S-57

S-S-03 ELECTROMAGNETIC INTERFERENCE (EMI) ACTIVITY HIGH

APPLICABILITY:

All generators

DESCRIPTION:

Electromagnetic interference (EMI) and radio frequency (RF) are both different names given to electromagnetic signals that can be picked up and measured. This is done by running the generator neutral cable that goes from the generator neutral point to the neutral grounding transformer,

FIGURE 4.7
Radio-frequency current transformer (RFCT) shown installed on a grounding cable.

through a *radio-frequency*-capable CT (current transformer) called a RFCT (radio-frequency current transformer—see Figure 4.7). The RFCT is then connected to an oscilloscope or spectrum analyzer, to measure the frequencies spectrum of the partial discharge activity embedded in the RF/EMI signal, as well as other electric discharges, such as sparking and arcing. A band-pass filter of about 1 MHz is also generally used to filter the output signal of the RFCT. Because of this wide frequency range and the location of the RFCT, this method of discharge activity monitoring generally detects all sources of RF and EMI activity. Further, this makes the RF/EMI measurements not only sensitive to partial discharge activity (PDA) in the winding, but to all of the sources of electromagnetic *noise* associated with the generator. The down side of this technique is that unless very severe discharging is occurring, it is difficult for nonspecialists to separate the PDA of interest, from the noise.

As shown in Figure 4.7, the RFCT also can be installed on a generator grounding cable, in lieu of the neutral; however, with some lesser efficacy in picking up generator internal signals.

Also refer to the symptom: RADIO-FREQUENCY (RF) MONITOR HIGH (S-S-13).

S-S-04 HYDROGEN IN STATOR COOLING WATER (SCW) HIGH

APPLICABILITY:

Stators with inner water-cooled stator windings

DESCRIPTION:

In machines with stators windings directly cooled by water, collection/test methods exist that they are aimed at measuring the rate of hydrogen

FIGURE 4.8
An indigenous stator hydrogen-in-cooling water detector made by a large utility for its own fleet of generators. Commercial instruments also exist for the same purpose.

ingress into the SCW system. In the case of some units, the measurements are done online (Figure 4.8), while in others they require special tests. In all cases, the OEMs supply information about maximum levels of hydrogen-into-stator water leakage rates. The concern is that excessive hydrogen leakage into the SCW can cause gas locking, which is a form of flow blockage. If this occurs, then the cooling of the affected stator bar(s) will be lost. At that point the bar(s) may overheat and fail. In many cases, stations develop guidelines for operator action when the hydrogen leak rate into SCW is approaching or exceeding high levels.

RELATED MALFUNCTIONS	NUMBER
BREACH OF THE FLUX-SHIELD COOLING PATH	M-C&F-03
BREACH OF THE STRANDS OR TUBES CARRYING COOLING WATER	M-S-06
CRACKED CONDUCTOR STRANDS	M-S-10
DAMAGE TO CONDUCTOR BAR INSULATION	M-S-15
DEGRADED/RUPTURED "O" RINGS OR OTHER SCW PIPING CONNECTIONS	M-SCW-07
GAS LOCKING IN FLUX-SHIELD COOLING PATH	M-C&F-08
GAS LOCKING IN WATER-CARRYING STRANDS OR TUBES	M-S-21
HYDROGEN ESCAPING THE GENERATOR	M-H-08
HYDROGEN INGRESSION INTO STATOR COOLING WATER	M-S-29

(Continued)

RELATED MALFUNCTIONS	NUMBER
HYDROGEN INGRESSION INTO STATOR COOLING WATER FROM STATOR TERMINALS	M-S-30
HYDROGEN INGRESSION INTO STATOR COOLING WATER (SCW) FROM FLUX SHIELDS	M-C&F-11
INSUFFICIENT FLOW OF COOLING MEDIUM IN A STATOR BAR	M-S-32
INSUFFICIENT FLOW OF PHASE COOLANT	M-S-33
LEAK OF HYDROGEN INTO H_2-COOLING WATER SYSTEM	M-H-13
LEAKY SCW HOSES	M-SCW-11
RUPTURED WATER MANIFOLD (HEADER)	M-S-47
RUPTURED WATER PARALLEL MANIFOLD (HEADER)	M-S-48
STATOR COOLING WATER (SCW) LEAK INTO GENERATOR FROM MANIFOLDS	M-S-50
STATOR COOLING WATER (SCW) LEAK INTO GENERATOR FROM PHASE PARALLEL	M-S-49

S-S-05 LARGE DECREASE IN PARTIAL DISCHARGE ACTIVITY (PDA)

APPLICABILITY:

All generators

DESCRIPTION:

This symptom relates to the few instances when suddenly or over a short time the PDA decreases significantly inside a generator experiencing elevated partial discharge activity, in particular if the PDA was trending upward. This can be the result large arcing replacing the low intensity PDA, indicating a possible impending failure of the stator insulation.

PDA can be measured online on real-time by a number of devices. Figure 4.7 shows a RFCT installed on a grounding cable. Other location for the installation of the RFCT is the neutral of the main generator, before the grounding transformer. Another device that picks up PDA is a slot coupler, as shown in Figure 4.9.

RELATED MALFUNCTIONS	NUMBER
DAMAGE TO CONDUCTOR BAR INSULATION	M-S-15
DAMAGE TO CONDUCTOR BAR INSULATION DUE TO A WATER LEAK	M-S-16
DEFECTIVE STATOR WINDING INSULATION	M-S-18
GROUND FAULT OF STATOR WINDING	M-S-26

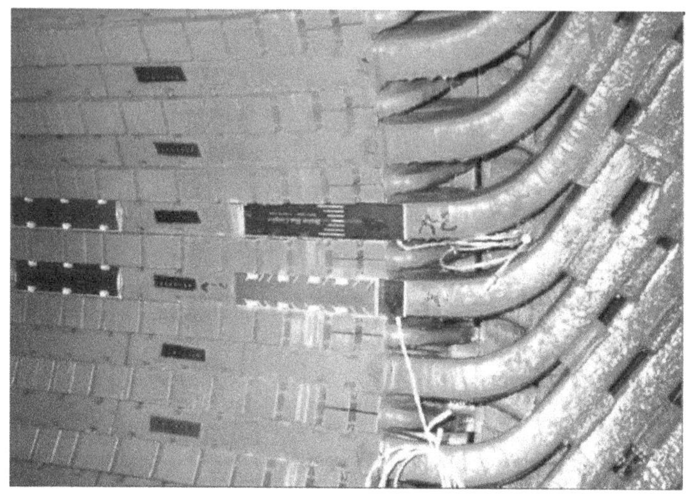

FIGURE 4.9
A slot coupler for measuring online partial discharge activity on a section of the stator winding.

S-S-06 NEGATIVE-SEQUENCE CURRENTS ALARM AND/OR TRIP

APPLICABILITY:

All generators

DESCRIPTION:

Negative-sequence events generally refer to negative-sequence currents on the generator rotor induced by an unbalance in the three stator phase currents. These currents flow on the surface of the rotor, in the forging and in the wedges, representing skin effect currents that occur at twice rated fundamental frequency. The unbalanced current condition may be due to a generator or grid problem, but it is often grid-related rather than a problem in the generator. Large unbalanced stator currents also can be generated due to a major short circuit on the step-up or auxiliary unit transformers, on in the switchyard.

There are two components of negative-sequence to consider. The first is the continuous negative-sequence current (I_2) component, which refers to the amount of phase unbalance the generator can tolerate for an infinite operating period. The second is the transient component called *I22t*, which refers to the level of transient negative-sequence current that the generator can withstand.

For large steam turbine generators (T-Gs) up to 350 MVA, a typical continuous I_2 value of 8% of rated stator current would be normal. This means that the generator could carry a continuous phase unbalance in the stator winding yielding a negative-sequence current component of 8% or 0.08 p.u. of the rated stator current, without damaging any of the generator components, specifically the rotor. The value of 8% is smaller for larger units (per IEEE C50.13-2005 or later). A typical transient value for I_2 (the *I22t* component) would be 10. This means that the generator could withstand 100% or 1 p.u. negative-sequence current for 10 s (i.e., 1 p.u. negative-sequence current times 10 s equals an *I22t* value of 10).

The actual values of continuous and transient negative-sequence current withstand capabilities depend on the rating of the generator and the type of cooling of the rotor, and they can be found in the IEEE C50.13 standard. Figure 4.10 shows an example of the withstand limits expressed in graphical form.

There is always a small natural degree of unbalance in the three phase currents, but it is not harmful below the continuous I_2 value. When the degree of unbalance becomes significant, it appears as 120 (100) Hz (i.e., the twice per fundamental frequency) currents flowing on the surface of the rotor body and wedges, which can overheat the rotor forging and/or wedges. One of the symptoms may be high vibration of the rotor due to asymmetric heating of the forging. It can lead to component degradation and eventual failure should serious overheating occur. Relay protection is generally provided to detect the level of negative-sequence currents and initiate a generator trip.

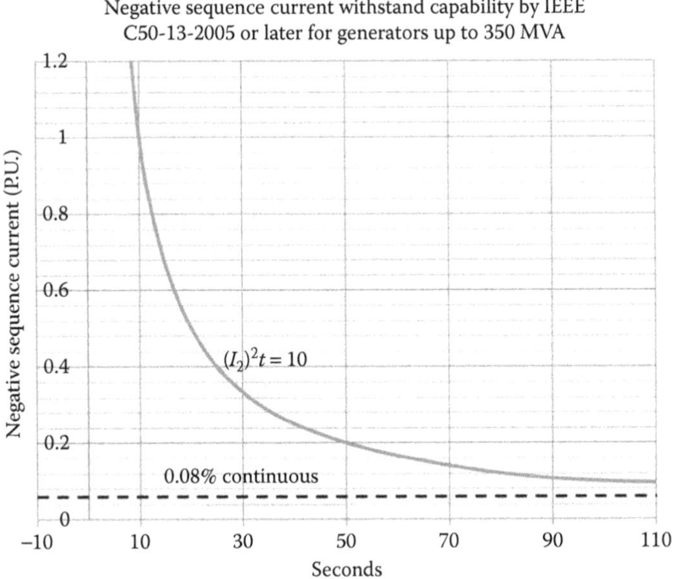

FIGURE 4.10
Continuous and sudden negative-sequence current withstand capability.

Figure 4.10 shows both the continuous withstand capability at 0.08 p.u. of negative-sequence current, with the transient one equal to 10, for a given machine. As stated above, the standard IEEE C50.13-2005 or later revisions have values for continuous capability and transient capability that depend on the rating of the machine and the type of cooling of the rotor. Hence, one must refer to the standard when addressing a particular machine. Also, some of these values are different in older versions of the standard from the values found in later versions. Machines designed to the older versions must meet the values found in the version of the standard in place at the time the machine was designed.

RELATED MALFUNCTIONS	NUMBER
CURRENT UNBALANCED BETWEEN PHASES	M-S-12

S-S-07 OUTLET TEMPERATURE OF STATOR CONDUCTOR-BAR COOLANT—ABOVE NORMAL (CASE 1)

APPLICABILITY:

CASE 1: Machine energized and synchronized carrying load

- Hydrogen inner-cooled stator windings
- Water inner-cooled stator windings

DESCRIPTION:

This symptom refers to the case where the temperature of the coolant medium (hydrogen or water) as it emerges from a stator bar, and measured by RTD or TC, is above *normal*. By *normal*, it means the range of temperatures typically measured on the same machine under the same load and ambient conditions, or the maximum temperature or range of allowable temperatures given by the OEM.

One must be careful in diagnosing this type of symptom because full bar blockage and partial bar blockage of coolant flow will show up differently on the temperature readings. With partial blockage, the coolant temperature will vary proportionally as load moves up or down. However, with full coolant flow blockage of a stator bar, the temperature measurement of the outlet TC/RTD will often decrease to the level of the local temperature surrounding the temperature sensor. That is most likely to be the hydrogen gas temperature in the local area of the measurement point.

RELATED MALFUNCTIONS	NUMBER
AIR OR HYDROGEN HEAT-EXCHANGER COOLING-WATER SUPPLY OFF	M-H-01
BLOCKAGE OF COOLING FLOW IN A STATOR BAR	M-S-02

(Continued)

RELATED MALFUNCTIONS	NUMBER
BLOCKAGE OF HYDROGEN COOLERS	M-H-03
BLOCKED COOLING VENT IN STATOR CORE	M-C&F-02
BREACH OF STRANDS OR TUBES CARRYING COOLING WATER	M-S-06
CURRENT OVERLOAD OF STATOR WINDING	M-S-11
FOREIGN BODY INTERACTING WITH STATOR CORE	M-C&F-07
GAS LOCKING IN WATER-CARRYING STRANDS OR TUBES	M-S-21
GAS LOCKING OF COOLING WATER IN HYDROGEN OR AIR HEAT EXCHANGERS	M-H-07
GLOBAL STATOR TEMPERATURE ABOVE NORMAL	M-S-25
GLOBAL TEMPERATURE OF CORE ABOVE NORMAL	M-C&F-09
HYDROGEN PRESSURE INSIDE THE GENERATOR BELOW NORMAL	M-H-09
INADEQUATE FLOW OF COOLING WATER TO AIR OR H_2-COOLERS	M-H-10
INLET TEMPERATURE OF AIR OR HYDROGEN COOLING WATER HIGH	M-H-11
INSUFFICIENT FLOW OF COOLANT DUE TO BUCKLED COOLING STRANDS OR TUBES	M-S-31
INSUFFICIENT FLOW OF COOLING MEDIUM IN A STATOR BAR	M-S-32
INSUFFICIENT PERFORMANCE OF AIR OR HYDROGEN HEAT EXCHANGERS	M-H-12
INTERLAMINAR INSULATION FAILURE	M-C&F-13
LOCAL CORE FAULT	M-C&F-15
LOCALIZED OVERHEATING OF STATOR CONDUCTOR	M-S-39
OPERATION OUTSIDE CAPABILITY CURVE—ONLINE	M-E-XX
PHASE CURRENT HIGH	M-S-45
PRESSURE OF HYDROGEN SUPPLY—LOW	M-H-16
RUPTURED WATER MANIFOLD (HEADER)	M-S-48
TEMPERATURE OF COLD GAS ABOVE NORMAL	M-H-17
TEMPERATURE OF CORE END ABOVE NORMAL	M-C&F-21
TEMPERATURE OF LOCAL CORE AREA ABOVE NORMAL	M-C&F-20
TEMPERATURE OF A SINGLE PHASE ABOVE NORMAL	M-S-52
TEMPERATURE OF STATOR CONDUCTOR ABOVE NORMAL	M-S-53

S-S-08 OUTLET TEMPERATURE OF STATOR CONDUCTOR-BAR COOLANT—ABOVE NORMAL (CASE 2)

APPLICABILITY:

CASE 2: Machine energized and NOT synchronized

- Hydrogen inner-cooled stator windings
- Water inner-cooled stator windings

DESCRIPTION:

This symptom refers to the case where the temperatures measured by RTD or TC of the coolant medium (hydrogen or water) as it emerges from a stator bars, is above *normal*. By *normal*, it means the range of temperatures typically measured in the same machine under the same load and ambient conditions, or the maximum temperature or range of allowable temperatures given by the OEM.

Under normal conditions, there should be very little increase in outlet coolant temperatures, because the only stator current flowing is the vector summation of the small currents required to magnetize (excite) the step-up transformer (if connected to the generator) and the unit auxiliary transformers up to rated voltage. This assumes the HV side of the SUT is open.

RELATED MALFUNCTIONS	NUMBER
AIR OR HYDROGEN HEAT-EXCHANGER COOLING-WATER SUPPLY OFF	M-H-01
BLOCKAGE OF COOLING FLOW IN A STATOR BAR	M-S-02
BLOCKAGE OF HYDROGEN COOLERS	M-H-03
BLOCKED COOLING VENT IN STATOR CORE	M-C&F-02
BREACH OF STRANDS OR TUBES CARRYING COOLING WATER	M-S-06
FOREIGN BODY INTERACTING WITH STATOR CORE	M-C&F-07
GAS LOCKING IN WATER-CARRYING STRANDS OR TUBES	M-S-21
GAS LOCKING OF COOLING WATER IN HYDROGEN OR AIR HEAT EXCHANGERS	M-H-07
GLOBAL STATOR TEMPERATURE ABOVE NORMAL	M-S-25
GLOBAL TEMPERATURE OF CORE ABOVE NORMAL	M-C&F-09
HYDROGEN PRESSURE INSIDE THE GENERATOR BELOW NORMAL	M-H-09
INADEQUATE FLOW OF COOLING WATER TO AIR OR H_2-COOLERS	M-H-10
INLET TEMPERATURE OF AIR OR HYDROGEN COOLING WATER HIGH	M-H-11
INSUFFICIENT FLOW OF COOLANT DUE TO BUCKLED COOLING STRANDS OR TUBES	M-S-31
INSUFFICIENT FLOW OF COOLING MEDIUM IN A STATOR BAR	M-S-32
INSUFFICIENT PERFORMANCE OF AIR OR HYDROGEN HEAT EXCHANGERS	M-H-12
INTERLAMINAR INSULATION FAILURE	M-C&F-13
LOCAL CORE FAULT	M-C&F-15
LOCALIZED OVERHEATING OF STATOR CONDUCTOR	M-S-39
OPERATION OUTSIDE CAPABILITY CURVE—ONLINE	M-E-11
PHASE CURRENT HIGH	M-S-45
PRESSURE OF HYDROGEN SUPPLY—LOW	M-H-16
RUPTURED WATER MANIFOLD (HEADER)	M-S-47
TEMPERATURE OF COLD GAS ABOVE NORMAL	M-H-17
TEMPERATURE OF CORE END ABOVE NORMAL	M-C&F-21
TEMPERATURE OF LOCAL CORE AREA ABOVE NORMAL	M-C&F-20
TEMPERATURE OF SINGLE PHASE ABOVE NORMAL	M-S-52
TEMPERATURE OF STATOR CONDUCTOR ABOVE NORMAL	M-S-53

S-S-09 OUTLET TEMPERATURE OF STATOR CONDUCTOR-BAR COOLANT BELOW NORMAL

APPLICABILITY:

All generators

DESCRIPTION:

A stator bar-coolant temperature decrease would be a rather rare type of occurrence. It would most likely be the result of a fully blocked flow of SCW, for example, due to gas locking, as explained in S-S-07. However, it could also be due to an open circuit in the stator winding resulting in no stator current flow, while the generator is connected to the system. For instance, a generator with an open stator parallel circuit will still carry current in each phase, but one section of the winding (the open one) will show a much lower temperature. This situation is basically impossible with inner-cooled-by-water windings, because an electrically open circuit will almost always be accompanied by a major rupture in the SCW path. On the other hand, this situation is clearly possible in the case of an inner cooled by H_2 winding.

An open parallel will most probably also result in additional symptoms, such as an increase in negative-sequence currents. Figure 4.11 schematically illustrates the case of a stator winding with two-parallel circuits per phase, with one parallel open in phase B.

In that case, the temperature measured by the outlet thermocouples on the affected circuit may be closer to the temperature of the hydrogen in the generator than that of the hot SCW flowing in other conducting bars.

RELATED MALFUNCTIONS	NUMBER
WINDING (STATOR) OPEN CIRCUIT	M-S-58
BLOCKAGE OF COOLANT FLOW IN A STATOR BAR	M-S-02

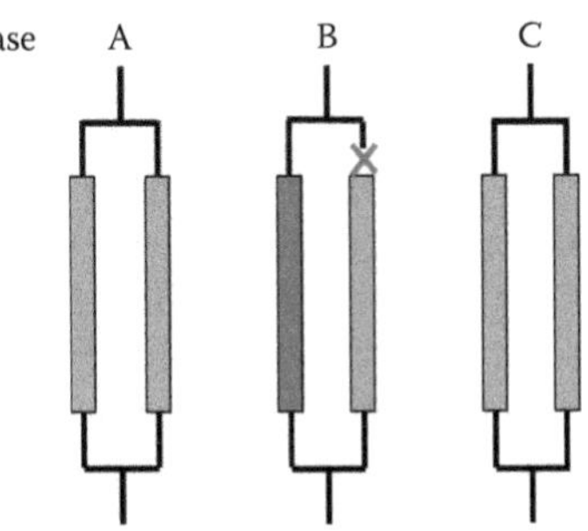

FIGURE 4.11
Example of an open stator circuit. In this case, two parallel circuits per phase are shown. One parallel circuit in phase B (the open one) is much cooler than the rest of the winding.

S-S-10 OZONE SMELL NEAR AIR-COOLED GENERATORS

APPLICABILITY:

Air-cooled generators, in particular those not totally enclosed

DESCRIPTION:

This symptom refers to those cases where near the generator one experiences a pungent ozone smell. The smell can be particularly intense near air-cooled generators that are not totally enclosed.

RELATED MALFUNCTIONS	NUMBER
PARTIAL DISCHARGE ACTIVITY (PDA) HIGH	M-S-44

S-S-11 PARTIAL DISCHARGE ACTIVITY (PDA) HIGH

APPLICABILITY:

All generators
Note: The generator must be energized such that the stator winding has voltage induced and this voltage is above the level of PDA and/or corona inception. And, there must be some type of partial discharge (PD) monitoring installed to be able to detect it, such as

- Capacitive coupling
- RFCT
- SSC (stator slot coupler)
- EMI

DESCRIPTION:

There are a number of different types of PDA that can be monitored inside the generator, including sources of discharge that are not classical *generator PD*, but rather discharges from the isophase bus (IPB), collector ring sparking, static exciter field forcing spikes, and so on. Some devices look at pure PDA in the higher frequency range for generator PDA, while others measure the discharge over a wider or lower frequency range. Therefore, devices such as these may require some type of pattern recognition to determine the type and source of the PDA in the generator. And some instruments mainly look at the frequency spectrum that can also *see* electrical discharges that are not necessarily from the generator itself.

RELATED MALFUNCTIONS	NUMBER
PARTIAL DISCHARGE ACTIVITY (PDA) HIGH	M-S-44

S-S-12 PHASE CURRENT UNBALANCE

APPLICABILITY:

All generators

DESCRIPTION:

All large generators are three phase machines, with the large majority being *star* (*wye*) connected, as opposed to being *delta* connected. This type of connection provides for the 3-phases and a *neutral* or *ground* point. The 3-phase currents, when balanced, are always very close in magnitude, and displaced by 120 electrical degrees. When the currents become unbalanced in magnitude, there is also usually a change in the displacement angle. During online operation, this condition will result in a negative-sequence current circulating in the stator winding. The negative-sequence current generates an airgap magnetic flux rotating at 60 (or 50) Hz in reverse direction. This flux induces 120 (or 100) Hz eddy currents on the surface of the rotor. Under certain conditions, these eddy currents can cause significant damage to wedges, retaining rings, the forging, and other rotor components. All machines generally have phase current unbalance protection in one form or another to prevent damage to the generator.

A situation where the phase currents are not balanced generally results from some unbalance in the grid voltages, a large fault in the grid or a severe internal fault inside the generator, such as a short circuit, or an open circuit on one phase.

RELATED MALFUNCTIONS	NUMBER
NEGATIVE-SEQUENCE CURRENTS	M-S-42
WINDING (STATOR) OPEN CIRCUIT	M-S-58

S-S-13 RADIO-FREQUENCY (RF) MONITOR HIGH

APPLICABILITY:

- Generator spinning at any speed with voltage present at its terminals (i.e., excitation on)
- Generator spinning at rated speed and at rated terminal voltage (i.e., open circuit condition)
- Generator connected and synchronized to the grid

PDA detectability by RF monitoring is confined to generators with RF monitors installed. Figure 4.12 exhibits a RFCT installed on the neutral cable between the grounding transformer and the neutral point of a 560 MVA generator.

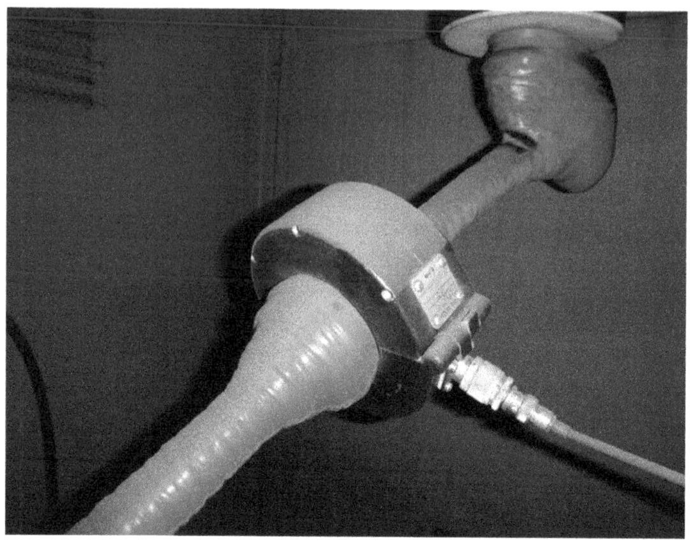

FIGURE 4.12
Radio-frequency current transformer (RFCT) installed on the neutral cable of a large generator.

DESCRIPTION:

This symptom applies to those cases when readings are higher than *normal*. By *normal* it means previous readings on the same unit under the same or similar operating conditions. RF techniques are best when used as a trending device and not as an absolute threshold value. This is because there is inherent noise in partial discharge signals monitored in this manner. In addition, the PD signals seen by an RF monitor can be from a number of sources. Therefore, a considerable amount of expertise and familiarity is required to recognize the RF signal patterns (generally in the frequency spectrum) to distinguish what the source of the PD is in the RF signal.

RELATED MALFUNCTIONS	NUMBER
ARCING BETWEEN CORE AND KEYBARS	M-C&F-01
CRACKED CONDUCTOR STRANDS	M-S-10
DAMAGE TO CONDUCTOR BAR INSULATION	M-S-15
FAILURE OF HYDROGEN SEALS	M-R-11
FAILURE OF OIL WIPER	M-R-12
HIGH RESISTANCE OF PHASE-CONNECTOR JOINT	M-S-27
LOOSE CORE LAMINATIONS	M-C&F-16
LOOSENING OF STATOR WINDING SLOT WEDGES	M-S-41
RADIO-FREQUENCY (RF) ACTIVITY	M-S-46
RUB OF OIL WIPER	M-R-22
VIBRATION OF SATATOR CONDUCTOR BAR IN SLOT	M-S-55

S-S-14 SLOT TEMPERATURE OF STATOR CONDUCTOR BAR—ABOVE NORMAL (CASE 1)

APPLICABILITY:

All generators

CASE 1: Machine energized and synchronized carrying load

DESCRIPTION:

Temperature of the top and bottom bars in the slot is measured by RTD or TC embedded between the conductor bars, inside the slot area. This symptom becomes relevant when the measured temperature is above *normal*. By *normal*, it means the range of temperatures normally measured in the same machine and at same or similar location, under same load and ambient conditions. Figure 4.13 shows the typical location for slot-embedded RTDs.

Full bar blockage and partial bar blockage of coolant flow (in inner-cooled stators) will show up differently on slot temperature measurements. With partial blockage, the coolant temperature will vary proportionally as load varies up or down. However, with full coolant flow blockage, the same is true, but the stator bar slot temperature in general will increase dramatically, due to uncooled I^2R losses.

FIGURE 4.13
The typical location of slot-embedded RTDs. They are in general found on indirectly cooled stator windings, but may be of great help in troubleshooting problems when installed on inner-cooled windings.

This is a critical piece of information, because with a fully blocked bar, the outlet temperature of the coolant most likely will go down toward the temperature of the hydrogen in the machine.

RELATED MALFUNCTIONS	NUMBER
AIR OR HYDROGEN HEAT-EXCHANGER COOLING-WATER SUPPLY OFF	M-H-01
BLOCKAGE OF COOLING FLOW IN A STATOR BAR	M-S-02
BLOCKAGE OF HYDROGEN COOLERS	M-H-03
BLOCKAGE OF PARALLEL-PHASE CONNECTOR COOLING PATH	M-S-03
BLOCKAGE OF PHASE-CONNECTOR COOLING PATH	M-S-05
BLOCKAGE OF PHASE-COOLANT FLOW	M-S-04
BLOCKED COOLING VENT IN STATOR CORE	M-C&F-02
BREACH OF STRANDS OR TUBES CARRYING COOLING WATER	M-S-06
CURRENT OVERLOAD OF STATOR WINDING	M-S-11
FOREIGN BODY INTERACTION WITH STATOR CORE	M-C&F-07
GAS LOCKING IN WATER-CARRYING STRANDS OR TUBES	M-S-21
GAS LOCKING OF COOLING WATER IN AIR OR HYDROGEN HEAT EXCHANGERS	M-H-07
GAS LOCKING OF PARALLEL-PHASE COOLANT	M-S-23
GAS LOCKING OF PHASE COOLANT	M-S-22
GLOBAL STATOR TEMPERATURE ABOVE NORMAL	M-S-25
GLOBAL TEMPERATURE OF CORE ABOVE NORMAL	M-C&F-09
HYDROGEN PRESSURE INSIDE THE GENERATOR BELOW NORMAL	M-H-09
INADEQUATE FLOW OF COOLING WATER TO AIR OR H_2-COOLERS	M-H-10
INLET TEMPERATURE OF AIR OR HYDROGEN COOLING WATER HIGH	M-H-11
INSUFFICIENT FLOW OF COOLANT DUE TO BUCKLED COOLING STRANDS OR TUBES	M-S-31
INSUFFICIENT FLOW OF COOLING MEDIUM IN A STATOR BAR	M-S-32
INSUFFICIENT FLOW OF PHASE-PARALLEL COOLANT	M-S-34
INSUFFICIENT PERFORMANCE OF AIR OR HYDROGEN HEAT EXCHANGERS	M-H-12
INTERLAMINAR INSULATION FAILURE	M-C&F-13
LOCAL CORE FAULT	M-C&F-15
LOCALIZED OVERHEATING OF STATOR CONDUCTOR	M-S-39
OPERATION OUTSIDE CAPABILITY CURVE—ONLINE	M-E-11
PHASE CURRENT HIGH	M-S-45
PRESSURE OF HYDROGEN SUPPLY—LOW	M-H-16
RUPTURED WATER MANIFOLD (HEADER)	M-S-47
TEMPERATURE OF A PARALLEL CIRCUIT ABOVE NORMAL	M-S-51
TEMPERATURE OF COLD GAS ABOVE NORMAL	M-H-17
TEMPERATURE OF CORE END ABOVE NORMAL	M-C&F-21
TEMPERATURE OF LOCAL CORE AREA ABOVE NORMAL	M-C&F-20
TEMPERATURE OF SINGLE PHASE ABOVE NORMAL	M-S-52
TEMPERATURE OF STATOR CONDUCTOR ABOVE NORMAL	M-S-53

S-S-15 SLOT TEMPERATURE OF STATOR CONDUCTOR BAR—ABOVE NORMAL (CASE 2)

APPLICABILITY:

All generators
CASE 2: Machine energized and NOT synchronized

DESCRIPTION:

Temperature of the top and bottom bars in the slot measured by RTD or TC embedded between the conductor bars, inside the slot area. This symptom becomes relevant when the measured temperature is above *normal*. By *normal*, it means the range of temperatures normally measured in the same machine, same or similar location, under same load and ambient conditions.

There ought to be very little noticeable increase in stator slot temperatures because the only stator current flowing will be that of the amount required to magnetize (i.e., excite) the unit auxiliary transformers and the step-up transformer, if it is connected to the generator, up to rated terminal voltage. This no-load current is very small in relation to the rated current of the generator. Also, if the SUT is connected to the generator (breaker closed or no breaker existing between generator and SUT), the HV side of the SUT is not connected to the grid (open breaker(s)).

RELATED MALFUNCTIONS	NUMBER
AIR OR HYDROGEN HEAT-EXCHANGER COOLING-WATER SUPPLY OFF	M-H-01
BLOCKAGE OF COOLING FLOW IN A STATOR BAR	M-S-02
BLOCKAGE OF HYDROGEN COOLERS	M-H-03
BLOCKAGE OF PARALLEL-PHASE CONNECTOR COOLING PATH	M-S-03
BLOCKAGE OF PHASE-CONNECTOR COOLING PATH	M-S-05
BLOCKAGE OF PHASE-COOLANT FLOW	M-S-04
BLOCKED COOLING VENT IN STATOR CORE	M-C&F-02
BREACH OF STRANDS OR TUBES CARRYING COOLING WATER	M-S-06
FOREIGN BODY INTERACTING WITH STATOR CORE	M-C&F-07
GAS LOCKING IN WATER-CARRYING STRANDS OR TUBES	M-S-21
GAS LOCKING OF COOLING WATER IN HYDROGEN HEAT EXCHANGERS	M-H-07
GAS LOCKING OF PARALLEL-PHASE COOLANT	M-S-23
GAS LOCKING OF PHASE COOLANT	M-S-22
GLOBAL TEMPERATURE OF CORE ABOVE NORMAL	M-C&F-09
HYDROGEN PRESSURE INSIDE THE GENERATOR BELOW NORMAL	M-H-09
INADEQUATE FLOW OF COOLING WATER TO AIR OR H_2-COOLERS	M-H-10

(Continud)

RELATED MALFUNCTIONS	NUMBER
INLET TEMPERATURE OF AIR OR HYDROGEN COOLING WATER HIGH	M-H-11
INSUFFICIENT FLOW OF COOLANT DUE TO BUCKLED COOLING STRANDS OR TUBES	M-S-31
INSUFFICIENT FLOW OF COOLING MEDIUM IN A BAR OR TUBE	M-S-32
INSUFFICIENT FLOW OF PHASE-PARALLEL COOLANT	M-S-34
INSUFFICIENT PERFORMANCE OF AIR OR HYDROGEN HEAT EXCHANGERS	M-H-12
INTERLAMINAR INSULATION FAILURE	M-C&F-13
LOCAL CORE FAULT	M-C&F-15
PRESSURE OF HYDROGEN SUPPLY—LOW	M-H-16
RUPTURED WATER MANIFOLD (HEADER)	M-S-47
TEMPERATURE OF A PARALLEL CIRCUIT ABOVE NORMAL	M-S-51
TEMPERATURE OF COLD GAS ABOVE NORMAL	M-H-17
TEMPERATURE OF CORE END ABOVE NORMAL	M-C&F-21
TEMPERATURE OF LOCAL CORE AREA ABOVE NORMAL	M-C&F-20
TEMPERATURE OF STATOR CONDUCTOR ABOVE NORMAL	M-S-53

S-S-16 STATOR CURRENT ABOVE NOMINAL

APPLICABILITY:

- All generators
- *Machine energized and synchronized*

DESCRIPTION:

Every generator has a defined nominal (rated) stator current value which should not be exceeded during the operation. This value may be found on the generator nameplate as well as on the capability and V curves. A machine may have more than one nominal stator current value, for instance, one for each H_2 pressure that appears on the capability curve(s). However, once the hydrogen pressure is set, the applicable nominal stator current value must not be exceeded. This value is defined as *maximum stator current*.

A generator may undergo transient current overloads during grid disturbances. In the vast majority of cases there is little or nothing the operators can do during the transient, given its short duration. Depending on the severity of the overload, a relatively short occurrence can have significant negative lasting effects on a number of major components.

This symptom applies to those short transients discussed above, as well as to long periods of operation with currents above nominal.

RELATED MALFUNCTIONS	NUMBER
CURRENT OVERLOAD OF STATOR WINDING	M-S-11
MVAR TRANSIENTS	M-E-09
MW TRANSIENTS	M-E-10
OPERATION OUTSIDE CAPABILITY CURVE—ONLINE	M-E-11
PHASE CURRENT HIGH	M-S-45
SUDDEN LARGE CHANGE IN GRID IMPEDANCE	M-E-18
TERMINAL VOLTAGE EXCURSION—ONLINE	M-E-23
VOLTAGE OF GRID—UNBALANCED	M-S-57

S-S-17 STATOR GROUND ALARM AND/OR TRIP

APPLICABILITY:

All generators

DESCRIPTION:

This symptom applies when a stator ground alarm is received. In some plants the alarm is concurrent with the tripping of the unit. Stator ground alarms/trips can also result from grounds outside the generator terminals (e.g., on cables, isophase busses, the generator-side windings of auxiliary or step-up transformers, and potential transformers connected to the generator output).

RELATED MALFUNCTIONS	NUMBER
DAMAGE FROM MAGNETIC TERMITES	M-S-14
DAMAGE TO CONDUCTOR BAR INSULATION	M-S-15
DEFECTIVE STATOR WINDING INSULATION	M-S-18
GROUND FAULT OF STATOR WINDING	M-S-26

S-S-18 TEMPERATURE OF PHASE CONNECTOR ABOVE NORMAL

APPLICABILITY:

All generators

DESCRIPTION:

This symptom applies whenever the temperature of the stator winding phase connectors is above *normal*. By *normal*, it means the range of temperatures measured in the same machine, at the same of similar location, under the same load and ambient conditions or as defined by the OEM. This temperature can be measured directly by embedded temperature sensors or indirectly by measuring the cooling medium by RTDs or TCs. Generally, however, the phase connectors are not usually equipped with TCs or RTDs. *Note:* The number and arrangement of phase connectors will depend on the stator winding arrangement; and this will vary greatly depending on the winding being 2- or 4-pole and either connected as a single circuit or with parallels.

RELATED MALFUNCTIONS	NUMBER
BLOCKAGE OF PHASE-CONNECTOR COOLANT PATH	M-S-05
BROKEN/CRACKED STATOR TERMINAL	M-S-07
CURRENT OVERLOAD OF STATOR WINDING	M-S-11
GAS LOCKING OF PHASE COOLANT	M-S-22
GAS LOCKING OF PHASE PARALLEL COOLANT	M-S-23
GLOBAL STATOR TEMPERATURE ABOVE NORMAL	M-S-25
INSUFFICIENT FLOW PHASE COOLANT	M-S-33
TEMPERATURE OF STATOR TERMINAL ABOVE NORMAL	M-S-54

S-S-19 TEMPERATURE OF STATOR TERMINAL ABOVE NORMAL

APPLICABILITY:

All generators

DESCRIPTION:

This symptom applies whenever the temperature of the stator *terminals* is above *normal*. By *normal*, it means the range of temperatures measured in the same machine and same or similar location, under same load and ambient conditions or as defined by the OEM. This temperature can be measured directly by embedded temperature sensors or indirectly by measuring the cooling medium by RTDs or TCs. Generally, however, terminals are not usually equipped with TCs or RTDs.
Note: By *terminals*, it means the end portion of the winding, including the bushings.

RELATED MALFUNCTIONS	NUMBER
BLOCKAGE IN THE FLOW OF STATOR TERMINAL COOLANT	M-S-01
BROKEN/CRACKED STATOR TERMINAL	M-S-07
CURRENT OVERLOAD OF STATOR WINDING	M-S-11
GAS LOCKING OF WATER-COOLED TERMINALS	M-S-24
GLOBAL STATOR TEMPERATURE ABOVE NORMAL	M-S-25
INSUFFICIENT FLOW OF TERMINAL COOLANT	M-S-35
TEMPERATURE OF STATOR TERMINAL ABOVE NORMAL	M-S-54

S-S-20 TERMINAL VOLTAGES UNBALANCED

APPLICABILITY:

All generators

DESCRIPTION:

Terminal voltage unbalance in the generator is possible by a fault on the grid, including equipment directly connected to the generator (e.g., UAT,[1] SUT, etc.) on one or two phases, or by a generator internal fault such as an open or short circuit on one or two phases.

RELATED MALFUNCTIONS	NUMBER
GROUND FAULT OF STATOR WINDING	M-S-26
VOLTAGE OF GRID—UNBALANCED	M-S-57

S-S-21 VIBRATION OF STATOR ENDWINDING—HIGH

APPLICABILITY:

All generators with stator endwinding vibration probes

DESCRIPTION:

All generators have some level of endwinding vibration present during operation. The two most prevalent modes of vibration are the power frequency mode of 60 (50) Hz, and the twice-power frequency mode of 120 (100) Hz. Of those machines that have endwinding vibration probes installed, some provide online real-time values of endwinding vibration and, some provide the information only when desired, by hooking up instruments to the probes' outputs. The measured values of vibration are compared against (typically) vendor-provided maxima, or other industry guidelines, and for trending purposes. When maximum values are provided, a *high* indication means that the measurements are above the maximum values recommended by the generator manufacturer (or other reputable source, such as relevant standards, etc.).

Endwinding vibrations are also generally load dependent—the most common relationship is that the higher the stator current, the higher the vibrations. It is possible, however, that vibration may go up at lower

[1] UAT—Unit auxiliaries transformer; this book also employs the equivalent term AUT, for auxiliary unit transformer.

loads, or behave nonlinearly vis-à-vis generator loading. The nature of the changes in vibration levels during load and VAR changes depend on the individual machine and its core-to-keybar coupling. Thus, it is important, when comparing vibration levels, to make comparisons at similar load points—both MW and MVAR output.

Figures 4.14 and 4.15 include graphs obtained from a 2-pole, 188 MVA, air-cooled generator exhibiting higher-than-normal vibrations of a radial phase connection. The graphs clearly show the dependency of the vibration on temperature and load. The temperature shifts the natural frequency of the vibrating component (due to thermal relaxation), and the load changes the level of the driving electromagnetic forces, which are proportional to the stator current squared.

The most common units for vibration used are peak-to-peak (p-t-p). However, vibration velocity and acceleration are also used. There are simple formulas for unit conversion among those three.

RELATED MALFUNCTIONS	NUMBER
LOOSENING OF STATOR ENDWINDING BLOCKING	M-S-40
VIBRATION OF STATOR ENDWINDING—HIGH	M-S-56

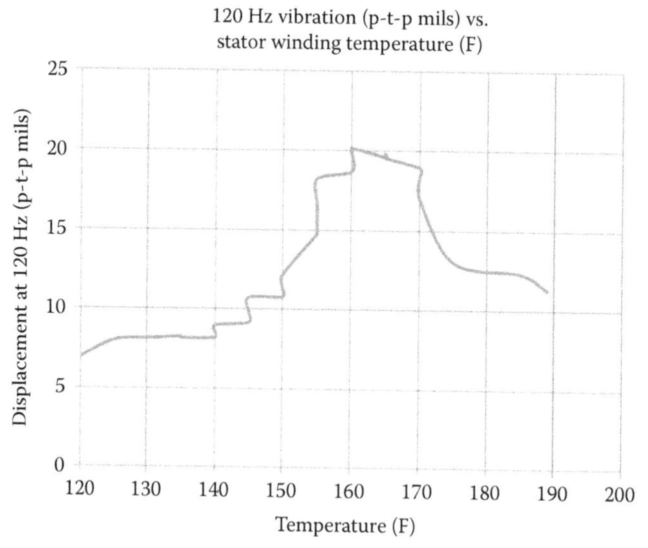

FIGURE 4.14
Vibration of a radial connection with natural frequency close to 120 Hz, versus the temperature of the winding.

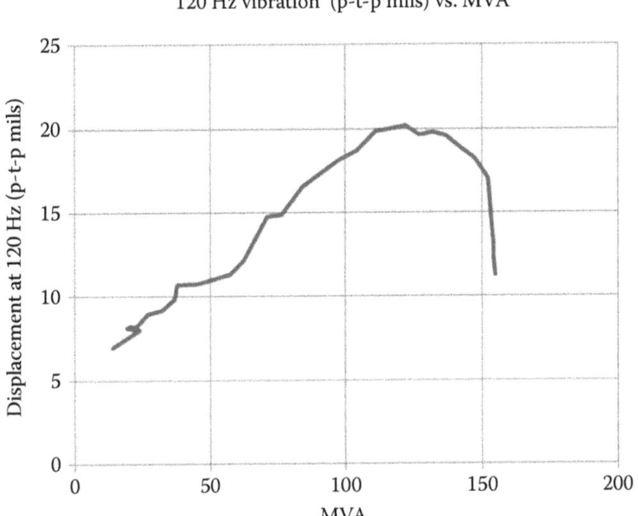

FIGURE 4.15
Same machine as in Figure 4.14. This graph shows the vibration at 120 Hz versus the load MVA.

4.2.3 Rotor Symptoms

S-R-01 GROUNDED FIELD-WINDING ALARM AND/OR TRIP

APPLICABILITY:

All generators

DESCRIPTION:

This symptom applies whenever a trip and/or alarm is received indicating the presence of a short circuit to ground on the rotor winding. (i.e., a breakdown of the rotor winding insulation).

A ground alarm/trip does not discriminate between ground faults on the rotor winding or outside the rotor. (e.g., DC cables, busses, rotating diodes, rotating armature [in the case of brushless exciters], other excitation system components, and brushgear).

In the case of the majority of machines with rotor-winding ground protection, the relay(s) only send an alarm signal on a ground fault condition. Automatic tripping the unit due to a rotor-winding ground fault is not too common.

RELATED MALFUNCTIONS	NUMBER
BURNING/PYROLYSIS OF ROTOR INSULATION	M-R-04
DEFECTIVE FIELD WINDING INSULATION	M-R-06
DEGRADED/DAMAGED FIELD-WINDING INSULATION	M-R-07
GROUNDED FIELD WINDING	M-R-13

S-R-02 INSULATION RESISTANCE LOW IN WATER-COOLED ROTORS

APPLICABILITY:

Generators with field windings inner cooled by water

DESCRIPTION:

This symptom applies to those few rotors with windings that are directly cooled by water.

RELATED MALFUNCTIONS	NUMBER
WATER LEAK IN A WATER-COOLED ROTOR	M-R-31

Figure 4.16 schematically shows the rotor cooling arrangement for a water inner-cooled rotor. Pressures of the cooling water in the order of 1000–1300 psig are common (Figure 4.17). The temperature of the rotor winding is obtained by measuring the temperature of the outlet water—similarly as done with water inner-cooled stators.

S-R-03 RESISTANCE OF SHAFT GROUNDING BRUSH—HIGH

APPLICABILITY:

All generators

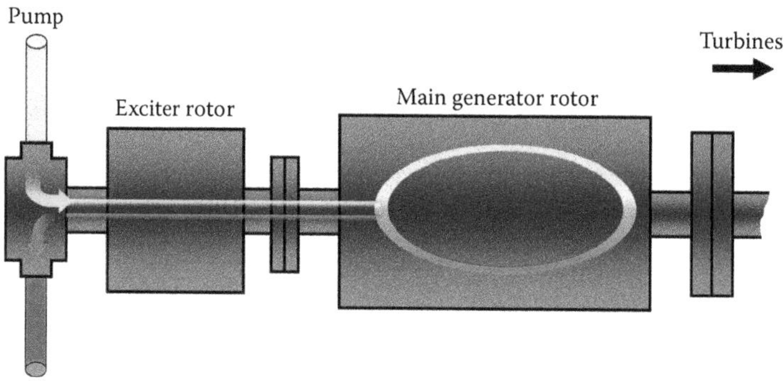

FIGURE 4.16
General arrangement of a water-cooled rotor winding.

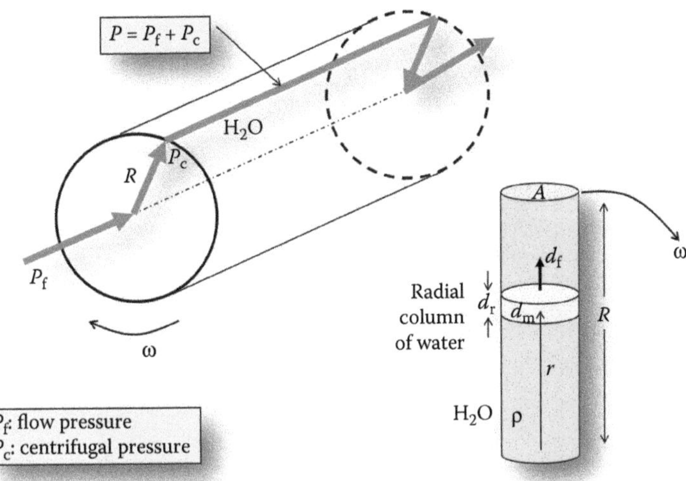

FIGURE 4.17
Pressure of the cooling water in a water inner-cooled field winding.

DESCRIPTION:

This symptom applies whenever an alarm is received from instruments monitoring the integrity of the shaft grounding circuit and shaft grounding brushes. Receiving this alarm is usually an indication that the grounding brushes are worn and require replacement, or they are fouled in some way and require maintenance.

There are several types of shaft grounding artifacts and monitoring systems. Figure 4.18 shows a double brush arrangement. This allows removing one brush for replacement while the other takes care of keeping effective grounding. Figure 4.19 shows a twin-braided strip shaft grounding arrangement.

RELATED MALFUNCTIONS	NUMBER
INEFFECTIVE SHAFT GROUNDING	M-R-15

S-R-04 SHAFT CURRENT ABOVE NORMAL

APPLICABILITY:

All generators

DESCRIPTION:

This symptom applies whenever an elevated (above *normal*) magnitude of shaft current is present, as measured with dedicated online real-time monitoring instrumentation installed on many generators. In other machines, these variables can, and sometimes are, measured periodically.

FIGURE 4.18
Twin shaft-grounding brush arrangement.

FIGURE 4.19
Twin shaft-grounding braided strip arrangement.

By *normal*, it means shaft currents measured in previous occasions under similar load conditions. This alarm is usually an indication that a shaft ground condition exists somewhere on the T-G shaft. Most likely locations are bearings, hydrogen seals, and oil piping or exciter-end pedestal instrumentation.

RELATED MALFUNCTIONS	NUMBER
FAILURE OF BEARING INSULATION	M-R-09
FAILURE OF H_2-SEAL INSULATION	M-R-10
GROUNDED FIELD WINDING	M-R-13
INEFFECTIVE SHAFT GROUNDING	M-R-15
INTERLAMINAR INSULATION FAILURE	M-C&F-13
LOCAL CORE FAULT	M-C&F-15
RUB BETWEEN BEARING AND JOURNAL	M-R-20
RUB OF HYDROGEN SEAL	M-R-21
SHAFT VOLTAGE—HIGH	M-R-23

S-R-05 SHAFT VOLTAGE ABOVE NORMAL

APPLICABILITY:

All generators

DESCRIPTION:

This symptom applies whenever an elevated (above *normal*) magnitude of shaft voltage is present, as measured with dedicated online real-time monitoring instrumentation installed on many generators. In other machines, these variables can, and sometimes are, measured periodically.

By *normal*, it means shaft voltage measured in previous occasions under similar load conditions. When there is no actual problem in progress this alarm is generally an indication of the shaft grounding brushes needing replacement or maintenance.

RELATED MALFUNCTIONS	NUMBER
GROUNDED FIELD-WINDING	M-R-13
INEFFECTIVE SHAFT GROUNDING	M-R-15
INTERLAMINAR INSULATION FAILURE	M-C&F-13
LOCAL CORE FAULT	M-C&F-15
SHAFT VOLTAGE—HIGH	M-R-23

S-R-06 SHAFT VOLTAGE BELOW NORMAL

APPLICABILITY:

All generators

DESCRIPTION:

This symptom applies whenever a reduced (below *normal*) magnitude of shaft voltage is present, as measured with dedicated online real-time monitoring instrumentation installed on many generators. In other machines, these variables can, and sometimes are, measured periodically.

By *normal*, it means shaft voltage measured in previous occasions under similar load conditions.

Low shaft voltage is most often found along with high shaft current and a shaft ground condition on the generator T-G[2] line somewhere.

RELATED MALFUNCTIONS	NUMBER
GROUNDED FIELD WINDING	M-R-13
RUB OF HYDROGEN SEAL	M-R-21
RUB BETWEEN BEARING AND JOURNAL	M-R-20

S-R-07 SHORTED TURNS IN FIELD WINDING

APPLICABILITY:

All generators

DESCRIPTION:

There are no commercial devices that alarm due to the presence of shorted turns in the field winding, or that alert to the development of new shorted turns in a rotor with a number of those shorts already present. However, there are systems that continuously monitor for the presence of shorted turns, mainly by the employment of rotor flux monitor (RFM) probes. Other systems attempt to discern the presence of shorted turns in the field winding by *looking* at the signature created by shaft voltages. Research in these topics is proceeding in a number of places.

This symptom applies whenever a new shorted turn(s) is (have been) identified by a direct method, such as traces obtained from an RFM, or other indirect methods, such as an RSO test or an impedance test.

Figure 4.20 shows a computer originated display of shorted turns on the rotor winding of a large nuclear-powered turbogenerator. The data are fed to the computer online real-time, and the calculated values can be displayed on any computer inside the plant with access to this program. Commercially developed instrumentation and software are moving in the direction of providing real-time feedback on the presence of shorted turns.

When rotor shorted turns occur they can cause rotor thermal asymmetry that is field current-dependent. Special *thermal sensitivity* tests are designed to investigate the nature of the unbalance should this occur. This involves introducing MW, MVAR, and temperature-induced changes to

[2] T-G—Turbine generator.

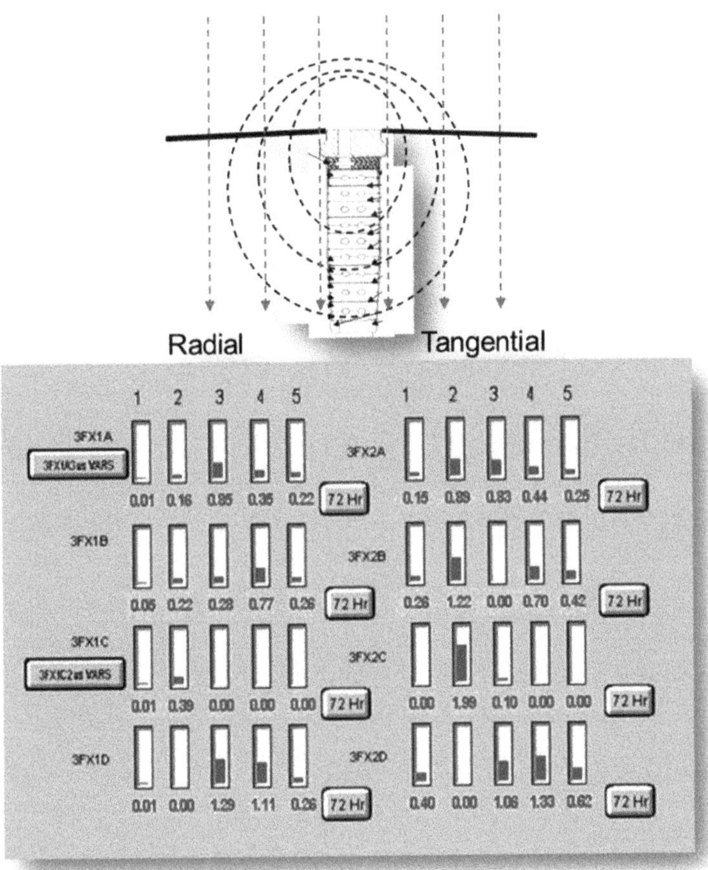

FIGURE 4.20
A real-time display of the number of shorted turns in a cylindrical rotor. Both radial flux and tangential flux are used in the calculations and are displayed to aid in obtaining a reliable indication of the actual number of shorted turns. By clicking on any of the *72 Hr* buttons, the screen displays traces of the last 72 h.

identify the nature of the rotor vibrations; looking for magnitude and phase changes, as well as reversible versus nonreversible vibration characteristics. See **THERMAL SENSITIVITY TEST** (M-R-29).

RELATED MALFUNCTIONS	NUMBER
DEGRADED/DAMAGED FIELD-WINDING INSULATION	M-R-07
SHORTED TURNS IN FIELD WINDING	M-R-24

S-R-08 TEMPERATURE OF BEARING ABOVE NORMAL

APPLICABILITY:

All generators

DESCRIPTION:

This symptom applies in situations where the metal temperature of a bearing(s) is above its normal range. By *normal,* it means temperatures that are higher than historically measured for the bearing in question, under similar load conditions, or higher than the allowable range, as provided by the OEM or other relevant Industry sources.

RELATED MALFUNCTIONS	NUMBER
ARCING ACROSS BEARINGS AND HYDROGEN SEALS	M-R-01
BEARING MISALIGNMENT	M-R-02
BLOCKAGE OF LUBE-OIL FLOW	M-LOS-01
DYNAMIC ROTOR IMBALANCE (MECHANICAL)	M-R-08
EFFICIENCY OF LUBE-OIL COOLERS DEGRADED	M-LOS-02
FAILURE OF LUBE-OIL PUMP	M-LOS-03
FLOW OF LUBE OIL TO BEARINGS BELOW NORMAL	M-LOS-06
INEFFECTIVE SHAFT GROUNDING	M-R-15
LEAK FROM LUBE-OIL SYSTEM	M-LOS-08
LUBE-OIL IMPURITIES	M-LOS-10
RUB BETWEEN BEARING AND JOURNAL	M-R-20
TEMPERATURE OF BEARING ABOVE NORMAL	M-R-25

S-R-09 TEMPERATURE OF FIELD-WINDING ABOVE NORMAL

APPLICABILITY:

All generators

DESCRIPTION:

This symptom applies in those situations when the rotor (field) winding reaches temperatures above *normal*. By *normal,* it means temperatures that were measured previously on the same machine under same operating conditions. Also, *normal* means temperatures within *Insulation Class* and *Class Rise* limits and/or as set by the OEM.

The temperature of the rotor winding cannot be measure directly. In rotors with external excitation (having sliprings), the temperature is derived from the value of the resistance of the field, measured continuously by dividing the field voltage by the field current. (Refer to IEEE Std 115-2009: *IEEE Guide: Test Procedures for Synchronous Machines Part*

I—Acceptance and Performance Testing Part II—Test Procedures and Parameter Determination for Dynamic Analysis.) Similar arrangement is made with rotors with shaft-mounted excitation, via auxiliary sliprings. In some brushless machines, measuring rotor winding temperature requires installing circuit components and instrumentation to the extent many stations do not monitor this parameter.

RELATED MALFUNCTIONS	NUMBER
AIR OR HYDROGEN HEAT-EXCHANGER COOLING-WATER SUPPLY OFF	M-H-01
BLOCKAGE OF AIR COOLERS	M-H-02
BLOCKAGE OF HYDROGEN COOLERS	M-H-03
BLOCKAGE OF ROTOR VENTILATION PATH	M-R-03
HYDROGEN PRESSURE INSIDE GENERATOR BELOW NORMAL	M-H-09
INADEQUATE FLOW OF COOLING WATER TO AIR OR H_2 COOLERS	M-H-10
INLET TEMPERATURE OF AIR OR HYDROGEN COOLING WATER HIGH	M-H-11
INSUFFICIENT PERFORMANCE OF AIR OR HYDROGEN HEAT EXCHANGERS	M-H-12
OPERATION OUTSIDE CAPABILITY CURVE—ONLINE	M-E-11
OVEREXCITATION ONLINE	M-E-13
SHORTED TURNS IN FIELD WINDING	M-R-24
TEMPERATURE OF COLD GAS ABOVE NORMAL	M-H-17
TEMPERATURE OF FIELD WINDING ABOVE NORMAL	M-R-26

S-R-10 VIBRATION OF ROTOR—HIGH

APPLICABILITY:

All generators

DESCRIPTION:

This symptom applies when the vibration of either the TE[3] or CE[4] generator bearings or both, is (are) above the normal operating range, as given by the OEM or Industry guides, such as:

- ISO 7919-2—2011 or later: *Mechanical vibration—Evaluation of machine vibration by measurements on rotating shafts/Part 2: Land-based steam turbines and generators in excess of 50 MW with normal operating speeds of 1500 r/min, 3000 r/min and 3600 r/min.* and ISO 7919–4.
- ISO 7919-4—2009 and later: *ditto—Part 4: Gas turbine sets with fluid-film bearings.*

[3] TE—Turbine end.
[4] CE—Connection end, also indicated as the exciter end (EE).

RELATED MALFUNCTIONS	NUMBER
BEARING MISALIGNMENT	M-R-02
DEGRADED/DAMAGED FIELD-WINDING INSULATION	M-R-07
DYNAMIC ROTOR IMBALANCE (MECHANICAL)	M-R-08
FAILURE OF HYDROGEN SEALS	M-R-11
FAILURE OF OIL WIPER	M-R-12
FAILURE OF SEAL-OIL PUMP	M-SOS-05
GROUNDED SHAFT (MOMENTARILY)	M-R-14
LOSS OF STABILITY	M-R-17
MECHANICAL FAILURE OF ROTOR COMPONENT	M-R-18
MISALIGNMENT OF HYDROGEN SEAL	M-R-19
RUB BETWEEN BEARING AND JOURNAL	M-R-20
RUB OF HYDROGEN SEAL	M-R-21
RUB OF OIL WIPER	M-R-22
THERMAL ASYMMETRY OF ROTOR	M-R-28
THERMAL SENSITIVITY TEST	M-R-29
VIBRATION OF ROTOR—HIGH	M-R-30

4.2.4 Excitation Symptoms

S-E-01 ANOMALOUS GENERATOR OUTPUT VOLTAGE

APPLICABILITY:

- All generators
- Machine energized but NOT synchronized

DESCRIPTION:

This symptom applies when the unit is at rated speed and not synchronized, with excitation applied, and the output voltage of the generator is

- Higher than 105% of nominal voltage, or
- Lower than 95% of nominal voltage, or
- The difference in the amplitude of the voltages between the three phases is too large

RELATED MALFUNCTIONS	NUMBER
DEGRADED/DAMAGED FIELD-WINDING INSULATION	M-R-07
FAILURE OF EXTERNAL STATIC EXCITATION	M-E-04
FAILURE OF RADIAL PHASE CONNECTOR	M-S-19
FAILURE OF SERIES CONNECTION	M-S-20
FIELD CURRENT TOO LOW OFFLINE	M-E-05
HIGH RESISTANCE OF PHASE CONNECTOR JOINT	M-S-27

(Continued)

RELATED MALFUNCTIONS	NUMBER
HIGH RESISTANCE OF PHASE-PARALLEL CONNECTOR JOINT	M-S-28
OVEREXCITATION OFFLINE	M-E-12
TERMINAL VOLTAGE EXCURSION—OFFLINE	M-E-22

S-E-02 ERRATIC BEHAVIOUR OF GENERATOR FIELD TEMPERATURE

APPLICABILITY:

All generators

DESCRIPTION:

This symptom applies to all machines where the temperature of the main generator rotor winding is monitored. The most common method for measuring and monitoring field winding temperature in machines having sliprings is indirectly done, by calculating the ratio between the *field voltage* and the *field current*; this ratio equals the resistance of the winding (*field resistance*), after some small allowance is done for a 1.5–2 volts drop at the sliprings.

$$\text{Field resistance} = \frac{\text{Field voltage at sliprings (less 1.5 to 2 VDC)}}{\text{Field current}}$$

The field resistance is related to the temperature of the winding by the following equation:

$$\text{Field temperature} = \frac{\text{Field resistance}}{\text{Reference resistance}}\left(\text{Reference temperature} + k\right) - k$$

In the equation above, the *reference resistance* is the resistance of the winding measured at factory referred to a given voltage, typically to 25°C or 40°C, or some other value. The *reference temperature* is the temperature of the winding when the *reference resistance* was measured and k is a constant which equals 234.5 for pure conductivity copper, and 225 for aluminum with an equivalent copper conductivity equal to 62% of pure copper.

In many generators—in particular brushless machines, there is no access for direct measurement of the field voltage and current. In those cases, the measurements are done indirectly by measuring exciter output voltage and current. In some cases there is no indication whatsoever in the control room of field temperature.

Field temperature may behave erratically under certain circumstances, and this symptom relates to those cases.

A related symptom is **SPARKING ON SLIPRINGS (S-E-14).**

RELATED MALFUNCTIONS	NUMBER
DAMAGE/DEGRADATION OF THE BRUSHGEAR	M-E-03
DEFECTIVE FIELD WINDING INSULATION	M-R-06
GROUNDED FIELD WINDING	M-R-13
MVAR TRANSIENTS	M-E-09
WATER LEAK IN A WATER-COOLED ROTOR	M-R-31
Some MVAR transients may be due to excitation malfunction	N/A
Some problems with the supply of water to a water-cooled rotor	N/A

S-E-03 FIELD CURRENT TOO HIGH OFFLINE

APPLICABILITY:

- Machine energized but not synchronized and its speed is anywhere from zero to rated rpm
- All generators

DESCRIPTION:

This symptom applies whenever a generator is spinning with its field applied and has not been synchronized to the grid, and the field current is such that the volts-per-hertz ratio is larger than the nominal volts-per-hertz.

Example

For a 50 Hz, 4-pole generator, with a nominal voltage equal to 22 kV, the nominal V/Hz = 22,000/50 = 440.

Now, let us assume that when the machine reaches the speed of 1000 rpm, and the stator voltage equals 18 kV. At this point the frequency in the machine's stator equals (1000/1500) × 50 Hz = 33.3 Hz. Thus, the V/Hz ratio at this condition equals 18,000/33.33 = 540.

The ratio of 540 is higher than the nominal V/Hz value of 440. Thus, it can be said that at this point, the field current is too high (and the machine is 22.7% overfluxed).

RELATED MALFUNCTIONS	NUMBER
OVEREXCITATION OFFLINE	M-E-12
OVERFLUXING OFFLINE	M-E-14
TERMINAL VOLTAGE EXCURSION OFFLINE	M-E-22

S-E-04 FIELD CURRENT TOO HIGH ONLINE

APPLICABILITY:

- Machine energized and synchronized
- All generators

DESCRIPTION:

A generator that is energized and synchronized to the grid has its rotor winding fed by its excitation system. The level of field current depends on the load point of the machine and its terminal voltage. This symptom applies whenever the field current is above its maximum value, as given by the OEM. One can find the maximum field current value in the operation manual for the unit, and/or the *V-curves* supplied with the machine. The other implication is that the generator is operating outside the capability curve in the *lagging* power factor range.

RELATED MALFUNCTIONS	NUMBER
OVEREXCITATION ONLINE	M-E-13
OVERFLUXING ONLINE	M-E-15
OVERHEATING OF SOLID-STATE EXCITATION POWER COMPONENT	M-E-16

S-E-05 FIELD CURRENT TOO LOW ONLINE

APPLICABILITY:

- Machine energized and synchronized
- All generators

DESCRIPTION:

A generator that is energized and synchronized to the grid has its rotor winding fed by its excitation system. The level of field current depends on the load point of the machine and its terminal voltage. This symptom applies whenever the field current is so low, that the generator is operating on or below its maximum *leading* reactive power or very close to the *steady-state stability limit* (*SSSL*).

The SSSL depends on additional factors such as the impedance of the MOT and the grid, and will be different for every machine. Pole slipping is a concern if the field current goes too low.

There is also a concern about core-end heating effects, and this will also be different on any given machine.

RELATED MALFUNCTIONS	NUMBER
FIELD CURRENT TOO LOW ONLINE	M-E-06

S-E-06 HIGH RATE OF BRUSH WEAR

APPLICABILITY:

All generators with sliprings

DESCRIPTION:

This symptom relates to the case where brushes are replaced too often due to excessive wear. An investigation of the originating cause for the high wear rate is required if this symptom arises. Figure 4.21 shows the result of brushes allowed to wear past the minimum recommended length.

RELATED MALFUNCTIONS	NUMBER
ACCELERATED BRUSH WEAR	M-E-01

S-E-07 INSUFFICIENT COOLING OF SOLID-STATE EXCITATION CABINETS

APPLICABILITY:

All generators excited by solid-state exciters through brushes and sliprings

DESCRIPTION:

This symptom applies to those situations where the cubicles containing the solid-state power devices of the excitation are not properly cooled (for instance, due to a failed cooling fan, plugged filters, etc.). Typical indication will be by a temperature sensor or alarm from a failed fan.

RELATED MALFUNCTIONS	NUMBER
INSUFFICIENT COOLING OF SOLID-STATE EXCITATION CABINETS	M-E-08

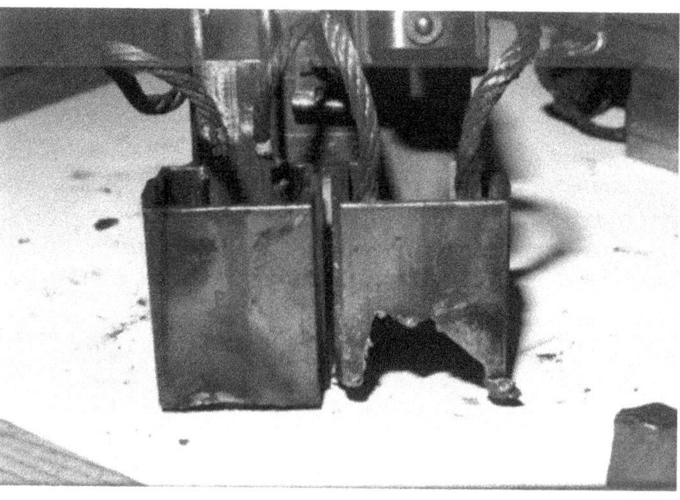

FIGURE 4.21
The results of brushes allowed to wear far beyond the recommended minimum length. The result was a full dc current arc developing across the sliprings with major damage to the machine.

S-E-08 MVAR TRANSIENT

APPLICABILITY:

All generators

DESCRIPTION:

This symptom applies whenever the main generator undergoes a reactive power output transient.

By *transient*, it means a fast swing of MVARs that is not the result of a planned load change, but the result of a grid disturbance (such as a sudden voltage change), or operator error, or malfunction of excitation system (including problems with the field winding).

Typically, this condition is detected in the control room by the operators. Usually, there is no alarm generated by this type of event.

RELATED MALFUNCTIONS	NUMBER
LOSS OF STABILITY	M-R-17
MVAR TRANSIENTS	M-E-09

S-E-09 MW TRANSIENT

APPLICABILITY:

All generators

DESCRIPTION:

This symptom applies to the case when the main generator undergoes an active (real) power output transient: either an increase or decrease in real power.

By *transient*, it means a fast swing of MW that is not the result of a planned operator control action.

Generally, there is no alarm associated with this type of event. Typically, this condition is detected in the control room by the operators, or it may go unnoticed, depending on its severity and duration.

Example

A fast MW swing would occur due to a trip of one of the coal mills in a fossil plant. For instance, if a boiler has 5 coal pulverizers and one is lost, that would account for a 20% reduction in fuel input and show up as a fairly fast reduction down to 80% power output.

RELATED MALFUNCTIONS	NUMBER
LOSS OF STABILITY	M-R-17
MW TRANSIENTS	M-E-10

S-E-10 OPERATION WITH *LEADING* POWER FACTOR BEYOND ALLOWABLE REGION

APPLICABILITY:

All generators

DESCRIPTION:

This situation, where the unit was/is operated beyond its allowable leading capability or stability limit, is also captured by the following symptom.

- FIELD CURRENT TOO LOW ONLINE (S-E-05)

RELATED MALFUNCTIONS	NUMBER
OVEREXCITATION ONLINE	M-E-13
OVERFLUXING ONLINE	M-E-15

Figure 4.22 shows the various regions of the capability curve. The leading operating region is indicated by the negative MVARs. One of the limits

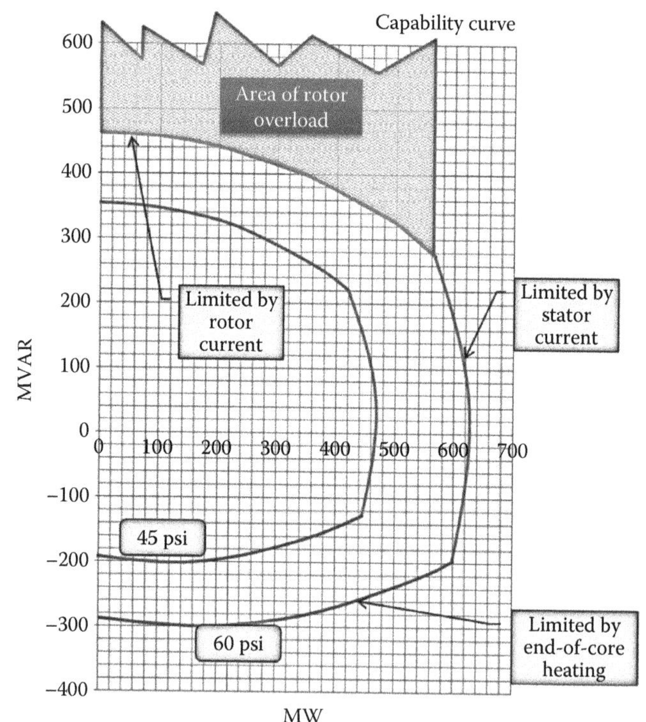

FIGURE 4.22
The capability curve shows the various limits and the overload area in the leading power factor while operating at 60 psi hydrogen pressure.

in leading power factor operation is the overheating of the core ends. Another is related to the steady and transient stability of the generator.

S-E-11 OVERFLUXING (VOLTS PER HERTZ) TOO HIGH OFFLINE

APPLICABILITY:

All generators

DESCRIPTION:

For an explanation, see the following symptom.

- FIELD CURRENT TOO HIGH OFFLINE (S-E-03)

RELATED MALFUNCTIONS	NUMBER
BURNING/PYROLYSIS OF CORE INSULATION	M-C&F-05
LOCAL CORE FAULT	M-C&F-15
OVERFLUXING OFFLINE	M-E-14
TERMINAL VOLTAGE EXCURSION—OFFLINE	M-E-22
SHORTED TURNS IN FIELD WINDING	M-R-24

Figure 4.23 shows how the application of excitation current higher than full open-circuit current (I_{FOC}) during starting can result in overfluxing of the generator.

S-E-12 PRESSURE DROP ACROSS BRUSHGEAR FILTER—HIGH

APPLICABILITY:

All generators with sliprings

DESCRIPTION:

This symptom relates to the situation where the filter (if applicable) at the inlet air path of the brushgear is partially or fully clogged, impeding the flow of cooling air to the brushgear.

RELATED MALFUNCTIONS	NUMBER
BLOCKAGE OF BRUSHGEAR COOLING AIR	M-E-02
TEMPERATURE OF SLIPRINGS/BRUSHGEAR ABOVE NORMAL	M-E-21

S-E-13 ROTOR SPEED BELOW NOMINAL—ONLINE

APPLICABILITY:

- Machine energized and synchronized
- All generators

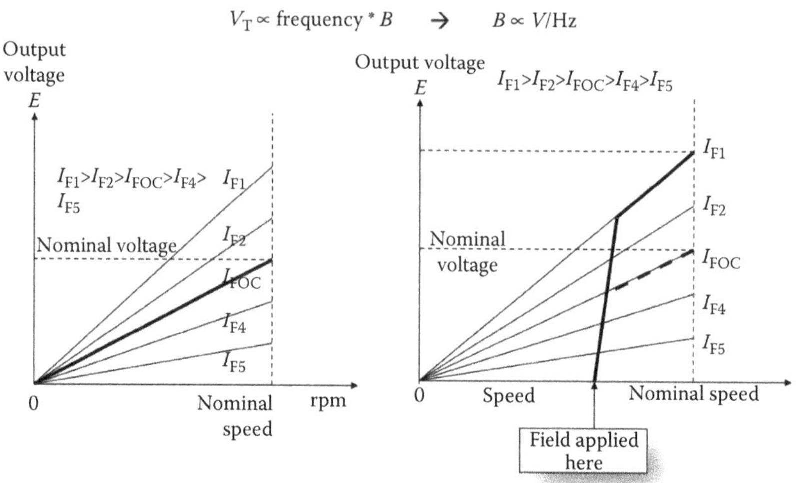

FIGURE 4.23
The graphs show the relationships between the speed, excitation current, and terminal voltage. The graph on the left of the reader is for when the excitation current is applied at zero speed (in general, not done for lack of rotor cooling at very low speeds); the graph on the right of the reader is for when the excitation is applied at speeds other than zero. In the graph on the right, application of excessive field current leads to overfluxing of the generator.

DESCRIPTION:

This symptom applies when the generator is synchronized to the grid, and the rpm are below the rated value. Unless the grid in question is a small system, in which the given generator represents a significant portion of the total generation, the frequency (speed) is, for all practical purposes, exclusively determined by the grid. In any case, the consequences of operating at lower-than-rated speeds are the same: the stator core will be in jeopardy of overfluxing. The degree of resulting core damage or degradation depends on how low has the frequency fallen below 60 (50) Hz. There is generally overfluxing protection available (see Chapter 2).

RELATED MALFUNCTIONS	NUMBER
FREQUENCY BELOW NOMINAL—ONLINE	M-E-07

S-E-14 SPARKING ON SLIPRINGS

APPLICABILITY:

All generators with sliprings

DESCRIPTION:

This symptom relates to the situation when there is sparking under the brushes. Sparking may be minor or intensive, which can lead to failure of the brushing operation, with damage to brushes, sliprings, brushgear structure, and in extreme cases, great damage to the rotor itself.

Sparking at the slip/collector rings can be detected by PDA or EMI monitoring. In the case of PDA monitoring, sparking between brushes and collector rings generally shows up as clearly defined and equally spaced spikes on the PD pattern. And for EMI, it shows up as clearly defined discharges at a specific frequency in the EMI signature. Most often, sparking will be detected by operators or electricians during their daily rounds (Figure 4.24).

RELATED MALFUNCTIONS	NUMBER
ACCELERATED BRUSH WEAR	M-E-01
DAMAGE/DEGRADATION OF THE BRUSHGEAR	M-E-03

S-E-15 TEMPERATURE IN SOLID-STATE EXCITATION CUBICLE—HIGH

APPLICABILITY:

All generators excited by solid-state exciters through brushes and sliprings

DESCRIPTION:

This symptom applies to those situations where the temperature in the excitation cubicle(s) containing the solid-state power devices of the

FIGURE 4.24
A snapshot of a video taken of sparking between brushes and sliprings. As shown in the figure, the best practice to visually capture the sparking is by keeping the lights out in the brushgear enclosure during this activity.

excitation is abnormally high (for instance, due to a failed cooling fan, plugged filters, etc.), or from a component or number of components that are overheating.

Typical indication will be by a temperature sensor that may or may not relay the information to the control room.

In this case, *high* means above a temperature range that guaranties proper operation of the excitation system over its design life-span. There is generally an alarm for this occurrence.

See Figure 4.25 for an example of an excitation system and its air filters requiring periodic replacement for maintaining adequate cooling of the cubicle internals.

RELATED MALFUNCTIONS	NUMBER
INSUFFICIENT COOLING OF SOLID-STATE EXCITATION CABINETS	M-E-08
OVERHEATING OF SOLID-STATE EXCITATION COMPONENT	M-E-16

S-E-16 TEMPERATURE OF SLIPRINGS/ BRUSHGEAR ABOVE NORMAL

APPLICABILITY:

All generators with external excitation (sliprings)

FIGURE 4.25
The figure shows the excitation cubicles of a large generator. Replacing filters during periodic preventive maintenance (PM) is important for the proper cooling of the exciter components.

DESCRIPTION:

This symptom applies to the situation when the temperature of the sliprings/brushgear is above its *normal* range. By *normal*, it means the range of temperatures that is typical for that component under similar load conditions, or above the values recommended by the OEM. In general, the method of measuring the temperature of the slipring/brushgear is by measuring the temperature of the outlet cooling air. Temperature of the sliprings can also be measured (periodically) with infrared probes.

RELATED MALFUNCTIONS	NUMBER
BLOCKAGE OF BRUSHGEAR COOLING AIR	M-E-02
DAMAGE/DEGRADATION OF THE BRUSHGEAR	M-E-03
TEMPERATURE OF INLET COOLING AIR TO BRUSHGEAR—HIGH	M-E-20
TEMPERATURE OF SLIPRINGS/BRUSHGEAR ABOVE NORMAL	M-E-21

4.2.5 Lubrication System Symptoms

S-LOS-01 DIFFERENTIAL PRESSURE ACROSS LUBE-OIL COOLER—ABOVE NORMAL

APPLICABILITY:

All generators

DESCRIPTION:

This symptom relates to the case when the differential pressure across the filter of the lube-oil system is above normal. *Normal* means the delta-pressure range recommended by the OEM or typical range of differential pressures for this unit.

If this symptom occurs, it is most often an indicator of some impediment to the flow of the lube oil in the coolers, for example, clogging or debris build up in the oil coolers.

RELATED MALFUNCTIONS	NUMBER
BLOCKAGE OF LUBE-OIL FLOW	M-LOS-01

S-LOS-02 FLOW OF COOLING WATER TO LUBE-OIL COOLERS—BELOW NORMAL

APPLICABILITY:

All generators

DESCRIPTION:

This symptom applies whenever the flow of cooling water to the lube-oil heat exchanger(s) is below normal. By *normal,* it means that the rate of flow required for maintaining bearing metal temperatures within allowable operating range.

If this symptom occurs, it is most often an indicator of some impediment to the flow of the cooling water to the coolers, for example, clogging or debris buildup in the water side of the lube-oil coolers, or a problem of insufficient water from the raw cooling water supply.

RELATED MALFUNCTIONS	NUMBER
FLOW OF COOLING WATER TO LUBE-OIL COOLERS BELOW NORMAL	M-LOS-05
FLOW OF COOLING WATER TO LUBE-OIL COOLERS BLOCKED	M-LOS-06

S-LOS-03 FLOW OF LUBE OIL BELOW NORMAL

APPLICABILITY:

All generators

DESCRIPTION:

This symptom relates to the case when the flow of lube oil to the bearings is below that required to maintain adequate lubrication, and/or below the flow recommended by the OEM.

RELATED MALFUNCTIONS	NUMBER
BLOCKAGE OF LUBE-OIL FLOW	M-LOS-01
FAILURE OF LUBE-OIL PUMP	M-LOS-03
FLOW OF LUBE OIL TO BEARINGS BELOW NORMAL	M-LOS-05
LEAK FROM LUBE-OIL SYSTEM	M-LOS-08

S-LOS-04 INLET TEMPERATURE OF LUBE-OIL COOLING WATER—HIGH

APPLICABILITY:

All generators

DESCRIPTION:

This symptom applies whenever the temperature of the cooling water entering the lube-oil heat exchanger(s) is high. By *high,* it means that the temperature is above the normal operating range.

This type of symptom is more likely to occur in summer months when the raw water supply from a river or lake is highest.

RELATED MALFUNCTIONS	NUMBER
INLET TEMPERATURE OF LUBE-OIL COOLING WATER HIGH	M-LOS-07

S-LOS-05 PRESSURE OF LUBE OIL BELOW NORMAL

APPLICABILITY:

All generators

DESCRIPTION:

This symptom applies whenever the pressure of the lubricating oil at the inlet of the bearings is below normal. *Below normal* infers a pressure that does not allow maintaining the bearing temperatures within allowable operating limits. If not investigated and corrected as soon as possible, it can also lead to operational issues with the bearings or mechanical damage such as a *rub* and/or electrical arcing between the rotor shaft and the bearings.

RELATED MALFUNCTIONS	NUMBER
FLOW OF LUBE OIL TO BEARINGS BELOW NORMAL	M-LOS-06

S-LOS-06 TEMPERATURE OF LUBE OIL AT BEARING OUTLET—HIGH

APPLICABILITY:

All generators

DESCRIPTION:

This symptom applies whenever the temperature of the lube oil at the bearing outlet is high (above range recommended by the OEM or industry practice).

Lubrication oil temperature must be maintained within an OEM-recommended margin. In certain circumstances, when there are operational problems with a bearing, the temperature of the oil as it exits the bearing is above the recommended range. If this is occurring, it is also likely that the embedded thermocouple (if applicable) in the bearing's white metal will also show temperatures above normal range.

RELATED MALFUNCTIONS	NUMBER
FLOW OF COOLING WATER TO LUBE-OIL COOLERS BELOW NORMAL	M-LOS-04
RUB BETWEEN BEARING AND JOURNAL	M-R-20
TEMPERATURE OF LUBE-OIL COOLER OUTLET ABOVE NORMAL	M-LOS-11

S-LOS-07 TEMPERATURE OF LUBE-OIL COOLER OUTLET ABOVE NORMAL

APPLICABILITY:

All generators

DESCRIPTION:

This symptom applies whenever the temperature of the lube oil, as it leaves the lube-oil heat exchanger(s) or enters the bearings, is above normal. By *normal,* it means the temperature that allows the bearings to remain within allowable temperature limits, as given by the OEM or other proven sources in the industry.

RELATED MALFUNCTIONS	NUMBER
BLOCKAGE OF LUBE-OIL FLOW	M-LOS-01
EFFICIENCY OF LUBE-OIL COOLERS DEGRADED	M-LOS-02
FLOW OF COOLING WATER TO LUBE-OIL COOLERS BELOW NORMAL	M-LOS-04
INLET TEMPERATURE OF LUBE-OIL COOLING WATER HIGH	M-LOS-07
LEAK OF COOLING WATER BEFORE INLET TO LUBE-OIL COOLERS	M-LOS-09
RUB BETWEEN BEARING AND JOURNAL	M-R-20
TEMPERATURE OF LUBE-OIL COOLER OUTLET ABOVE NORMAL	M-LOS-11

4.2.6 Seal-Oil System Symptoms

S-SOS-01 COOLING-WATER PRESSURE AT INLET TO SEAL-OIL HEAT EXCHANGER—LOW

APPLICABILITY:

All hydrogen-cooled generators

DESCRIPTION:

This symptom applies when the pressure of the hydrogen seal-oil cooling water at the inlet side of the heat exchanger(s) is below normal. By *low,* it means it is not capable of performing its cooling functions satisfactorily under normal operating conditions and rated load.

RELATED MALFUNCTIONS	NUMBER
FLOW OF COOLING WATER TO SEAL-OIL COOLERS BLOCKED	M-SOS-07
FLOW OF PRIMARY (RAW) COOLING WATER TO SEAL-OIL COOLERS BELOW NORMAL	M-SOS-08

S-SOS-02 DIFFERENTIAL PRESSURE ACROSS SEAL-OIL FILTER ABOVE NORMAL

APPLICABILITY:

All hydrogen-cooled generators

DESCRIPTION:

This symptom relates to the case where the differential pressure across the filter in the seal-oil system is above *normal*. By *normal*, it means the normal delta-pressure operational range, as given by the OEM, or as indicated by the previous operational history of the machine. Typical reason for this situation is a clogged filter, requiring maintenance. Additional insight can be obtained from the following malfunctions:

RELATED MALFUNCTIONS	NUMBER
BLOCKAGE OF SEAL-OIL FLOW	M-SOS-01
PRESSURE OF SEAL-OIL ABOVE NORMAL	M-SOS-13

Figure 4.26 shows typical seal-oil filters.

FIGURE 4.26
A couple of seal-oil filters after undergoing maintenance cleaning.

S-SOS-03 HEAT-EXCHANGER OUTLET TEMPERATURE OF SEAL OIL ABOVE NORMAL

APPLICABILITY:

Hydrogen-cooled generators

DESCRIPTION:

This symptom applies to those cases where the temperature of the seal oil exiting the seal-oil coolers (or the temperature of the seal oil as it enters the hydrogen seals) is above normal. By *normal*, it means the allowable range of temperatures specified by the manufacturer, or the range of temperatures that results in the proper functioning of the hydrogen seals under normal operating conditions.

RELATED MALFUNCTIONS	NUMBER
BLOCKAGE OF SEAL-OIL FLOW	M-SOS-01
EFFICIENCY OF SEAL-OIL COOLERS REDUCED	M-SOS-04
FLOW OF PRIMARY (RAW) COOLING WATER TO SEAL-OIL COOLERS BELOW NORMAL	M-SOS-08
LEAK FROM SEAL-OIL SYSTEM	M-SOS-10
LEAK OF COOLING WATER BEFORE INLET TO SEAL-OIL COOLERS	M-SOS-11
SEAL-OIL COOLING-WATER TERMPERATURE HIGH	M-SOS-14
TEMPERATURE OF SEAL-OIL COOLER OUTLET ABOVE NORMAL	M-SOS-15

S-SOS-04 INLET PRESSURE OF SEAL OIL BELOW NORMAL

APPLICABILITY:

Hydrogen-cooled generators

DESCRIPTION:

This symptom applies to those cases when the pressure of the hydrogen seal oil at the inlet of the hydrogen seals is below normal. By *normal*, it means the allowable range of pressures specified by the manufacturer, or the range of pressures that result in the proper functioning of the hydrogen seals under normal operating conditions.

RELATED MALFUNCTIONS	NUMBER
BLOCKAGE OF SEAL-OIL FLOW	M-SOS-01
BLOCKAGE OF SEAL-OIL HEAT EXCHANGER ON OIL SIDE	M-SOS-02
FAILURE OF HYDROGEN SEALS	M-R-11
FAILURE OF SEAL-OIL PUMP	M-SOS-05
LEAK FROM SEAL-OIL SYSTEM	M-SOS-10

S-SOS-05 PRESSURE IN SEAL-OIL DETRAINING TANK—ABOVE NORMAL

APPLICABILITY:

Hydrogen-cooled generators

DESCRIPTION:

This symptom applies to those cases when the pressure in the seal-oil detraining tank is above normal. By *above normal*, it means that there is not enough vacuum to carry out an efficient detraining of the seal oil.

Oil foaming in the seal-oil tank is one of the most common additional indicators. There is usually a sight-glass on the seal-oil tank to observe this condition (Figure 4.27).

RELATED MALFUNCTIONS	NUMBER
FAILURE OF SEAL-OIL VACUUM PUMP	M-SOS-06
INEFFECTIVE SEAL-OIL DETRAINING	M-SOS-09

S-SOS-06 PRESSURE OF SEAL OIL—ABOVE NORMAL

APPLICABILITY:

All hydrogen-cooled generators

FIGURE 4.27
Oil foaming can be seen in the sight-glass window on the seal-oil tank.

DESCRIPTION:

This symptom applies whenever the pressure of the seal oil is above normal. By *normal,* it means the allowable range of pressures specified by the manufacturer, or a range of pressures that results in the proper functioning of the hydrogen seals under normal operating conditions.

RELATED MALFUNCTIONS	NUMBER
BLOCKAGE OF SEAL-OIL HEAT EXCHANGER ON OIL SIDE	M-SOS-02
PRESSURE OF SEAL OIL ABOVE NORMAL	M-SOS-13

S-SOS-07 SEAL-OIL COOLING-WATER INLET TEMPERATURE—HIGH

APPLICABILITY:

All hydrogen-cooled generators

DESCRIPTION:

This symptom applies to those cases where the temperature of the cooling water at the inlet side of the seal oil heat exchangers is *high.* By *high,* it means to be above OEM-recommended values or a temperature range that does not allow sufficient cooling of the hydrogen seals under normal operating conditions and rated load.

The most likely occurrence for this is during summer months when the raw cooling water supply from the river or lake is highest. This symptom is similar to S-SOS-09.

RELATED MALFUNCTIONS	NUMBER
SEAL-OIL COOLING-WATER INLET TEMPERATURE HIGH	M-SOS-14

S-SOS-08 TEMPERATURE OF HYDROGEN SEAL ABOVE NORMAL

APPLICABILITY:

Hydrogen-cooled generators

DESCRIPTION:

This symptom applies whenever the temperature of the seal-oil metal is higher than normal. By *normal,* it means a temperature that is above the allowable operating range as given by the OEM or industry standards, or higher than temperatures measured at the same seal in previous occasions under similar load condition. When this occurs, the outlet oil from the hydrogen seals will likely also be high.

RELATED MALFUNCTIONS	NUMBER
DYNAMIC ROTOR IMBALANCE (MECHANICAL)	M-R-08
FAILURE OF HYDROGEN SEALS	M-R-11
FAILURE OF SEAL-OIL PUMP	M-SOS-05
GROUNDED SHAFT (MOMENTARILY)	M-R-14
MISALIGNMENT OF HYDROGEN SEAL	M-R-19
RUB OF HYDROGEN SEAL	M-R-21
TEMPERATURE OF HYDROGEN SEAL ABOVE NORMAL	M-R-27

S-SOS-09 TEMPERATURE OF RAW WATER AT SEAL-OIL COOLER INLET ABOVE NORMAL

APPLICABILITY:

All hydrogen-cooled generators

DESCRIPTION:

This symptom applies to those cases where the temperature of the cooling water at the inlet of the seal-oil system heat exchanger is above normal.

The most likely occurrence for this is during summer months when the raw cooling water supply from the river or lake is highest.

RELATED MALFUNCTIONS	NUMBER
FLOW OF COOLING WATER TO SEAL-OIL COOLERS BLOCKED	M-SOS-07
FLOW OF PRIMARY (RAW) COOLING WATER TO SEAL-OIL COOLERS BELOW NORMAL	M-SOS-08

S-SOS-10 TEMPERATURE OF SEAL OIL ABOVE NORMAL

APPLICABILITY:

All hydrogen-cooled generators

DESCRIPTION:

This symptom applies to those cases when the temperature of the seal oil is above normal range.

RELATED MALFUNCTIONS	NUMBER
BLOCKAGE OF SEAL-OIL FLOW	M-SOS-01
DEGRADED CONDTION OF THE SEAL OIL	M-SOS-03
EFFICIENCY OF SEAL-OIL COOLERS REDUCED	M-SOS-04
FAILURE OF SEAL-OIL PUMP	M-SOS-05

(Continued)

RELATED MALFUNCTIONS	NUMBER
FLOW OF COOLING WATER TO SEAL-OIL COOLERS BLOCKED	M-SOS-07
FLOW OF PRIMARY (RAW) COOLING WATER TO SEAL-OIL COOLERS BELOW NORMAL	M-SOS-08
LEAK OF COOLING WATER BEFORE INLET TO SEAL-OIL COOLERS	M-SOS-11
TEMPERATURE OF SEAL-OIL COOLER OUTLET ABOVE NORMAL	M-SOS-15

4.2.7 Hydrogen System Symptoms

S-H-01 COOLING-GAS HEAT-EXCHANGER DIFFERENTIAL TEMPERATURE—HIGH

APPLICABILITY:

- Hydrogen-cooled generators
- Air-cooled generators

DESCRIPTION:

This symptom applies to those machines where the differential temperature of the *hot gas* minus the temperature of *cold gas* across the heat exchanger is monitored, and it is higher than the OEM-recommended value. This value is typically a function of the load of the generator. Thus, instead of a single value, a curve may be given, where the maximum differential temperature is plotted against the load MVA. However, if the cooling water to the heat exchangers has a recirculation temperature control valve, then it is more likely to have a constant differential temperature between the hot and cold cooling gas.

RELATED MALFUNCTIONS	NUMBER
AIR OR HYDROGEN HEAT-EXCHANGER COOLING-WATER SUPPLY OFF	M-H-01
BLOCKAGE OF AIR COOLERS	M-H-02
BLOCKAGE OF HYDROGEN COOLERS	M-H-03
GAS LOCKING OF COOLING WATER IN HYDROGEN OR AIR HEAT EXCHANGERS	M-H-07
INADEQUATE FLOW OF COOLING WATER TO AIR OR H2 COOLERS	M-H-10
INSUFFICIENT PERFORMANCE OF AIR OR HYDROGEN HEAT EXCHANGERS	M-H-12
OPERATION OUTSIDE CAPABILITY CURVE—ONLINE	M-E-11
PHYSICAL BREACH OF HYDROGEN OR AIR HEAT EXCHANGER	M-H-15
FAILURE OF THE COOLING WATER RECIRCULATION SYSTEM CONTROL VALVE	N/A

S-H-02 DEGRADED HYDROGEN PURITY

APPLICABILITY:

Hydrogen-cooled generators

DESCRIPTION:

This symptom applies to those situations where the purity of the hydrogen in the casing of the machine is below industry-accepted values or/and OEM recommendations.

Figure 4.28 shows typical hydrogen control cubicles for a large turbogenerator. The panels include instrumentation to monitor purity, dew point, pressure and temperature, and pressure regulators.

RELATED MALFUNCTIONS	NUMBER
DEGRADED HYDROGEN PURITY	M-H-04
FAILURE OF HYDROGEN DRYER	M-H-06
FAILURE OF OIL WIPER	M-R-12
FAILURE OF SEAL-OIL PUMP	M-SOS-05
INEFFECTIVE SEAL-OIL DETRAINING	M-SOS-09
MISALIGNMENT OF HYDROGEN SEAL	M-R-19
MOISTURE IN HYDROGEN—HIGH	M-H-14
OIL & WATER LEAKS INSIDE THE GENERATOR	M-C&F-18
WATER LEAK INTO GENERATOR FROM AIR- OR H_2-COOLER	M-H-18

FIGURE 4.28
Hydrogen control panel.

S-H-03 ELEVATED HYDROGEN DEW POINT

APPLICABILITY:

Hydrogen-cooled generators

DESCRIPTION:

This symptom applies whenever the dew point of the hydrogen inside the generator is higher than normal. Most manufactures recommend that the dew point be kept below –10°C. Dew point excursions above 0°C ought to be investigated and remedied.

RELATED MALFUNCTIONS	NUMBER
DEGRADED CONDITION OF THE SEAL OIL	M-SOS-03
ELEVATED HYDROGEN DEW POINT	M-H-05
FAILURE OF HYDROGEN DRYER	M-H-06
FAILURE OF SEAL-OIL PUMP	M-SOS-05
FAILURE OF SEAL-OIL VACUUM PUMP	M-SOS-06
HYDROGEN INGRESSION INTO STATOR COOLING WATER	M-S-29
WATER LEAK INTO GENERATOR FROM AIR OR HYDROGEN COOLER	M-H-18

S-H-04 HYDROGEN LEAK INTO THE ISOPHASE BUS (IPB)—ALARM

APPLICABILITY:

All hydrogen-cooled generators connected to an IPB system

DESCRIPTION:

Most large generators have their terminals connected directly to an IPB. In some of these machines, the generator terminal box is directly attached to the IPB, in which case, a hydrogen leak through the high-voltage bushings allows the hydrogen to flow directly from the generator to the IPB.

Under certain conditions (specifically, IBPs that are not forced cooled), significant hydrogen leaks can result in hydrogen accumulation under the generator in the IPB transition compartment and the IPB itself. Within certain ranges, the hydrogen–air mixture is explosive; hence, some machines have hydrogen detectors inside the IPB under the generator that will alarm when certain level of hydrogen content is measured.

RELATED MALFUNCTIONS	NUMBER
HYDROGEN ESCAPING THE GENERATOR	M-H-08

S-H-05 HYDROGEN MAKE-UP RATE ABOVE NORMAL

APPLICABILITY:

Hydrogen-cooled generators

DESCRIPTION:

This symptom applies whenever the hydrogen's make-up rate is above its normal range and implies a leak of hydrogen from the generator. By *normal,* it means an OEM-provided range of makeup that is considered adequate for a machine of a given size and hydrogen pressure. There are also industry-accepted values for what is considered *normal* rate of hydrogen makeup. One of the most widely used criteria is that hydrogen-cooled generators should be able to maintain a leakrate below 500 ft^3/day. And if the leakrate reaches 1500 ft^3/day, then serious investigation of the leak location should be undertaken.
Note: Hydrogen leaks from the generator casing can be extremely serious because hydrogen leaking from a pressure vessel also has the tendency to self-ignite. The flame is invisible to the naked eye, it is very hot, and the byproducts of the leak are heat and water.

RELATED MALFUNCTIONS	NUMBER
FAILURE OF HYDROGEN SEALS	M-R-11
FAILURE OF SEAL-OIL PUMP	M-SOS-05
HYDROGEN ESCAPING THE GENERATOR	M-H-08
HYDROGEN INGRESSION INTO STATOR COOLING WATER	M-S-29
LEAK OF HYDROGEN INTO H2 COOLING WATER SYSTEM	M-H-13
MISALIGNMENT OF HYDROGEN SEAL	M-R-19

S-H-06 HYDROGEN PRESSURE INSIDE GENERATOR BELOW NORMAL

APPLICABILITY:

Hydrogen-cooled generators

DESCRIPTION:

This symptom applies when the hydrogen pressure inside the casing is below its *normal* value.

Most generators are rated for operation at one or more zones of the capability curve, according to the hydrogen pressure in the machine. Instead of a formula allowing operation at any hydrogen pressure up to a maximum, the OEM provides a set of curves on the capability curve for a number of hydrogen pressures, for example, such as 15, 30, 45, 60, and

75 psig.[5] Hence, the minimum required value of hydrogen pressure depends also on the operating regime of the machine.

In machines with water inner-cooled stator windings, the pressure of the hydrogen ought to always remain above that of the SCW by at least 5 psi at the inlet manifold, even when the unit is idle, but with the SCW in operation. This is to ensure that water does not leak into the generator, if there is a leak in the SCW system. Most generators have a detraining system to remove hydrogen from the SCW and so it is preferable to have hydrogen leaking into the SCW rather than water leaking into the generator. Figure 4.29 shows a typical case where the capability curve is drawn for more than a single pressure.

RELATED MALFUNCTIONS	NUMBER
FAILURE OF HYDROGEN SEALS	M-R-11
FAILURE OF SEAL-OIL PUMP	M-SOS-05
HYDROGEN PRESSURE INSIDE GENERATOR BELOW NORMAL	M-H-09
MISALIGNMENT OF HYDROGEN SEAL	M-R-19

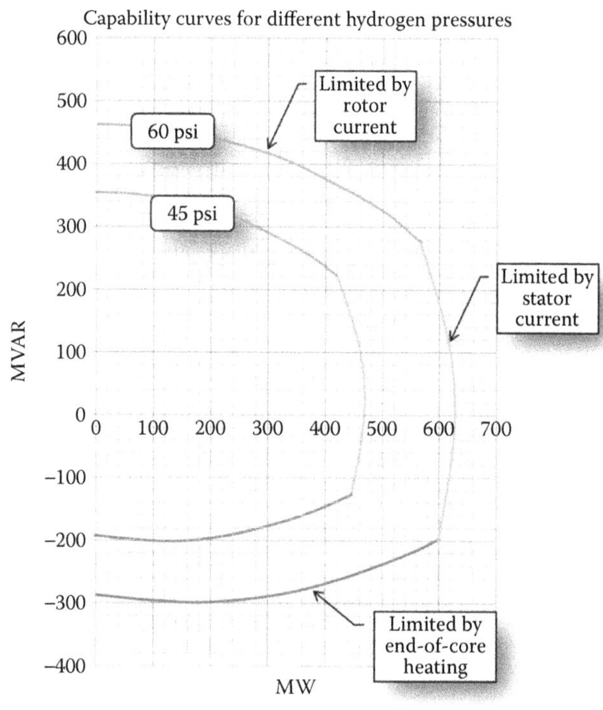

FIGURE 4.29
Output as a function of hydrogen pressure.

[5] 1 psi = 6.895 kPa.

S-H-07 INADEQUATE FLOW OF COOLING WATER TO AIR OR H_2 COOLERS

APPLICABILITY:

- Air-cooled generators with heat exchangers
- Hydrogen-cooled generators

DESCRIPTION:

This symptom applies to those cases where the flow of cooling water to the air or hydrogen heat exchangers is below that required to maintain effective cooling of the generator. See Figure 4.30 for an example of hydrogen cooler fouling limiting the flow of cooling water.

RELATED MALFUNCTIONS	NUMBER
AIR OR HYDROGEN HEAT-EXCHANGER COOLING-WATER SUPPLY OFF	M-H-01
BLOCKAGE OF AIR COOLERS	M-H-02
BLOCKAGE OF HYDROGEN COOLERS	M-H-03
GAS LOCKING OF COOLING WATER IN HYDROGEN OR AIR HEAT EXCHANGERS	M-H-07
GAS LOCKING OF PHASE COOLANT	M-S-22
GAS LOCKING OF PHASE-PARALLEL COOLANT	M-S-23

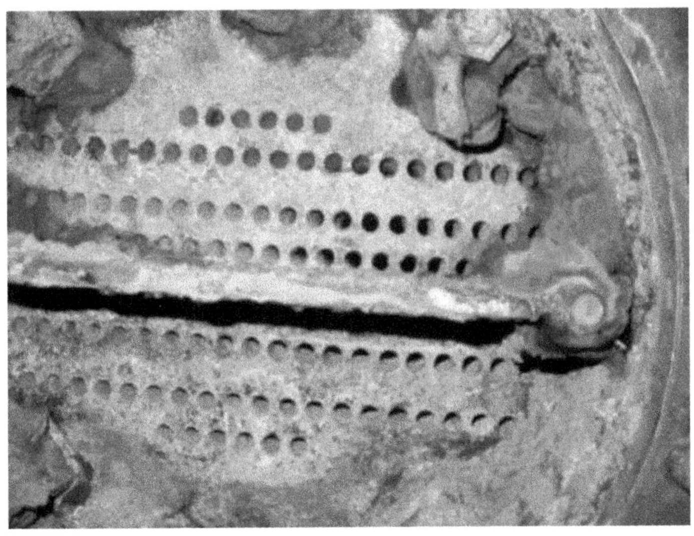

FIGURE 4.30
Severe fouling of a hydrogen heat exchanger.

RELATED MALFUNCTIONS	NUMBER
INADEQUATE FLOW OF COOLING WATER TO AIR OR H_2 COOLERS	M-H-10
PHYSICAL BREACH OF HYDROGEN OR AIR HEAT EXCHANGER	M-H-15

S-H-08 INLET TEMPERATURE OF AIR OR HYDROGEN COOLING WATER—HIGH

APPLICABILITY:

- Air-cooled generators with heat exchangers
- Hydrogen-cooled generators

DESCRIPTION:

This symptom applies to those cases where the temperature of the cooling water at the inlet of the air or hydrogen heat exchangers is too high to allow operation of the machine at rated load, within its designed temperature rise limits. This is generally more of an issue in summer months when the raw water supply to the heat exchangers is highest.

RELATED MALFUNCTIONS	NUMBER
INLET TEMPERATURE OF AIR OR HYDROGEN COOLING WATER—HIGH	M-H-11

S-H-09 PRESSURE OF HYDROGEN SUPPLY—LOW

APPLICABILITY:

Hydrogen-cooled generators

DESCRIPTION:

This symptom applies to those situations where the hydrogen makeup supply's pressure is dropping and cannot maintain the operational pressure of the hydrogen in the casing. The most likely cause of this is that the H_2 supply requires replenishing.

RELATED MALFUNCTIONS	NUMBER
PRESSURE OF HYDROGEN SUPPLY—LOW	M-H-16

S-H-10 TEMPERATURE OF COLD GAS ABOVE NORMAL

APPLICABILITY:

- Hydrogen-cooled generators
- Air-cooled generators

DESCRIPTION:

This symptom applies whenever the temperature of the *cold gas* (hydrogen or air leaving the heat exchanger after being cooled), is above *normal.* By *above normal,* it means temperatures that are substantially higher than in previous occasions under very similar load and ambient temperature conditions, or higher than recommended by the manufacturer or/and the pertinent standards and/or industry practice.

The significance of this symptom is that there is a problem with the air or hydrogen coolers or the raw water supply to the coolers.

Figure 4.31 shows a cold gas (hydrogen) temperature sensor in the cooling path of a large turbogenerator.

RELATED MALFUNCTIONS	NUMBER
AIR OR HYDROGEN HEAT-EXCHANGER COOLING-WATER SUPPLY OFF	M-H-01
BLOCKAGE OF AIR COOLERS	M-H-02
BLOCKAGE OF HYDROGEN COOLERS	M-H-03
GAS LOCKING OF COOLING WATER IN HYDROGEN OR AIR HEAT EXCHANGERS	M-H-07
INADEQUATE FLOW OF COOLING WATER TO AIR OR H_2 COOLERS	M-H-10
INLET TEMPERATURE OF AIR OR HYDROGEN COOLING WATER HIGH	M-H-11
TEMPERATURE OF COLD GAS ABOVE NORMAL	M-H-17

FIGURE 4.31
Cold-gas temperature sensor inside a generator.

4.2.8 Stator Cooling Water System Symptoms

S-SCW-01 BLOCKAGE IN THE FLOW OF STATOR COOLING WATER

APPLICABILITY:

All generators with water inner-cooled stator windings

DESCRIPTION:

This symptom applies to whenever there is a partial or total blockage in the normal flow of SCW.

RELATED MALFUNCTIONS	NUMBER
BLOCKAGE IN THE FLOW OF STATOR TERMINAL COOLANT	M-S-01
BLOCKAGE OF COOLANT FLOW IN A STATOR BAR	M-S-02
BLOCKAGE OF PARALLEL-PHASE-CONNECTOR COOLING PATH	M-S-03
BLOCKAGE OF PHASE COOLANT FLOW	M-S-04
BLOCKAGE OF PHASE-CONNECTOR COOLING PATH	M-S-05
GAS LOCKING IN WATER-CARRYING STRANDS OR TUBES	M-S-21
GAS LOCKING OF PHASE COOLANT	M-S-22
GAS LOCKING OF PHASE-PARALLEL COOLANT	M-S-23
GAS LOCKING OF WATER-COOLED TERMINALS	M-S-24
INSUFFICIENT FLOW OF COOLANT DUE TO BUCKLED COOLING STRANDS OR TUBES	M-S-31
INSUFFICIENT FLOW OF COOLING MEDIUM IN A STATOR BAR	M-S-32
INSUFFICIENT FLOW OF PHASE COOLANT	M-S-33
INSUFFICIENT FLOW OF PHASE-PARALLEL COOLANT	M-S-34
INSUFFICIENT FLOW OF TERMINAL COOLANT	M-S-35

S-SCW-02 BLOCKAGE OF STATOR COOLING WATER (SCW) HEAT EXCHANGER

APPLICABILITY:

All generators with water inner-cooled stator windings

DESCRIPTION:

This symptom applies when there is a blockage of the flow of SCW in the SCW heat exchanger. The likelihood of this happening is very low, given the good quality of the SCW. If any plugging of the cooler happens, it is much more likely that it will be in the piping carrying the raw cooling water to the exchanger. In the case where the plugging affects the flow of the SCW, the reader can refer to the following symptom.

- BLOCKAGE IN THE FLOW OF STATOR COOLING WATER (SCW) (S-SCW-01)

S-SCW-03 CONDUCTIVITY OF STATOR COOLING WATER ABOVE NORMAL

APPLICABILITY:

All generators with water inner-cooled stator windings

DESCRIPTION:

This symptom applies whenever the conductivity of the SCW is above *normal*. By *normal*, it means the widely accepted value of 0.1 μS/cm, or a value provided by the OEM.

RELATED MALFUNCTIONS	NUMBER
CONDUCTIVITY OF STATOR COOLING WATER ABOVE NORMAL	M-SCW-04

To understand the significance of the water conductivity, let us look at an example of how it may affect the integrity of the SCW system. Figure 4.32 shows a typical water inner-cooled winding.

Example

Calculation of the water resistance to ground along the Teflon hoses.

A typical large machine may have 72 stator slots. So now let us calculate the current leaking from the conductors to each manifold. For that, we may use the following typical data:

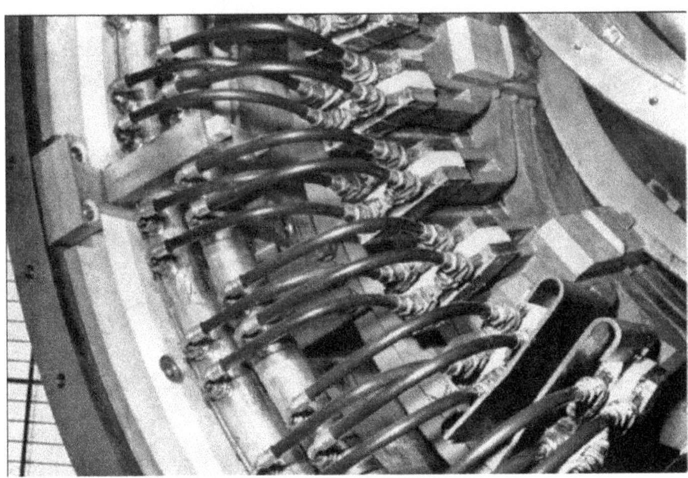

FIGURE 4.32
Water inner-cooled stator bars. Each bar has two stacks of strands; each stack is fed by one Teflon hose, insulating the conductor from the grounded manifold (header). Both inlet and outlet headers are at the same side of the machine.

Hose length: 30 cm
Hose diameter: 2 cm
Water conductivity: 0.1 μS/cm
Line nominal voltage: 24 kV
Figure 4.33 helps to explain the electrical and cooling water circuits for each bar.

From the data, the following is obtained

$$\text{Conductivity of the water inside a single hose} = \frac{0.1 * 10^{-6} * 2^2}{30}$$

$$= 13 * 10^{-9}(\text{Siemens})$$

$$\text{Leakage current through a single hose} = \frac{24{,}000}{\sqrt{3}} * 13 * 10^{-9}$$

$$= 0.18(\text{mA})$$

Leakage current into the header from 72 hoses = 0.013 amperes

$$\text{Enegy dissipated as heat in each hose} = \frac{24\,\text{kV}}{\sqrt{3}} * 0.18\,\text{mA} = 2.5\text{ W}$$

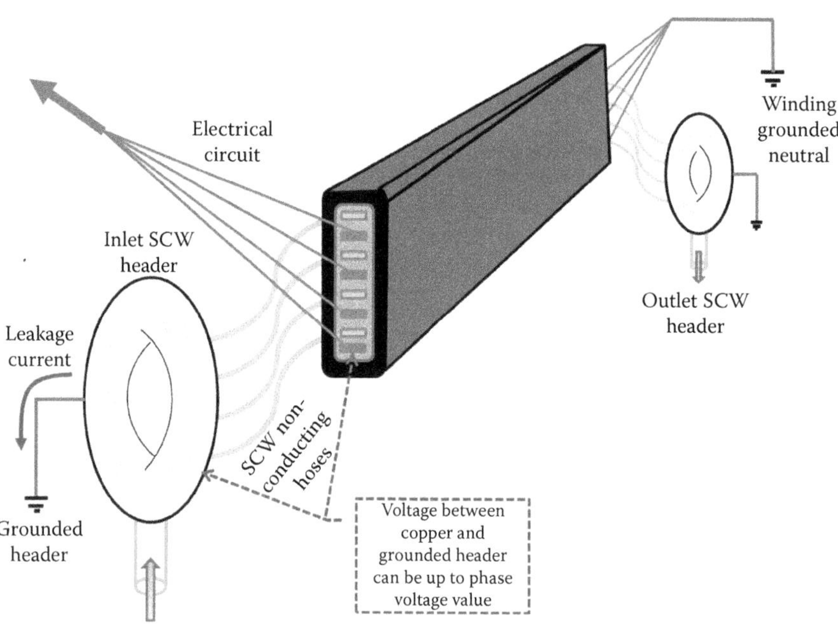

FIGURE 4.33
Stator cooling water and electric circuits shown for water inner-cooled bar.

Now imagine the water conductivity increases to 10 μS/cm, in which case each hose will dissipate 250 W. As the conductivity increases, it reaches a point when damage to the hoses may occur, leading to a ground fault.

S-SCW-04 COOLER OUTLET TEMPERATURE OF SCW ABOVE NORMAL

APPLICABILITY:

All generators with water inner-cooled stator windings

DESCRIPTION:

This symptom applies whenever the temperature of the SCW, as it leaves the heat exchanger, is above *normal.*

By *normal,* it means the maximum temperature, that if exceeded, the SCW system may not be able to remove all the heat it must remove to keep the stator temperature within normal range, with the unit at rated load.

RELATED MALFUNCTIONS	NUMBER
COOLER OUTLET TEMPERATURE OF SCW ABOVE NORMAL	M-SCW-05
PRIMARY WATER FLOW TO SCW COOLERS BELOW NORMAL	M-SCW-12
TEMPERATURE OF PRIMARY (RAW) WATER AT INLET OF SCW COOLERS—ABOVE NORMAL	M-SCW-16

S-SCW-05 DIFFERENTIAL PRESSURE ACROSS SCW HEAT EXCHANGER ABOVE NORMAL

APPLICABILITY:

All generators with water inner-cooled stator windings

DESCRIPTION:

This symptom applies whenever the pressure drop across the SCW heat exchanger(s) is above *normal*. By *normal,* it means a pressure higher than typically encountered in the same unit under similar load and ambient conditions, or/and a pressure that renders the SCW coolers (Figure 4.34) incapable to keeping the stator winding temperatures within allowable operating range.

RELATED MALFUNCTIONS	NUMBER
BLOCKAGE OF STATOR COOLING WATER (SCW) HEAT EXCHANGERS	M-SCW-02

FIGURE 4.34
A typical skid arrangement for a stator cooling water system (SCW). The horizontal twin heat exchangers are shown.

S-SCW-06 DIFFERENTIAL PRESSURE OF SCW ACROSS STATOR WINDING ABOVE NORMAL

APPLICABILITY:

All generators with water inner-cooled stator windings

DESCRIPTION:

This symptom applies whenever the pressure drop across the stator winding is above *normal*. By *normal*, it means the differential pressure of the SCW across the stator when operating at rated load, and/or the maximum delta pressure recommended by the OEM.

RELATED MALFUNCTIONS	NUMBER
BLOCKAGE IN THE FLOW OF STATOR COOLING WATER (SCW)	M-SCW-01
GAS LOCKING OF STATOR COOLING WATER (SCW) SYSTEM	M-SCW-08

S-SCW-07 DIFFERENTIAL PRESSURE OF SCW ACROSS STATOR WINDING BELOW NORMAL

APPLICABILITY:

All generators with water inner-cooled stator windings

DESCRIPTION:

This symptom applies whenever the pressure drop across the stator winding is below *normal*. By *normal*, it means the differential pressure of the SCW across the stator when operating at rated load, and/or the minimum delta pressure recommended by the OEM.

RELATED MALFUNCTIONS	NUMBER
GAS LOCKING OF STATOR COOLING WATER (SCW) SYSTEM	M-SCW-08
RUPTURED WATER MANIFOLD (HEADER)	M-S-47
RUPTURED WATER PARALLEL MANIFOLD (HEADER)	M-S-48
STATOR COOLING WATER FLOW BELOW NORMAL	M-SCW-14

S-SCW-08 INLET PRESSURE TO PRIMARY SIDE OF SCW HEAT EXCHANGER BELOW NORMAL

APPLICABILITY:

All generators with water inner-cooled stator windings

DESCRIPTION:

This symptom applies whenever the pressure of the raw water at the inlet of the SCW heat exchanger(s) is below *normal*. By *normal*, it means a pressure that is below that pressure typically measured for this unit under similar operating conditions, or that result in the SCWs underperforming.

RELATED MALFUNCTIONS	NUMBER
LEAK OF SCW PRIMARY (RAW) COOLING WATER	M-SCW-10
PRIMARY WATER FLOW TO SCW COOLERS BELOW NORMAL	M-SCW-12

S-SCW-09 LOW LEVEL OF SCW IN THE H_2-DETRAINING TANK

APPLICABILITY:

All generators with water inner-cooled stator windings

DESCRIPTION:

This symptom applies whenever the level of SCW in the hydrogen detraining tank is lower than a vendor-recommended value or, as typical for the same unit at similar operating conditions.

First of all, check that there is no instrument error or mechanical malfunction with the tank valves or related components.

Also check to ensure that there is SCW makeup water available. For additional information to operators, see the following symptoms.

RELATED MALFUNCTIONS	NUMBER
GAS LOCKING OF STATOR COOLING WATER (SCW) SYSTEM	M-SCW-08
HYDROGEN INGRESSION INTO STATOR COOLING WATER	M-S-29

S-SCW-10 LOW LEVEL OF SCW IN THE MAKE-UP TANK

APPLICABILITY:

All generators with water inner-cooled stator windings

DESCRIPTION:

This symptom applies whenever a *low level* alarm is displayed indicating the level of treated water in the make-up tank is low.

First of all check that the instrumentation is in working order. Also check that there is a supply of treated water from the station's water treatment plant. For additional information see the following symptoms.

RELATED MALFUNCTIONS	NUMBER
LEAK IN THE FLOW PATH OF THE STATOR COOLING WATER (SCW)	M-SCW-09
Problem with the supply of treated water to the SCW make-up tank	N/A

S-SCW-11 PRIMARY WATER FLOW TO SCW COOLERS BELOW NORMAL

APPLICABILITY:

All generators with water inner-cooled stator windings

DESCRIPTION:

This symptom applies in the case where the flow of primary cooling water to the SCW system is *below normal*. By *below normal,* it means a flow that cannot sustain adequate cooling of the stator winding at rated output. Zero flow is also covered by this symptom.

RELATED MALFUNCTIONS	NUMBER
PRIMARY WATER FLOW TO SCW COOLERS BELOW NORMAL	M-SCW-12

S-SCW-12 STATOR COOLING WATER FLOW BELOW NORMAL

APPLICABILITY:

All generators with water inner-cooled stator windings

DESCRIPTION:

This symptom applies whenever the flow of SCW is *below normal*. By *below normal*, it means a flow that is below the flow rate required to keeping the winding temperatures within their allowable operating range, while the unit is at rated load.

RELATED MALFUNCTIONS	NUMBER
BLOCKAGE IN THE FLOW OF STATOR COOLING WATER (SCW)	M-SCW-01
GAS LOCKING OF PHASE COOLANT	M-S-22
GAS LOCKING OF PHASE-PARALLEL COOLANT	M-S-23
STATOR COOLING WATER FLOW BELOW NORMAL	M-SCW-14
STATOR COOLING WATER HEAT EXCHANGER UNDERPERFORMING	M-SCW-15

S-SCW-13 STATOR COOLING WATER OUTLET PRESSURE BELOW NORMAL

APPLICABILITY:

All generators with water inner-cooled stator windings

DESCRIPTION:

This symptom applies whenever the pressure of the SCW is below *normal*. By *normal*, it means a pressure that is required to maintain efficient cooling of the stator winding at rated load, and/or as a minimum recommended value by the OEM. (Outlet pressure means outlet from the pump/motor or inlet to the generator.)

RELATED MALFUNCTIONS	NUMBER
BLOCKAGE IN THE FLOW OF STATOR COOLING WATER (SCW)	M-SCW-01

S-SCW-14 STATOR COOLING WATER PRESSURE DROP ACROSS SCW FILTER/STRAINER ABOVE NORMAL

APPLICABILITY:

All generators with water inner-cooled stator windings

DESCRIPTION:

This symptom applies to those cases where the differential pressure across the SCW filter/strainer is *above normal*. By *above normal*, it means a drop in pressure that is higher than normally found for the same ambient and load conditions, and/or that renders the SCW system unable to keep the temperature of the stator winding within its allowable operating range.

FIGURE 4.35
A typical SCW organic filter (1–3 μm).

Figure 4.35 shows a typical filter. Copper corrosion byproducts and/or other abnormal deposits may clog these filters increasing the pressure drop across the filter. Similar clogging can happen to the strainer(s).

RELATED MALFUNCTIONS	NUMBER
STATOR COOLING WATER (SCW) FILTER/STRAINER RESTRICTION	M-SCW-13

S-SCW-15 TEMPERATURE OF GLOBAL OUTLET STATOR COOLING WATER ABOVE NORMAL

APPLICABILITY:

All generators with water inner-cooled stator windings

DESCRIPTION:

This symptom applies whenever the temperature of the water (typically measured by TCs) collected from all conductor bars at the outlet manifold/header is above *normal*. By *normal*, it means to be above the range of temperatures commonly measured in the same machine under similar load and ambient temperature conditions, or as defined by the OEM.

RELATED MALFUNCTIONS	NUMBER
PRIMARY WATER FLOW TO SCW COOLERS BELOW NORMAL	M-SCW-12
STATOR COOLING WATER FLOW BELOW NORMAL	M-SCW-14
STATOR COOLING WATER HEAT-EXCHANGER UNDER PERFORMING	M-SCW-15
TEMPERATURE OF A PARALLEL CIRCUITABOVE NORMAL	M-S-51
TEMPERATURE OF A SINGLE PHASE ABOVE NORMAL	M-S-52
TEMPERATURE OF PRIMARY (SCW) WATER AT INLET OF SCW COOLERS ABOVE NORMAL	M-SCW-16

S-SCW-16 TEMPERATURE OF PRIMARY (RAW) WATER AT INLET OF COOLERS ABOVE NORMAL

APPLICABILITY:

All generators with water inner-cooled stator windings

DESCRIPTION:

This symptom applies to those situations where the inlet temperature of the primary (raw) water to the SCW heat exchanger is *above normal*. By *above normal*, it means a temperature that renders the SCW system incapable of removing enough heat during operation at rated load, to maintaining all stator temperatures within their allowable range.

Note: This can be a problem during the hottest days of the summer months if the raw water supply is abnormally high.

RELATED MALFUNCTIONS	NUMBER
TEMPERATURE OF PRIMARY (RAW) WATER AT INLET OF COOLERS ABOVE NORMAL	M-SCW-16

Figure 4.36 shows the inlet and outlet cooling water feeding the air heat exchangers on a 188 MVA 2-pole generator.

FIGURE 4.36
Cooling water piping to and from the air coolers of a 188 MVA, 2-pole turbogenerator. The coolers are installed horizontally under the core.

5

List of Malfunctions

This chapter, describing 171 malfunctions (Table 5.1), contains a comprehensive but not exhaustive list of malfunctions that may represent one or another developing problem within the generator. Each one of the malfunctions listed in this chapter has one or more related symptoms listed in Chapter 4. The malfunctions are listed under different major components or systems of the generator, and are shown with a short numerical identifier to help identify the component or system.

Component of System	Numerical Identifier Key
Core and frame	M-C&F-xx
Stator	M-S-xx
Rotor	M-R-xx
Excitation	M-E-xx
Lube-oil system	M-LOS-xx
Seal-oil system	M-SOS-xx
Hydrogen	M-H-xx
Stator cooling water system	M-SCW-xx

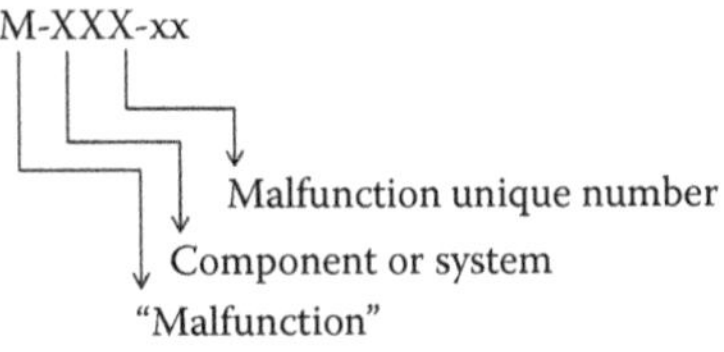

Table 5.1 lists all malfunctions with their corresponding explanations and each malfunction includes a list of related symptoms.

Note 1: Significant overspeed is a rear occurrence that normally results in serious degradation or even destruction of the generator, turbine, and/or other rotating components. This topic belongs in the realm of generator protection and it is not covered in this book. The reader may refer to Chapter 4 of the *Handbook of Large Turbo-Generator Operation and Maintenance* by G. Klempner and I. Kerszenbaum (2008).

Note 2: The *Thermal Sensitivity Test (M-R-29)* is actually *not* a malfunction. It is for convenience sake listed under malfunctions because it is the only test described in this book. It is designed to identify sources of vibration in the machine by devising a number of tests that check how the vibrations respond to certain parameter changes in the machine.

TABLE 5.1

List of Malfunctions

Number	MALFUNCTION DESCRIPTION BY SYSTEM
	CORE AND FRAME
M-C&F-01	ARCING BETWEEN CORE AND KEYBARS
M-C&F-02	BLOCKED COOLING VENT IN STATOR CORE
M-C&F-03	BREACH OF THE FLUX-SHIELD COOLING PATH
M-C&F-04	BURNING/PYROLYSIS OF FLUX-SHIELD INSULATION
M-C&F-05	BURNING/PYROLYSIS OF CORE INSULATION
M-C&F-06	BYPRODUCTS FROM BURNING INSULATION
M-C&F-07	FOREIGN BODY INTERACTION WITH STATOR CORE
M-C&F-08	GAS LOCKING IN FLUX-SHIELD COOLING PATH
M-C&F-09	GLOBAL TEMPERATURE OF CORE ABOVE NORMAL
M-C&F-10	HIGH VIBRATION OF CORE AND/OR FRAME
M-C&F-11	HYDROGEN INGRESSION INTO STATOR COOLING WATER (SCW) FROM FLUX SHIELD
M-C&F-12	INSUFFICIENT COOLING OF THE FLUX SHIELD
M-C&F-13	INTERLAMINAR INSULATION FAILURE
M-C&F-14	LEAK OF STATOR COOLING WATER (SCW) FROM A FLUX SHIELD
M-C&F-15	LOCAL CORE FAULT
M-C&F-16	LOOSE CORE LAMINATIONS
M-C&F-17	LOOSE GENERATOR FOOTING
M-C&F-18	OIL AND WATER LEAKS INSIDE THE GENERATOR
M-C&F-19	OVERVOLTAGE
M-C&F-20	TEMPERATURE OF A LOCAL CORE AREA ABOVE NORMAL
M-C&F-21	TEMPERATURE OF CORE END ABOVE NORMAL
M-C&F-22	TEMPERATURE OF FLUX SHIELD ABOVE NORMAL
	STATOR
M-S-01	BLOCKAGE IN THE FLOW OF STATOR TERMINAL COOLANT
M-S-02	BLOCKAGE OF COOLING FLOW IN A STATOR BAR
M-S-03	BLOCKAGE OF PARALLEL-PHASE-CONNECTOR COOLING PATH
M-S-04	BLOCKAGE OF PHASE COOLANT FLOW
M-S-05	BLOCKAGE OF PHASE-CONNECTOR COOLING PATH
M-S-06	BREACH OF THE STRANDS OR TUBES CARRYING COOLING WATER
M-S-07	BROKEN/CRACKED STATOR TERMINAL
M-S-08	BURNING/PYROLYSIS OF STATOR WINDING INSULATION
M-S-09	COPPER OXIDE BUILDUP IN WATER-CARRYING STRANDS
M-S-10	CRACKED CONDUCTOR STRANDS
M-S-11	CURRENT OVERLOAD OF STATOR WINDING
M-S-12	CURRENT UNBALANCE BETWEEN PHASES
M-S-13	CURRENT UNBALANCE BETWEEN PHASE PARALLELS
M-S-14	DAMAGE FROM MAGNETIC TERMITES

(Continued)

TABLE 5.1 (*Continued*)
List of Malfunctions

Number	MALFUNCTION DESCRIPTION BY SYSTEM
M-S-15	DAMAGE TO CONDUCTOR BAR INSULATION
M-S-16	DAMAGE TO CONDUCTOR WALL INSULATION DUE TO A WATER LEAK
M-S-17	DAMAGE TO HV BUSHINGS
M-S-18	DEFECTIVE STATOR WINDING INSULATION
M-S-19	FAILURE OF RADIAL PHASE CONNECTOR
M-S-20	FAILURE OF SERIES CONNECTION
M-S-21	GAS LOCKING IN WATER-CARRYING STRANDS OR TUBES
M-S-22	GAS LOCKING OF PHASE COOLANT
M-S-23	GAS LOCKING OF PHASE-PARALLEL COOLANT
M-S-24	GAS LOCKING OF WATER-COOLED TERMINALS
M-S-25	GLOBAL STATOR TEMPERATURE ABOVE NORMAL
M-S-26	GROUND FAULT OF STATOR WINDING
M-S-27	HIGH RESISTANCE OF PHASE-CONNECTOR JOINT
M-S-28	HIGH RESISTANCE OF PHASE-PARALLEL-CONNECTOR JOINT
M-S-29	HYDROGEN INGRESSION INTO STATOR COOLING WATER
M-S-30	HYDROGEN INGRESSION INTO STATOR COOLING WATER FROM STATOR TERMINALS
M-S-31	INSUFFICIENT FLOW OF COOLANT DUE TO BUCKLED COOLING STRANDS OR TUBES
M-S-32	INSUFFICIENT FLOW OF COOLING MEDIUM IN A STATOR BAR
M-S-33	INSUFFICIENT FLOW OF PHASE COOLANT
M-S-34	INSUFFICIENT FLOW OF PHASE-PARALLEL COOLANT
M-S-35	INSUFFICIENT FLOW OF TERMINAL COOLANT
M-S-36	LEAK OF CONDUCTING BAR COOLING WATER INTO GENERATOR
M-S-37	LEAK OF STATOR COOLING WATER (SCW) INTO GENERATOR
M-S-38	LEAK OF STATOR COOLING WATER (SCW) INTO GENERATOR FROM TERMINALS
M-S-39	LOCALIZED OVERHEATING OF STATOR CONDUCTOR
M-S-40	LOOSENING OF STATOR ENDWINDING BLOCKING
M-S-41	LOOSENING OF STATOR WINDING SLOT WEDGES
M-S-42	NEGATIVE-SEQUENCE CURRENTS
M-S-43	OVERHEATING OF PHASE CONNECTOR
M-S-44	PARTIAL DISCHARGE ACTIVITY (PDA) HIGH
M-S-45	PHASE CURRENT HIGH
M-S-46	RADIO FREQUENCY (RF) ACTIVITY
M-S-47	RUPTURED WATER MANIFOLD (HEADER)
M-S-48	RUPTURED WATER PARALLEL MANIFOLD (HEADER)
M-S-49	STATOR COOLING WATER (SCW) LEAK INTO GENERATOR FROM PHASE PARALLEL

(*Continued*)

TABLE 5.1 (*Continued*)

List of Malfunctions

Number	MALFUNCTION DESCRIPTION BY SYSTEM
M-S-50	STATOR COOLING WATER (SCW) LEAK INTO GENERATOR FROM MANIFOLDS
M-S-51	TEMPERATURE OF A PARALLEL CIRCUIT ABOVE NORMAL
M-S-52	TEMPERATURE OF A SINGLE PHASE ABOVE NORMAL
M-S-53	TEMPERATURE OF STATOR CONDUCTOR ABOVE NORMAL
M-S-54	TEMPERATURE OF STATOR TERMINAL ABOVE NORMAL
M-S-55	VIBRATION OF STATOR CONDUCTOR BAR IN SLOT
M-S-56	VIBRATION OF STATOR ENDWINDING—HIGH
M-S-57	VOLTAGE OF GRID—UNBALANCED
M-S-58	WINDING (STATOR) OPEN CIRCUIT
	ROTOR
M-R-01	ARCING ACROSS BEARINGS AND HYDROGEN SEALS
M-R-02	BEARING MISALIGNMENT
M-R-03	BLOCKAGE OF ROTOR VENTILATION PATH
M-R-04	BURNING/PYROLYSIS OF ROTOR INSULATION
M-R-05	CRACKED SHAFT
M-R-06	DEFECTIVE FIELD WINDING INSULATION
M-R-07	DEGRADED/DAMAGED FIELD WINDING INSULATION
M-R-08	DYNAMIC ROTOR UNBALANCE (MECHANICAL)
M-R-09	FAILURE OF BEARING INSULATION
M-R-10	FAILURE OF H_2 SEAL INSULATION
M-R-11	FAILURE OF HYDROGEN SEALS
M-R-12	FAILURE OF OIL WIPER
M-R-13	GROUNDED FIELD WINDING
M-R-14	GROUNDED SHAFT (MOMENTARILY)
M-R-15	INEFFECTIVE SHAFT GROUNDING
M-R-16	INGRESSION OF OIL INTO GENERATOR
M-R-17	LOSS OF STABILITY
M-R-18	MECHANICAL FAILURE OF ROTOR COMPONENT
M-R-19	MISALIGNMENT OF HYDROGEN SEAL
M-R-20	RUB BETWEEN BEARING AND JOURNAL
M-R-21	RUB OF HYDROGEN SEAL
M-R-22	RUB OF OIL WIPER
M-R-23	SHAFT VOLTAGE—HIGH
M-R-24	SHORTED TURNS IN FIELD WINDING
M-R-25	TEMPERATURE OF BEARING ABOVE NORMAL
M-R-26	TEMPERATURE OF FIELD WINDING ABOVE NORMAL
M-R-27	TEMPERATURE OF HYDROGEN SEAL ABOVE NORMAL
M-R-28	THERMAL ASYMMETRY OF ROTOR
M-R-29	THERMAL SENSITIVITY TEST (Note 2)

(*Continued*)

TABLE 5.1 *(Continued)*

List of Malfunctions

Number	MALFUNCTION DESCRIPTION BY SYSTEM
M-R-30	VIBRATION OF ROTOR—HIGH
M-R-31	WATER LEAK IN A WATER-COOLED ROTOR
	OVERSPEED (Note 1)
	EXCITATION
M-E-01	ACCELERATED BRUSH WEAR
M-E-02	BLOCKAGE OF BRUSHGEAR COOLING AIR
M-E-03	DAMAGE/DEGRADATION OF THE BRUSHGEAR
M-E-04	FAILURE OF THE EXTERNAL STATIC EXCITATION
M-E-05	FIELD CURRENT TOO LOW OFFLINE
M-E-06	FIELD CURRENT TOO LOW ONLINE
M-E-07	FREQUENCY BELOW NOMINAL—ONLINE
M-E-08	INSUFFICIENT COOLING OF SOLID-STATE EXCITATION CABINETS
M-E-09	MVAR TRANSIENTS
M-E-10	MW TRANSIENTS
M-E-11	OPERATION OUTSIDE CAPABILITY CURVE—ONLINE
M-E-12	OVEREXCITATION OFFLINE
M-E-13	OVEREXCITATION ONLINE
M-E-14	OVERFLUXING OFFLINE
M-E-15	OVERFLUXING ONLINE
M-E-16	OVERHEATING OF SOLID-STATE EXCITATION POWER COMPONENT
M-E-17	ROTOR SPEED BELOW NORMAL—ONLINE
M-E-18	SUDDEN LARGE CHANGE IN GRID IMPEDANCE
M-E-19	SYNCHRONIZATION OUT OF PHASE
M-E-20	TEMPERATURE OF INLET COOLING AIR TO BRUSHGEAR—HIGH
M-E-21	TEMPERATURE OF SLIPRINGS/BRUSHGEAR ABOVE NORMAL
M-E-22	TERMINAL VOLTAGE EXCURSION—OFFLINE
M-E-23	TERMINAL VOLTAGE EXCURSION—ONLINE
	HYDROGEN SYSTEM
M-H-01	AIR OR HYDROGEN HEAT-EXCHANGER COOLING-WATER SUPPLY OFF
M-H-02	BLOCKAGE OF AIR COOLERS
M-H-03	BLOCKAGE OF HYDROGEN COOLERS
M-H-04	DEGRADED HYDROGEN PURITY
M-H-05	ELEVATED HYDROGEN DEW POINT
M-H-06	FAILURE OF HYDROGEN DRYER
M-H-07	GAS LOCKING OF COOLING WATER IN HYDROGEN OR AIR HEAT EXCHANGERS
M-H-08	HYDROGEN ESCAPING THE GENERATOR

(Continued)

TABLE 5.1 *(Continued)*

List of Malfunctions

Number	MALFUNCTION DESCRIPTION BY SYSTEM
M-H-09	HYDROGEN PRESSURE INSIDE THE GENERATOR BELOW NORMAL
M-H-10	INADEQUATE FLOW OF COOLING WATER TO AIR OR H_2 COOLERS
M-H-11	INLET TEMPERATURE OF AIR OR HYDROGEN COOLING WATER HIGH
M-H-12	INSUFFICIENT PERFORMANCE OF AIR OR HYDROGEN HEAT EXCHANGERS
M-H-13	LEAK OF HYDROGEN INTO H_2-COOLING WATER SYSTEM
M-H-14	MOISTURE IN HYDROGEN—HIGH
M-H-15	PHYSICAL BREACH OF HYDROGEN OR AIR HEAT EXCHANGER
M-H-16	PRESSURE OF HYDROGEN SUPPLY—LOW
M-H-17	TEMPERATURE OF COLD GAS ABOVE NORMAL
M-H-18	WATER LEAK INTO GENERATOR FROM AIR OR HYDROGEN COOLER
	STATOR COOLING WATER (SCW) SYSTEM
M-SCW-01	BLOCKAGE IN THE FLOW OF STATOR COOLING WATER (SCW)
M-SCW-02	BLOCKAGE OF STATOR COOLING WATER (SCW) HEAT EXCHANGERS
M-SCW-03	BLOCKAGE OF STATOR COOLING WATER (SCW) STRAINERS/FILTERS
M-SCW-04	CONDUCTIVITY OF STATOR COOLING WATER ABOVE NORMAL
M-SCW-05	COOLER OUTLET TEMPERATURE OF SCW ABOVE NORMAL
M-SCW-06	COPPER OXIDE BUILDUP IN WATER-CARRYING STRANDS
M-SCW-07	DEGRADED/RUPTURED "O" RINGS OR OTHER SCW PIPING CONNECTIONS
M-SCW-08	GAS LOCKING OF STATOR COOLING WATER (SCW) SYSTEM
M-SCW-09	LEAK IN THE FLOW PATH OF THE STATOR COOLING WATER (SCW)
M-SCW-10	LEAK OF SCW PRIMARY (RAW) COOLING WATER
M-SCW-11	LEAKY SCW HOSES
M-SCW-12	PRIMARY WATER FLOW TO SCW COOLERS BELOW NORMAL
M-SCW-13	STATOR COOLING WATER (SCW) FILTER/STRAINER RESTRICTION
M-SCW-14	STATOR COOLING WATER FLOW BELOW NORMAL
M-SCW-15	STATOR COOLING WATER HEAT EXCHANGER UNDERPERFORMING
M-SCW-16	TEMPERATURE OF PRIMARY (RAW) WATER AT INLET OF SCW COOLERS ABOVE NORMAL
	SEAL-OIL SYSTEM (SOS)
M-SOS-01	BLOCKAGE OF SEAL-OIL FLOW
M-SOS-02	BLOCKAGE OF SEAL-OIL HEAT EXCHANGERS ON OIL SIDE
M-SOS-03	DEGRADED CONDITION OF THE SEAL OIL
M-SOS-04	EFFICIENCY OF SEAL-OIL COOLERS REDUCED

(Continued)

TABLE 5.1 *(Continued)*

List of Malfunctions

Number	MALFUNCTION DESCRIPTION BY SYSTEM
M-SOS-05	FAILURE OF SEAL-OIL PUMP
M-SOS-06	FAILURE OF SEAL-OIL VACUUM PUMP
M-SOS-07	FLOW OF COOLING WATER TO SEAL-OIL COOLERS BLOCKED
M-SOS-08	FLOW OF PRIMARY (RAW) COOLING WATER TO SEAL-OIL COOLERS BELOW NORMAL
M-SOS-09	INEFFECTIVE SEAL-OIL DETRAINING
M-SOS-10	LEAK FROM SEAL-OIL SYSTEM
M-SOS-11	LEAK OF COOLING WATER BEFORE INLET TO SEAL-OIL COOLERS
M-SOS-12	LEAK OF SEAL OIL INTO GENERATOR
M-SOS-13	PRESSURE OF SEAL OIL ABOVE NORMAL
M-SOS-14	SEAL-OIL COOLING-WATER INLET TEMPERATURE—HIGH
M-SOS-15	TEMPERATURE OF SEAL-OIL COOLER OUTLET ABOVE NORMAL
	LUBE-OIL SYSTEM (LOS)
M-LOS-01	BLOCKAGE OF LUBE-OIL FLOW
M-LOS-02	EFFICIENCY OF LUBE-OIL COOLERS DEGRADED
M-LOS-03	FAILURE OF LUBE-OIL PUMP
M-LOS-04	FLOW OF COOLING WATER TO LUBE-OIL COOLERS BELOW NORMAL
M-LOS-05	FLOW OF COOLING WATER TO LUBE-OIL COOLERS BLOCKED
M-LOS-06	FLOW OF LUBE OIL TO BEARINGS BELOW NORMAL
M-LOS-07	INLET TEMPERATURE OF LUBE-OIL COOLING WATER—HIGH
M-LOS-08	LEAK FROM LUBE-OIL SYSTEM
M-LOS-09	LEAK OF COOLING WATER BEFORE INLET TO LUBE-OIL COOLERS
M-LOS-10	LUBE-OIL IMPURITIES
M-LOS-11	TEMPERATURE OF LUBE-OIL COOLER OUTLET ABOVE NORMAL

Some clarifications about the description of the malfunctions:

Possible causes, possible consequences, and possible symptoms: In most malfunction descriptions, the reader will see an entry titled *Possible Causes*, another titled *Possible Consequences*, and a third one titled *Possible Symptoms*. Under possible causes, a number of malfunctions are listed that are or might be, by themselves, sources or factors in the development of the malfunction being discussed. For example, *High Vibration of Core and/or Frame (M-C&F-10)* may be one of the factors (malfunction) conducing to *Arcing between Core and Keybars (M-C&F-01)*. Also appearing under possible causes one may find other factors not described in the book as a malfunction, because they become obvious by their description.

Under possible consequences, the reader will find a list of malfunctions that can be of collateral damage to the malfunction being described. Also in this case, the list may include a number of consequences that are not directly covered in the book as malfunctions, but their meaning is clearly understood from the description in the text.

Finally, under *possible symptoms*, the reader may find a list of symptoms listed in the book that may be originated by the malfunction being discussed, as well as other symptoms that are not listed, but whose meanings are readily understood from the written description.

Operators may consider: Most malfunction descriptions have an entry under the title *Operators May Consider.* Obviously, the division of responsibilities between operator, technical staff, and management is different among the myriad of power plants in existence. Also, it differs depending on the type of problem encountered, the promptness of the required remedial action, and the criticality of the unit. Thus, when the authors state *Operators* they refer primarily to the operators, but also to anyone that has a decision making function on the operation of the unit.

5.1 Types of Malfunctions

5.1.1 Component Failure

Component-failure malfunctions describe those events that result in complete failure of a component. Depending on the component and its function, such a failure may or may not result in a trip, or require immediately removing the unit from service for repairs. For example, failure of a temperature probe may be of little significance to the overall integrity of the machine. It may require the making of some temporary or permanent changes to a low voltage circuit (carrying the probe's signal), and/or deleting a visual display of the probe's temperature. On the other hand, failure of a retaining ring will result in major damage with the unit coming offline instantaneously.

5.1.2 Component Degradation

Component-degradation malfunctions describe those events that are the result of degradations that have not, as yet, resulted in total failure of a component. For example, a slipring may have developed rough areas and maybe some eccentricity that shows itself as sparking between the ring and brushes. This type of degradation must be addressed, but may not require immediate action. Another example is the existence of shorted turns in the rotor field winding. Many machines are operating well while having one or more such shorts.

A key factor in assessing component degradation is the likelihood for developing into a component failure. For instance, in the case of the *sparking* slipring, what might be the likelihood it will result in a major short-circuit on the brushgear causing major damage to the equipment and a sudden trip of the generator offline.

In the pages ahead in this chapter, malfunctions describe either failures or degradation. However, the reader must keep in mind that in reading a malfunction, there is a level of severity, according to which a problem can be taken to represent component degradation or a component failure.

There are cases in which a symptom not necessarily indicates a component problem, but it is related to some type of abnormal operational event, such as a grid unbalance or frequency disturbance. These can result, if severe enough, in component malfunction (degradation and/or failure).

5.2 Core and Frame Malfunctions

M-C&F-01: ARCING BETWEEN CORE AND KEYBARS

APPLICABILITY:

All generators

MALFUNCTION DESCRIPTION:

All machines may at some time or another experience some degree of arcing between the back of core and the keybars. There are a number of possible contributing factors to this phenomenon. One of the primary factors is relative vibration between the core and frame allowing sparking to occur as the keybar makes and breaks contact with the stator core back. The other is the level of magnetic saturation of the stator core and the resulting leakage flux that causes current to flow in the keybar structure. Generally, both are required for significant arcing to occur. Another factor is how many keybars are in contact with each lamination (one, two, or three). One more is the manner the keybars are attached to the compression plate: bolted or welded, and if there are electrical-bonding strips of copper or aluminum between keybars. Finally, are all keybars grounded to the core, or only one?

Arcing between core and keybars is commonly known as *back of core burning* (BCB). It occurs due to current transfer between the stator core frame keybars and the stator iron core, at the back of the core where the keybars make contact with the core in the dovetail area. The currents induced can be anywhere from negligible to extremely damaging. The main influences for the current transfer mechanism are the degree of mechanical coupling in the radial and tangential direction between the core and keybars (i.e., vibration) and the magnetic saturation effects in the core, especially at the ends.

First, consider that the keybar structure forms a *squirrel cage*, which inherently carries currents due to leakage flux at the back of the core. This leakage flux links the squirrel cage and causes substantial current to flow axially down the keybars. The current then transfers from keybar to keybar at the core ends via a suitable path, which generally has the lowest impedance. The path can be the core itself or some part of the core-end clamping structure. It most likely occurs in the core, when noninsulated keybars are employed, but the current levels are generally not harmful in a machine that has a magnetically well designed core-end region and good contact between the core and the keybars. Magnetically well designed core end does not oversaturate during any mode of operation. In addition, the cage also acts as a damper to minimize or even negate flux linkage with the remainder of the core frame structure.

BCB generally occurs at the core ends. It has been reported that cases of BCB have been seen away from the ends but this has undoubtedly been on machines where a core failure has occurred. And, the BCB had occurred directly behind the location of the fault.

A major contributor to BCB, in many cases, has been oversaturation of the core ends. Therefore, it is essential to prevent it. If this cannot be prevented, then current transfer from the keybars to the core must be short-circuited by an alternative and lower impedance path. Having said this, BCB also requires an intermittent contact or high resistance joint between the core and the keybars for burning to occur. This has everything to do with the mechanical coupling between the core and the keybars, in the radial and tangential directions. Therefore, BCB can occur without oversaturation of the core ends, but it will be worse in a magnetically poorly designed core end.

Saturation of the core ends is most likely to occur from operation in the leading power factor range. This is due to the stator end region effects, which produce a higher level of axial flux impingement on the core ends in the leading power factor range. The net effect is that if the core is not protected against, or designed properly for this, the end iron will oversaturate and cause excessive leakage flux at the core back, on the ends.

Oversaturation can also occur on a well-designed core end, in the leading power factor range, if the particular machine employs magnetic retaining rings. Electromagnetic finite element analysis on generator end regions has shown a significant difference in effect between the use of magnetic versus nonmagnetic retaining rings, in some machines. In fact, up to a 40% reduction in core-end flux density at 0.95 power factor leading has been observed when nonmagnetic rings were used.

When there is core-end oversaturation, and hence higher leakage flux, there is a tendency for the interlaminar voltages at the core ends to increase and of course higher currents to occur in the keybar system due to this localized effect. It is believed that the increased interlaminar voltage at the core ends and the higher currents can create an arcing and sparking effect, which causes burning to occur at the core-to-keybar contact areas. In addition, make and break action from stator vibration is also suspected to influence the initiation of sparking due to the vibration effects. And further, if the core ends are oversaturated due to increased axial flux, the circulating currents in the individual laminations will be higher. (It also raises the core-end temperatures due to higher losses, which increases the chances of interlaminar core failures in the ends.) Therefore, there is also a possible contribution to the BCB simply by the eddy currents transferring and conducting through the keybar-to-core contact area.

One other interesting fact is that generally not all of the keybars actually carry the torsional load, and hence not all the keybars will have good contact with the core. Core–frame to core mechanical coupling analysis by finite elements has been done in the past and shown that in some cases, only about 20% of the keybars actually take up the torsional load. This may explain why BCB is not usually seen on all keybars, even though the leakage flux will be evenly distributed around the back of core circumference.

Some machines with significant BCB have been in operation for many years without any noticeable damage to the core and frame, or impact on production (see Figure 5.1). Nevertheless, these machines should undergo close scrutiny of their core backs, to ascertain the situation is not developing into a major problem.

FIGURE 5.1
The photo shows a mild back-of-core burning situation.

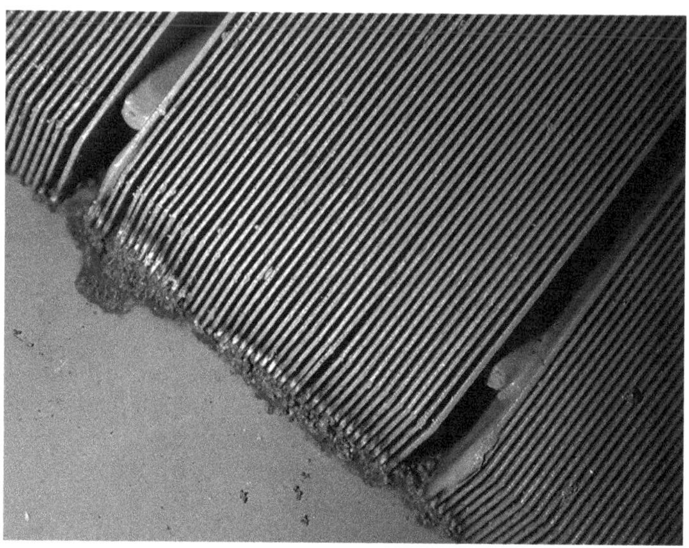

FIGURE 5.2
The photo shows a heavy back-of-core burning situation.

In some instances, otherwise mild BCB may result in annoying alarms from the generator condition monitor (GCM).

When the arcing is of such magnitude that significant core-burning occurs, the otherwise normal situation may become problematic, by burning interlaminar insulation, and releasing arcing by-products that may be harmful to the machine (in the form of metallic particles that may become *magnetic termites*) (see Figure 5.2).

PROBABLE CAUSES:

- Looseness of the core–frame fit due to inadequate design and/or manufacturing deficiencies
- HIGH VIBRATION OF CORE AND/OR FRAME (M-C&F-10)
- OVERFLUXING ONLINE (M-E-15)
- OVERFLUXING OFFLINE (M-E-14)
- TERMINAL VOLTAGE EXCURSION—ONLINE (M-E-23)
- TERMINAL VOLTAGE EXCURSION—OFFLINE (M-E-22)

POSSIBLE CONSEQUENCES:

- Damage to the interlaminar insulation at the back of the core
- Burning of the core laminates resulting in metallic particles that may, in extreme cases, be transported by the moving cooling gas onto the stator winding becoming *magnetic termites* (see Figure 5.2)

- Iron oxide deposits in the cooling vents and on the ID of the core
- Ozone smells in air-cooled units may be present

POSSIBLE SYMPTOMS:

- GENERATOR CONDITION MONITOR ALARM **(S-C&F-01)**
- RADIO FREQUENCY (RF) MONITOR HIGH **(S-S-13)**
- STATOR GROUND ALARM AND/OR TRIP **(S-S-17)** due to magnetic termites from excessive BCB
- If (1) and/or (2) above occur, and the machine has been experiencing high vibrations, it may be an indication of core–keybar arcing
- If (1) and/or (2) above occur, and they were preceded by an overfluxing event, or an overvoltage event, it may be an indication of core–keybar arcing

OPERATORS MAY CONSIDER:

1. In general, arcing between core and keybars does not usually represent an immediate problem. It is also difficult to diagnose without a visual inspection of the back of the core, which certainly cannot be done on most generators without a degree of disassembly. In general, BCB occurs on the last few packets of laminations at both ends of the core. Therefore, visual access is important mainly in those regions, and in most cases, some access for performing visual inspection of the end regions of the core is available after a degree of disassembly, usually by removing the hydrogen coolers.
2. If core–keybar arcing activity is determined to be significant (i.e., it is causing BCB that can be visually seen during inspections, and/or is generating pyrolysis byproducts that trigger the GCM or generating magnetic particles), an investigation should be undertaken to assess the severity of the problem and what, if any, measures should be taken to minimize risks to the machine. Bear in mind that if core–keybar arcing exists during normal operation (not overvoltages or overfluxing), there is little, if anything, that can be done to reduce the effect—most often the only viable action is to monitor the situation over time. In a few cases, OEMs were able to introduce changes to the existing machine that resulted in some improvement of the situation.

TEST AND VISUAL INDICATIONS:

- Visual BCB marks/deposits.
- Magnetic termites or loose metallic particles that chemical analysis shows are made of core-lamination material.

REFERENCES

1. G. Klempner and I. Kerszenbaum. *Handbook of Large Turbo-Generator Operation and Maintenance*. Piscataway, NJ: IEEE Press, 2008, chapters 4 and 8.
2. G. Klempner. Back of Core Burning. In: *GEC Generators Users Group Meeting, Ontario Power Generation*, Southern California Edison, San Onofre—Nuclear Generating Station, January 10–11, 2000.
3. D.R. Bertenshaw and A.C. Smith. Lamination Joint Reluctance in Stator Cores—A Forgotten Problem? In: *PEDM 2014, The 7th IET International Conference on Power Electronics, Machines and Drives*, Manchester, England, April 8–10, 2014.

M-C&F-02: BLOCKED COOLING VENT IN STATOR CORE

APPLICABILITY:

All generators

MALFUNCTION DESCRIPTION:

This malfunction relates to the partial or total blockage of a radial or axial core-cooling vent or a number of vents (see Figure 5.3).

Turbogenerator stator cores are made with radial and/or axial paths for the air or hydrogen to flow through, for the purpose of cooling it.

If radial vents are present, they take the form of gaps between packets of laminations separated by an "I" shaped nonmagnetic spacer. Therefore,

FIGURE 5.3
Blocked radial vents due to debris from a failed rotor.

a single vent extends 360° around the circumference of the core. Axial vents, when present, take the form of holes running from one side of the core to the other. Cores may have radial vents or axial vents.

Blocked vents effectively reduce the cooling of a section of the core. Depending on the degree of blocking, its location, and how extensive it is, the results could be global or local degradation of the interlaminar core insulation. In the case of indirect-cooled stator windings the core acts as a heat sink, removing the heat generated by the slot-embedded conductor bars. Therefore, degradation of the core cooling can affect the integrity of the conductor bar insulation, in particular if the resulting temperatures are high.

PROBABLE CAUSES/AREA AFFECTED:

- Axial vents may become plugged by debris, such as slot packing, or endwinding packing that has broken loose and found its way to the vent. The affected area is along the machine, limited to the vicinity of the blocked vents.
- Radial vents can become plugged by debris, such as slot packing, or endwinding packing that has broken loose and found its way to the vent. The affected area is local to the region of the blockage.
- Radial and axial vents can get plugged by rags or plastic bags left inadvertently during a maintenance or inspection activity. The affected areas are as noted above in previous two bullets, respectively.
- Radial vents can be partially blocked by deformation of the laminations at one or both sides of the vents. The affected area can be global, if the deformations are the result of wrongly designed *vent spacer* (the element that keeps the packets of laminations apart). It can also be local if the deformation is local due to a nondesign issue.
- Radial vents may become partially or completely plugged by extremely high production of oxidized copper dust in air-cooled machines with a very loose core, and/or if the core is loose within the frame. The affected area can be global, with different degrees of intensity along the core.

POSSIBLE CONSEQUENCES:

- Global or localized overheating of the interlaminar insulation with consequent degradation of the insulation
- Overheating of the stator coils with consequent damage or degradation of conducting-bar insulation, including a possible ground fault:
    - BURNING/PYROLYSIS OF CORE INSULATION **(M-C&F-05)**
    - BURNING/PYROLYSIS OF STATOR WINDING INSULATION **(M-S-08)**
    - BYPRODUCTS FROM BURNING INSULATION **(M-C&F-06)**

- DAMAGE TO CONDUCTOR BAR INSULATION **(M-S-15)**
- GLOBAL STATOR TEMPERATURE ABOVE NORMAL **(M-S-25)**
- GLOBAL TEMPERATURE OF CORE ABOVE NORMAL **(M-C&F-09)**
- GROUND FAULT OF STATOR WINDING **(M-S-26)**
- INTERLAMINAR INSULATION FAILURE **(M-C&F-13)**
- LOCAL CORE FAULT **(M-C&F-15)**
- LOCALIZED OVERHEATING OF STATOR CONDUCTOR **(M-S-39)**
- TEMPERATURE OF A LOCAL CORE AREA ABOVE NORMAL **(M-C&F-20)**

POSSIBLE SYMPTOMS:

Winding temperatures above normal, as indicated by several *symptoms*:

- GLOBAL TEMPERATURE OF CORE ABOVE NORMAL **(S-C&F-02)**
- GENERATOR CORE CONDITION—ALARM **(S-C&F-01)**

OPERATORS MAY CONSIDER:

1. If a core defect is suspected, merely reducing the load may not suffice, unless the defect is relatively small. Note should be taken of the fact that a localized core blockage may be somewhat removed from the RTD closest to it. Thus, the actual hot-spot temperature could be much higher than the one being measured. Damage to core insulation will result in a very long outage and loss of production. Therefore, any indication of a serious core problem is better addressed by bringing the unit offline as soon as possible.
2. Record all relevant temperatures versus load.

TEST AND VISUAL INDICATIONS:

- Visual inspection with rotor out or via airgap crawling robots may uncover clogged vents.
- EL-CID and/or full flux tests may be required to evaluate damage to core insulation.
- Electric tests and visual inspection may afford an estimation of how much the conducting bars insulation was affected by the hot core.

REFERENCES

1. G. Klempner and I. Kerszenbaum. *Handbook of Large Turbo-Generator Operation and Maintenance*. Piscataway, NJ: IEEE Press, 2008, chapters 2 and 8.
2. EPRI 1008378. Generator Core Overheating Risk Assessment: J.T. Deely Core Investigation, Technical Results, September 2003.

M-C&F-03: BREACH OF THE FLUX-SHIELD COOLING PATH

APPLICABILITY:

Machines with water-cooled flux shields

MALFUNCTION DESCRIPTION:

This malfunction describes a breach of the cooling path in a core-end flux shield (CEFS), also called *core-end flux screen*.

Many large machines have flux shields (almost always made of copper) installed on both ends of the core (see Figures 5.4 and 5.5). The purpose of these shields or screens is to ameliorate the impact that axial fluxes have on the ends of the core (called *core-end effects*), in particular when the generator operates at leading power factors.

CEFSs are usually water cooled when rotors with magnetic retaining rings are employed. This is because a magnetic retaining ring has the undesirable property of directing axial stray flux into the core ends in a more intense manner than nonmagnetic retaining rings. The airgap is reduced significantly in machines with this arrangement. Furthermore, the intensity is far greater in the leading power factor range and gets worse as the machine is operated more and more *leading*. Therefore, the flux shield becomes extremely hot if not force-cooled. Water tends to be better than hydrogen in this instance.

FIGURE 5.4
The water cooling of this flux shield is accomplished by a circular pipe or series of pipes brazed to the surface of the flux shield as shown (two in this case). The hottest part of the flux shield will be the bore side of the copper shield.

FIGURE 5.5
The cooling water is delivered from the stator cooling water system to the circular pipes brazed onto the flux shield by two pig tail-type pipes, isolated from ground (i.e., the stator core and frame) as shown above.

Another complication with water-cooled flux shields is that the use of water requires that it be demineralized. Hence, the source of the flux shield cooling water is the same stator cooling water (SCW) used for cooling the stator bars, and the cooling of the flux shield becomes part of the SCW system. And because of this, the flux shield must also be electrically isolated from the stator core. Thus, a failure of the cooling medium to fully meet its designed requirement can result in serious damage to the shield and to any insulation indirectly heated by the hot shield.

PROBABLE CAUSES:

Cracked/broken water-carrying hoses. This can be the result of wrong maintenance, inadequate design, and substandard manufacturing. It can also be the result of excessive machine vibration.

POSSIBLE CONSEQUENCES:

- Severe overheating of the flux shield, leading to deformation and failure of the support system, and burning of its insulation
- Overheating of core and winding insulation on and in the vicinity of the affected shield
- Gas locking by the hydrogen leaking into the CEFS may aggravate the lack of cooling

- Leak of SCW into the generator casing may adversely change the hydrogen's dew point and contaminate critical components
- BURNING/PYROLYSIS OF CORE INSULATION **(M-C&F-05)**
- BURNING/PYROLYSIS OF STATOR WINDING INSULATION **(M-S-08)**
- HYDROGEN INGRESSION INTO THE STATOR COOLING WATER **(M-S-29)**
- LEAK OF STATOR COOLING WATER (SCW) INTO GENERATOR **(M-S-37)**
- TEMPERATURE OF FLUX SHIELD ABOVE NORMAL **(M-C&F-22)**

POSSIBLE SYMPTOMS:

- HYDROGEN-IN-STATOR COOLING WATER (SCW) HIGH **(S-S-04)**
- LIQUID-IN-GENERATOR ALARM **(S-C&F-04)**
- TEMPERATURE OF CORE END ABOVE NORMAL **(S-C&F-06)**
- TEMPERATURE OF FLUX SHIELD ABOVE NORMAL **(S-C&F-07)** (By *normal*, it means temperatures measured on the same shield at previous occasions under similar load conditions).
- Some increase of stator winding temperatures above *normal* (see relevant symptoms).

OPERATORS MAY CONSIDER:

1. First thing after ascertaining instrumentation is working correctly is to assess the magnitude of the problem. Comparing temperatures with previous readings may offer a starting point.
2. A confirmed leak of hydrogen into the SCW system, without a subsequent *liquid-in-generator* alarm may indicate the breach is small, and if the temperature increase is mild, plans can be made to repair the unit at a convenient time.
3. If the hydrogen into SCW system leak is followed by *liquid-in-generator* alarm, or determination is made by other means that SCW is leaking into the generator, operators may consider taking the unit off-load for repairs ASAP, or in consultation with the OEM.
4. Keep record of pertinent temperatures and load points.
5. In all cases, evaluate not operating the unit in a leading power factor until the problem is resolved.

TEST AND VISUAL INDICATIONS:

- Pressure/vacuum tests
- Water in casing
- Excess hydrogen in the SCW system

REFERENCES

1. G. Klempner and I. Kerszenbaum. *Handbook of Large Turbo-Generator Operation and Maintenance*. Piscataway, NJ: IEEE Press, 2008, chapters 2, 8, and 11.
2. Generator End-Tooth Flux Shield, US patent 3731127 A, May 1973.
3. G. Vogt, C. Demian, R. Romary, G. Parent, and V. Costan. Experimental Model for Study of Electro-Magnetic Phenomena in Stator Core-End Laminations of Large Generators. *Progress in Electromagnetics Research B*, 56, 89–107, 2013.
4. P.J. Tavner and A.F. Anderson. Core Faults in Large Generators. *IEEE Proceedings—Electric Power Applications*, 152(6), 1427–1439, 2005.

M-C&F-04: BURNING/PYROLYSIS OF FLUX-SHIELD INSULATION

APPLICABILITY:

Generators with insulated flux shields at their core ends

MALFUNCTION DESCRIPTION:

To protect the core ends from overheating by eddy currents created by axial stray fluxes, mainly during operation with leading power factor, *CEFSs* are installed in many generators. In some designs, the CEFSs are insulated from the core and frame to avoid circulating currents and/or keeping the heat generated at the flux shield from impacting the core end. In particular, CEFSs are usually only insulated when they are water-cooled, to isolate the shield and the cooling water from ground (i.e., the stator core and frame). They may be only grounded in one point.

This malfunction relates to the case when for whatever reason the flux-shield overheats, damaging its insulation.

PROBABLE CAUSES:

In machines with water-cooled CEFSs:

- BREACH OF THE FLUX-SHIELD COOLING PATH (M-C&F-03)
- GAS LOCKING IN FLUX-SHIELD COOLING PATH (M-C&F-08)
- INSUFFICIENT COOLING OF THE FLUX SHIELD (M-C&F-12)
- Operation in leading power factors beyond the capability of the machine

POSSIBLE CONSEQUENCES:

- TEMPERATURE OF FLUX SHIELD ABOVE NORMAL (M-C&F-22)
- Severe overheating of the flux shield itself (normally made of copper), leading to deformation
- Damage to its insulation and/or insulated support systems

POSSIBLE SYMPTOMS:

- TEMPERATURE OF FLUX SHIELD ABOVE NORMAL **(M-C&F-07)**

(Meaning above the temperature that was normally measured in this machine under similar loads), or above maximum temperature recommended by OEM

- GENERATOR CORE CONDITION MONITOR—ALARM **(M-C&F-01)**

OPERATORS MAY CONSIDER:

1. If the only indication is temperature of the flux shield above normal, the urgency for taking action and the type of action will depend on the actual value of the temperature being measured. It is important the temperature remains below maximum values provided by the manufacturer, or in lieu of that, below temperature rise class for the machine in question.
2. If the increase in temperature precedes or comes together with an alarm from the GCM, consideration must be given to taking the unit offline ASAP to correct the problem, or in consultation with the OEM. Keep in mind that pyrolysis byproducts picked up by the GCM indicates the insulation is degrading quickly, and a loss of flux shield insulation may introduce circulating currents in the flux shield, creating additional losses and consequential temperature rise.

TEST AND VISUAL INDICATIONS:

- Discoloration of the flux shield and/or flux-shield insulation.
- Megger tests of flux-shield insulation (if grounded in one point, the ground must be lifted before insulation resistance [IR] tests are carried out).

REFERENCES

1. G. Klempner and I. Kerszenbaum. *Handbook of Large Turbo-Generator Operation and Maintenance*. Piscataway, NJ: IEEE Press, 2008, chapters 2, 8, and 11.
2. R. Singleton, P. Marshall, and J. Steel: Axial Magnetic Flux in Synchronous Machines: The Effect of Operating Conditions. *IEEE Transactions on Power Apparatus and Systems*, 100(3), 1226–1233, 1981.

M-C&F-05: BURNING/PYROLYSIS OF CORE INSULATION

APPLICABILITY:

All generators

MALFUNCTION DESCRIPTION:

This malfunction addresses the situation where core insulation has reached such high temperatures, that it is degrading much faster than following the Arrhenius progression. At this stage, it can be said that the insulation is burning or subject to pyrolysis.

Burning is the physical process of a fuel undergoing combustion, which is a rapid chemical oxidation creating byproducts and heat. Combustion requires the presence of oxygen, and the subjection of the fuel to a temperature above certain value, specific for each fuel. The fire triangle shows the conditions required for the presence of fire (Figure 5.6):

Let us look in the generator for the sources of the three components shown on the fire triangle:

Fuel: All insulation is rarely made entirely of organic materials; it is commonly made from a combination of organic and inorganic components (e.g., mica or glass as inorganic material and resins and epoxies as organic materials). The organic materials will burn in the presence of oxygen and high temperatures.

Oxygen: In air-cooled generators, the stator windings are in the presence of air, which is the source of the oxygen required for the initiation and sustenance of burning. In H_2-cooled generators, there is no air (and oxygen) present during the normal operation of the unit. Therefore, burning in the presence of fires can only exist if the casing has been severely breached and air is mixed with hydrogen. Depending on the H_2/air ratio, this by itself can produce high-combustion temperatures, lighting up stator insulation, which will then continue to burn.

Heat: The heat required to initiate burning of the insulation may have a number of sources. These are enumerated under *Probable Causes* below.

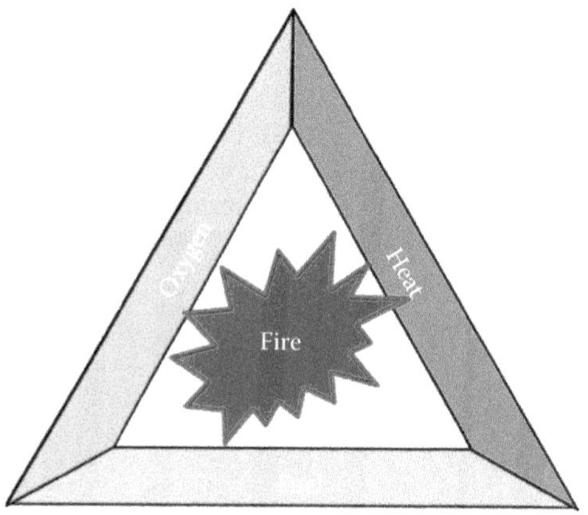

FIGURE 5.6
Fire triangle.

Pyrolysis: It is the chemical decomposition of organic materials in the absence of oxygen. Pyrolysis is the mechanism by which stator insulation deteriorates if a large amount of heat is applied in hydrogen-cooled generators.

For the purpose of investigating the development of rapid degradation of stator core insulation due to abnormal high temperatures, burning, and pyrolysis can be used as a single common factor.

Burning/pyrolysis may cause pyrolization of the interlaminar (core) insulation resulting in a conductive path for eddy currents, so that the eddy losses in the affected core area are magnified to a great extent.

PROBABLE CAUSES/AREAS AFFECTED:

Operation with leading power factor beyond capability curve:

- FIELD CURRENT TOO LOW ONLINE **(M-E-06)**
- OPERATION OUTSIDE CAPABILITY CURVE ONLINE **(M-E-11)**

Overfluxing of the core during a volts-per-hertz (V/Hz) event:

- OVERFLUXING OFFLINE **(M-E-14)**
- OVERFLUXING ONLINE **(M-E-15)**

Foreign metallic object shortening laminations:

- FOREIGN BODY INTERACTION WITH STATOR CORE **(M-C&F-07)**
- ARCING BETWEEN CORE AND KEYBARS **(M-C&F-01)**

Global loss of cooling:

- GLOBAL TEMPERATURE OF CORE ABOVE NORMAL **(M-C&F-09)**
- BLOCKED COOLING VENT IN STATOR CORE **(M-C&F-02)** (Localized)

Significant overload:

- OPERATION OUTSIDE CAPABILITY CURVE ONLINE **(M-E-11)**
- Damage to core laminations during a maintenance activity (like pulling out or in a rotor)
- A winding short-circuit that resulted in high currents flowing through the core
- Internal short-circuit, for example, between phases via core iron/section of the core
- Melting of core or significant core defect/localized

POSSIBLE CONSEQUENCES:

- Damage to the interlaminar core insulation to the extent that the condition will continue to deteriorate after the offending cause has been removed

- Partial meltdown of core material leading to
  - Major core repair
  - Ground faults
  - Stator rewind
  - Damage to the rotor from core material solidifying in the gas gap (airgap)

POSSIBLE SYMPTOMS:

- GENERATOR CORE CONDITION MONITOR—ALARM **(S-C&F-01)**
- OVERFLUXING (VOLTS per HERTZ) TOO HIGH OFFLINE **(S-E-11)**
- OPERATION WITH LEADING POWER FACTOR BEYOND ALLOWABLE REGION **(S-E-10)**

OPERATORS MAY CONSIDER:

Burning/pyrolysis of core insulation is difficult to determine while in operation, unless a recent V/Hz event occurred, or the unit was recently operated or is being operated with a significant leading power factor. Other causes will probably require taking the unit offline for testing and inspection, before they can be identified.

1. Confirming that the readings by the instrumentation (temperature and/or GCM alarm) are not due to instrumentation error.
2. Core burning may be deduced mainly from GCM (Figures 5.7 through 5.10) alarming. Higher-than-*normal* core temperatures

FIGURE 5.7
Typical old generator core/condition monitor (GCM).

FIGURE 5.8
Typical old instrument designed for the collection of pyrolysis byproducts. Normally installed together with the GCM of Figure 5.7.

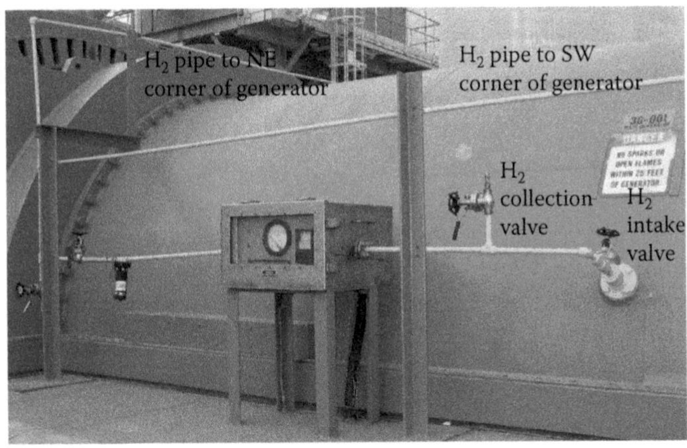

FIGURE 5.9
GCM that includes collection port for byproducts of pyrolysis installed on a four-pole nuclear-fueled generator.

without GCM information may only reveal a problem with the core, but not necessarily that temperatures are so high that burning/pyrolysis is occurring.

3. If a GCM alarm is received, analyze a hydrogen sample to ascertain the nature of the problem. (If tagged compounds were applied to the stator, analysis of the hydrogen sample may clearly indicate the core as the source of the problem, or may indicate a different source.)

FIGURE 5.10
Modern GCM for hydrogen-cooled generators. (From Environment One Corporation, Niskayuna, NY.)

4. Consider reducing the power output of the generator or removing it from service, according to the severity of the indications.

TEST AND VISUAL INDICATIONS:

- Analysis of GCM-trapped hydrogen sample
- Visual inspection of core
- EL-CID and/or offline flux tests

REFERENCES

1. G. Klempner and I. Kerszenbaum. *Handbook of Large Turbo-Generator Operation and Maintenance*. Piscataway, NJ: IEEE Press, 2008, chapter 4.
2. EPRI Technical Report 1007441. Repair and Testing Guide for Generator Cores Grounded at the Outside Diameter, December 2002.
3. P.J. Tavner and A.F. Anderson. Core Faults in Large Generators. *IEEE Proceedings—Electric Power Applications*, 152(6), 1427–1439, 2005.

M-C&F-06: BY-PRODUCTS FROM BURNING INSULATION

APPLICABILITY:

All generators

MALFUNCTION DESCRIPTION:

This malfunction addresses the situation where byproducts of burning insulation are captured by the GCM and generate an alarm.

Any insulation burning or undergoing pyrolysis inside the machine, be it from the stator winding, rotor winding, or stator laminations, indicates a very serious condition that must be given immediate attention.

If the GCM alarm is confirmed as *true*, then one avenue that may lead toward a better understanding of the developing problem is to remove a hydrogen sample and send it for analysis. Modern GCMs have the facility to do that within their system. Older instruments have one device just dedicated to the sampling of hydrogen (Figure 5.8); while in other cases a hydrogen collection valve is available (Figure 5.9).

FOR DISCUSSION OF THIS PROBLEM, REFER TO:

- BURNING/PYROLYSIS OF CORE INSULATION **(M-C&F-05)**
- BURNING/PYROLYSIS OF FLUX-SHIELD INSULATION **(M-C&F-04)**
- BURNING/PYROLYSIS OF ROTOR INSULATION **(M-R-04)**
- BURNING/PYROLYSIS OF STATOR WINDING INSULATION **(M-S-08)**

M-C&F-07: FOREIGN BODY INTERACTION WITH STATOR CORE

APPLICABILITY:

All generators

MALFUNCTION DESCRIPTION:

This malfunction addresses the presence of a foreign solid object interacting with the core.

Solid bodies may be nonmetallic, metallic, nonferromagnetic, and ferromagnetic. Depending on the type of material, the effects may be less or more severe, and the mechanisms involved different. For instance, nonmetallic objects will affect the core by impacting it with force (ejected by the rotor), or by blocking a cooling vent. Metallic objects may *weld* a number of continuous laminations creating a runaway core defect due to progressive deterioration of the interlaminar insulation. Large metallic objects may overheat due to eddy currents and if positioned next to insulation, damage it (see Figure 5.11). Magnetic particles may drill themselves into the coils producing ground faults or leaks, in water innercooled windings.

FIGURE 5.11
Shorted laminations on stator tooth top from loose foreign debris.

PROBABLE CAUSES:

- Foreign material exclusion (FME) violations resulting in foreign objects left inside the machine:
    - Washers/nuts/bolts
    - Pieces of insulation
    - Pieces of wiring
    - Coins/pens/screwdrivers/wrenches/other tools
    - And so on
- Component from inside the generator that broke loose
    - Washers and nuts
    - Insulation
    - Wiring
    - And so on

POSSIBLE CONSEQUENCES:

Local damage to core due to impact:

- LOCAL CORE FAULT (M-C&F-15)

Local damage to core due to short-circuiting of a number of laminations by a metallic object:

- INTERLAMINAR INSULATION FAILURE (M-C&F-13)

Local overheating due to blockage of a cooling vent by a foreign object:

- BLOCKED COOLING VENT IN STATOR CORE (M-C&F-02)

POSSIBLE SYMPTOMS:

- LOCAL CORE TEMPERATURE ABOVE NORMAL **(S-C&F-05)**
- GENERATOR CORE CONDITION MONITOR—ALARM **(S-C&F-01)**

OPERATORS MAY CONSIDER:

There are no direct indicators, with exception of a core temperature RTD picking up the resulting temperature rise from a core fault, or a GCM alarm. By then the fault may be well advanced. Another source of a core problem resulting in an elevated temperature rise is from the blockage of cooling vent(s). In this case, the situation can be reversed by removing the blockage. There is no way of knowing which one of the above, is the one creating the temperature rise. Therefore,

1. If the condition is mild, the temperature should be monitored to ascertain it is not trending higher. However, the location of the temperature detector may be removed from the actual fault location, erroneously giving the impression of a mild situation.
2. Core problems have the potential to cause severe economic losses—thus the most prudent action may be to remove the unit from operation and troubleshoot it if there is a concern or indication that core insulation may be in the process of being damaged.

TEST AND VISUAL INDICATIONS:

- Foreign object (or traces of it if is disintegrated by rotor action)
- EL-CID and/or flux tests
- Visual indications of core indentation and/or other type of damage caused by solid objects

REFERENCES

1. G. Klempner and I. Kerszenbaum. *Handbook of Large Turbo-Generator Operation and Maintenance*. Piscataway, NJ: IEEE Press, 2008, chapter 8.
2. EPRI Technical Report 1007441. Repair and Testing Guide for Generator Laminated Cores Grounded at the Core Outside Diameter, December 2002.

M-C&F-08: GAS LOCKING IN FLUX-SHIELD COOLING PATH

APPLICABILITY:

All generators with water-cooled end-core flux shields

MALFUNCTION DESCRIPTION:

This malfunction addresses the case of a partial or total loss of cooling to a flux shield cooled by SCW, when enough hydrogen leaks into the SCW system to create a flow restriction. A partial or total flow restriction to one or both CEFSs will have the effect of overheating the flux shield and its insulation. The result may extend to the point of deforming the flux shield and burning its insulation.

Many large machines have flux shields installed on both ends of the core. The purpose of these shields or screens is to ameliorate the impact that axial stray fluxes have on the ends of the core (called *core-end effects*), in particular when the generator operates at leading power factors.

Under certain load conditions (leading power factor operation), flux shields may become extremely hot if not force-cooled. Thus, a failure of the cooling medium to fully meet its designed requirement can result in serious damage to the shield, and any insulation that may be indirectly heated by the hot shield.

PROBABLE CAUSES:

- BREACH OF THE FLUX-SHIELD COOLING PATH **(M-C&F-03)**
- Hydrogen ingression into the SCW at other location than at the flux shield, but directly affecting one or both shields

POSSIBLE CONSEQUENCES:

- Severe overheating of the flux shield, leading to deformation and failure of support system, and burning of its insulation
- Leak of SCW into the generator casing may adversely change the hydrogen's dew point and contaminate critical components
- Overheating of core and winding insulation on and in the vicinity of the affected shield
- BURNING/PYROLYSIS OF STATOR WINDING INSULATION **(M-S-08)**
- BURNING/PYROLYSIS OF CORE INSULATION **(M-C&F-05)**
- HYROGEN INGRESSION INTO THE STATOR COOLING WATER **(M-S-29)**
- LEAK OF STATOR COOLING WATER (SCW) INTO GENERATOR **(M-S-37)**
- TEMPERATURE OF FLUX SHIELD ABOVE NORMAL **(M-C&F-22)**

POSSIBLE SYMPTOMS:

- GENERATOR CORE CONDITION MONITOR—ALARM **(S-C&F-01)**
- HYDROGEN-IN-STATOR COOLING WATER (SCW) HIGH **(S-S-04)**
- LIQUID-IN-GENERATOR ALARM **(S-C&F-04)**
- TEMPERATURE OF CORE END ABOVE NORMAL **(S-C&F-06)**
- TEMPERATURE OF FLUX SHIELD ABOVE NORMAL **(S-C&F-07)** (by *normal*, it means temperatures measured on the same shield at previous occasions under similar load conditions)
- Some increase of stator winding temperatures above *normal*

OPERATORS MAY CONSIDER:

1. First thing after ascertaining instrumentation is working correctly to assess the intensity of the problem. Comparing temperatures with previous readings may offer a starting point.

2. Keeping the flux-shield temperatures below the maximum limit provided by the manufacturer. This may require reducing load or bringing the unit offline.
3. Keeping records of pertinent temperatures and load points.
4. In all cases, evaluating operating the unit in power factors other than leading, until the problem is fixed.

TEST AND VISUAL INDICATIONS:

- Pressure/vacuum tests
- Water in casing
- Excess hydrogen in SCW system
- Hydrogen sample from GCM sent to analysis

REFERENCES

1. G. Klempner and I. Kerszenbaum. *Handbook of Large Turbo-Generator Operation and Maintenance*. Piscataway, NJ: IEEE Press, 2008, chapter 8.
2. EPRI Technical Report 1020275. Generator Stator Core Condition Monitoring by Tracking Shaft Voltage and Grounding Current, February 2010.

M-C&F-09: GLOBAL TEMPERATURE OF CORE—ABOVE NORMAL

APPLICABILITY:

All generators

MALFUNCTION DESCRIPTION:

This malfunction addresses the case where the entire core is exhibiting temperatures above *normal*. By *normal*, it means temperatures that have previously been taken for similar terminal voltage and load conditions, or within a range given by the OEM as permissive, or within a range given by the industry as common practice for this type of machine.

Figure 5.12 shows an RTD embedded in the back of the core in the middle of a laminations packet.

PROBABLE CAUSES:

- BLOCKAGE IN THE FLOW OF STATOR COOLING WATER (SCW) **(M-SCW-01)**
- HYDROGEN PRESSURE INSIDE GENERATOR BELOW NORMAL **(M-H-09)**
- INSUFFICIENT PERFORMANCE OF AIR OR HYDROGEN HEAT EXCHANGERS **(M-H-12)**
- OPERATION OUTSIDE CAPABILITY CURVE—ONLINE **(M-E-11)**
- OVEREXCITATION ONLINE **(M-E-13)**
- TERMINAL VOLTAGE EXCURSION—ONLINE **(M-E-23)**

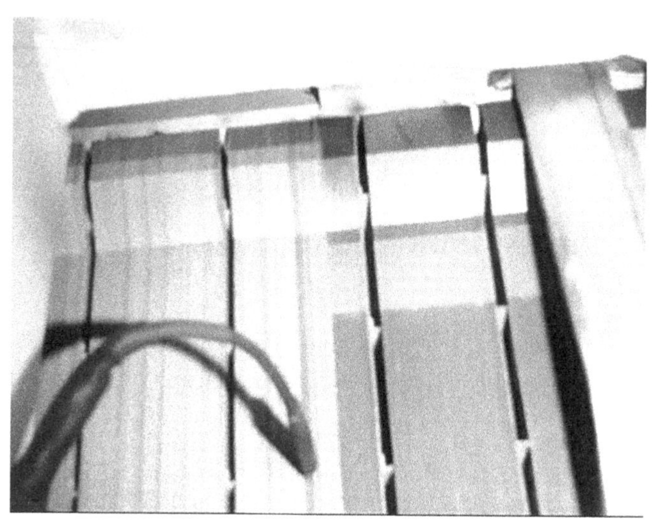

FIGURE 5.12
RTD embedded in back of the core.

POSSIBLE CONSEQUENCES:

- DAMAGE TO CONDUCTOR BAR INSULATION (M-S-15)
- GROUND FAULT OF STATOR WINDING (M-S-26)
- INTERLAMINAR INSULATION FAILURE (M-C&F-13)

POSSIBLE SYMPTOMS:

- Mainly temperature indicators embedded in the core
- Temperature probes embedded between top and bottom bars
- BYPRODUCTS FROM STATOR INSULATION PYROLYSIS (S-S-01)

OPERATORS MAY CONSIDER:

1. Reducing output until the temperatures return to the normal condition, if the instrumentation is verified as providing accurate readings.
2. Investigating the source of the problem and keeping monitoring the temperatures closely, if those temperatures are somewhat elevated but within the normal range, and not trending up quickly.
3. If the temperatures are back in the normal range, investigate source of the problem. If it is just an issue of adjusting valves, or similar activities, correct the situation and reload the unit.
4. If the temperature cannot be brought back to the normal range, consider taking the unit offline for troubleshooting.
5. At all times keep track of the GCM output. If burning/pyrolysis byproducts are detected, it could be a sign of a very deleterious situation developing requiring the unit to be brought offline immediately.

TEST AND VISUAL INDICATIONS:

Discoloration of insulation

REFERENCES

1. G. Klempner and I. Kerszenbaum. *Handbook of Large Turbo-Generator Operation and Maintenance*. Piscataway, NJ: IEEE Press, 2008, chapters 1, 3, 4, 5, and 9.
2. EPRI Technical Report 3002003590. Maintenance Guidelines for Generators Used in Simple and Combined Cycle Plants, December 2014.

M-C&F-10: HIGH VIBRATION OF CORE AND/OR FRAME

APPLICABILITY:

All generators

MALFUNCTION DESCRIPTION:

This malfunction is concerned with a situation where the core–frame of a generator is experiencing *high* vibrations. In the context of this problem, the term *high* means vibrations well above normally measured in the same machine under similar load conditions, or higher than recommended by the OEM or pertinent industry guidelines.

Vibrations in generators are typically within a narrow band, and change relatively little over the lifetime of the unit, unless some underlying problem causes them to do so. Vibrations can increase over time (trend up), or they can increase suddenly. Trending up of vibration may be related to a slow-progressing loosening of components (particularly, any anti-vibration system being employed, especially on two-pole machines), but can also be the result of growing cracks, or developing electrical problems (e.g., rotor shorted turns). In either instance, there is generally sufficient time to investigate and plan for a repair, if one is needed. On the other hand, a step increase may be the result of a problem that needs immediate attention.

PROBABLE CAUSES:

There are numerous causes that can result in elevated vibrations, among them:

- *Unbalanced magnetic pull*—primarily in four-pole machines, due to shorted turns in the rotor
- *Asymmetric heating and bowing of the rotor*—primarily in two-pole machines
- Loose fit between stator core and supporting keybars
- Cracked frame (see Figure 5.13)
- Cracked core supports
- Vibrations inducing in the frame–core assembly from an unbalanced rotor
- LOOSE CORE LAMINATIONS (M-C&F-16)

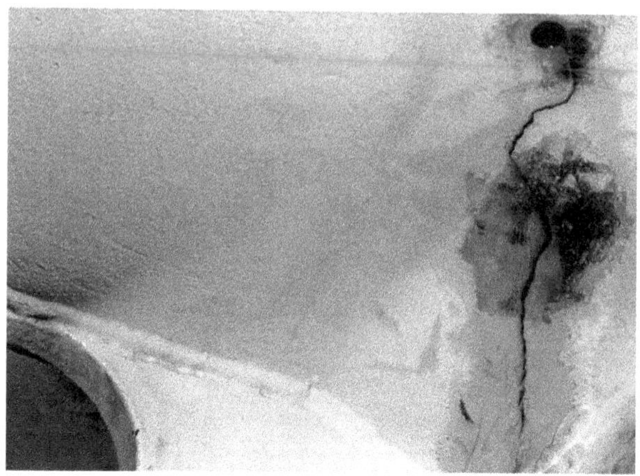

FIGURE 5.13
Cracked frame due to resonance issues.

POSSIBLE CONSEQUENCES:

Abrasion/degradation of the interlaminar insulation:

- INTERLAMINAR INSULATION FAILURE (M-C&F-13)
- Fretting of core against frame resulting in iron-dust deposits. Excessive fretting can result in enough iron-dust deposits to block core cooling vents.
- Cracking of the frame.
- Loosening of the core.
- Loosening of the endwindings.
- SCW leaks into the generator and hydrogen leaks into the SCW.
- ARCING BETWEEN CORE AND KEYBARS (M-C&F-01)
- HYDROGEN INGRESSION INTO STATOR COOLING WATER (M-S-29)
- LEAK OF STATOR COOLING WATER (SCW) INTO GENERATOR (M-S-37)
- LOOSENING OF STATOR WINDING SLOT WEDGES (M-S-41)
- RUPTURED WATER MANIFOLD (HEADER) (M-S-47)
- RUPTURED WATER PARALLEL MAINIFOLD (HEADER) (M-S-48)

Increase loss of hydrogen through the hydrogen seals:

- HYDROGEN ESCAPING THE GENERATOR (M-H-08)
- Loss of hydrogen through sealed casing joints (e.g., between casing and end doors)
- Accelerated aging of components due to high frequency metal fatigue

POSSIBLE SYMPTOMS:

- Audible sound increase in the vicinity of the generator
- HIGH VIBRATION OF CORE AND/OR FRAME **(S-C&F-03)**
- HYDROGEN MAKE-UP RATE ABOVE NORMAL **(S-H-05)**
- RADIO FREQUENCY (RF) MONITOR HIGH **(S-S-13)**

OPERATORS MAY CONSIDER:

1. Finding the origin of vibrations may be a very elusive goal. Operators may consider, first, trying to reduce high vibrations by reducing load (watts and/or vars). In most cases, this may not be of much help, and if the vibrations are high enough, operators may consider removing the unit from service for troubleshooting efforts.
2. Before the unit is brought offline, if the vibrations are still sustainable, but the root cause is unknown, an attempt should be made to perform a *thermal sensitivity test* (TST):
   - THERMAL SENSITIVITY TEST **(M-R-29)**

TEST AND VISUAL INDICATIONS:

- Iron-dust deposits
- *Greasing* of endwindings, endwinding support system, wedges, and so on
- Cracks (see Figure 5.13)
- Loose fasteners
- Loose core
- BCB
- Liquids in generator

REFERENCES

1. G. Klempner and I. Kerszenbaum. *Handbook of Large Turbo-Generator Operation and Maintenance*. Piscataway, NJ: IEEE Press, 2008, chapters 2 and 5.
2. J.J. Fleischmann and G.W. Staats. Present-Day Large Steam-Turbine-Generator Design Practice. In: *AIEE-ASME National Power Conference*, Boston, MA, September 1958, pp. 58–1109.
3. S.C. Barton, J.A. Massingill, and H.D. Barton. Design Features and Characteristics of Large Steam Turbine Generators. In: *AIEE-ASME National Power Conference*, Boston, MA, September 1958, pp. 58–1108.
4. CIGRE. Magnetic Vibration in Turbo-Generator Stators, Study Committee No. 11, Session 11-07, 1970.
5. L.T. Rosenberg. Abnormal Vibration Problems in Large Turbine-Driven Generators and their Solutions. *IEEE Transactions on Power Apparatus and Systems*, PAS-101(10), 4131–4135, October 1982.
6. EPRI Tutorial Guide. Generator Stator End Winding Vibration Guide. TR 1021774, August 2011.
7. IEEE Std 1129–2014. IEEE Guide for Online Monitoring of Large Synchronous Generators (10 MVA and Above).
8. IEEE Std 1665–2009. IEEE Guide for the Rewind of Synchronous Generators, 50 Hz and 60 Hz, Rated 1 MVA and Above.

M-C&F-11: HYDROGEN INGRESSION INTO STATOR COOLING WATER (SCW) FROM FLUX SHIELD

APPLICABILITY:

All machines with water-cooled flux shields

MALFUNCTION DESCRIPTION:

This malfunction relates to the situation where a leak of hydrogen into the SCW system occurred inside the generator, from one of the CEFSs.

Many large machines have flux shields installed on both ends of the core. The purpose of these shields or screens is to ameliorate the impact that axial fluxes have on the ends of the core (called *core-end effects*), in particular when the generator operates at leading power factors.

Flux shields under certain load conditions (leading power factor operation) may become extremely hot if not force-cooled. Thus, a failure of the cooling medium to fully meet its designed requirement can result in serious damage to the shield, and to any insulation that may indirectly be heated by the hot shield.

PROBABLE CAUSES:

- Cracked water hoses
- Joint failure
- Wrong maintenance
- Crack of water manifold
- BREACH OF THE FLUX-SHIELD COOLING PATH **(M-C&F-03)**

POSSIBLE CONSEQUENCES:

- BREACH OF FLUX-SHIELD COOLING PATH **(M-C&F-03)**
- HYDROGEN INGRESSION INTO STATOR COOLING WATER (SCW) **(M-S-29)**
- INSUFFICIENT COOLING OF THE FLUX SHIELD **(M-C&F-12)**

POSSIBLE SYMPTOMS:

- GENERATOR CORE CONDITION MONITOR—ALARM **(S-C&F-01)**
- HYDROGEN-IN-STATOR COOLING WATER (SCW) HIGH **(S-S-04)**
- TEMPERATURE OF FLUX SHIELD ABOVE NORMAL **(S-C&F-07)**

OPERATORS MAY CONSIDER:

1. Establishing and trending the rate of hydrogen ingression into SCW.
2. If the rate of H_2 ingression into SCW reaches maximum allowable limits, consideration should be given to unit shutdown to allow repairs.
3. Trending the temperature of the flux shield(s) if it is directly monitored (by RTD mounted on the shields). If this temperature becomes too elevated, consideration should be given to reducing load and if necessary, removing the unit from service for repairs.
4. Other operator activities can be found in:
   a. BREACH OF FLUX-SHIELD COOLING PATH **(M-C&F-03)**
   b. GAS LOCKING IN FLUX-SHIELD COOLING PATH **(M-C&F-08)**

TEST AND VISUAL INDICATIONS:

- BREACH OF FLUX-SHIELD COOLING PATH **(M-C&F-03)**
- GAS LOCKING IN FLUX-SHIELD COOLING PATH **(M-C&F-08)**
- HYDROGEN INGRESSION INTO STATOR COOLING WATER (SCW) **(M-S-29)**

M-C&F-12: INSUFFICIENT COOLING OF THE FLUX SHIELD

APPLICABILITY:

All generators with water-cooled flux shields

MALFUNCTION DESCRIPTION:

This malfunction addresses a situation where insufficient cooling of one or both CEFSs, also called core-end flux screens, exists. There are a number of reasons for this development, as noted under *Probable Causes*, below. Figure 5.14 shows a typical water-cooled flux shield made of copper.

Many large machines have flux shields installed on both ends of the core. The purpose of these shields or screens is to ameliorate the impact that axial fluxes have on the ends of the core (called *core-end effects*), in particular when the generator operates at leading power factor.

Flux shields, under certain load conditions (leading power factor operation) may become extremely hot if not force-cooled. Thus, a failure of the cooling medium to fully meet its designed requirement can result in serious damage to the shield, and to any insulation that may indirectly be heated by the hot shield.

The cooling of the flux shield is carried out by the SCW.

FIGURE 5.14
Water-cooled flux shield made of copper.

PROBABLE CAUSES/AREAS AFFECTED:

Note: This malfunction addresses insufficient cooling that is *not* derived from a breach in the cooling path. Breaches in the cooling path are discussed in:

- BREACH OF THE FLUX-SHIELD COOLING PATH (M-C&F-03)
- A partial or full blockage of the SCW path in or near the flux shield. The resulting overheating may affect only or mostly one flux shield.
- A global partial or full blockage of the flow of SCW. The entire stator winding is affected.
- Partial or total loss of hydrogen pressure in hydrogen-cooled generators.
- Problems with the hydrogen coolers (*cold* hydrogen temperature too high).
- Problems with the SCW system (*cold* SCW temperature too high).
- Causes for partial or total flow stoppage could be
  - Failure of the SCW system.
  - Debris found inside the SCW system.
  - Copper oxide buildup.
  - Inadvertent blocking during maintenance activity.

POSSIBLE CONSEQUENCES:

Severe overheating of the flux shield, leading to deformation and failure of support system, and damage to the flux-shield insulation.

- TEMPERATURE OF FLUX SHIELD ABOVE NORMAL (M-C&F-22)

Overheating of core and winding insulation on and in the vicinity of the affected shield:

- BURNING/PYROLYSIS OF STATOR WINDING INSULATION **(M-S-08)**
- BURNING/PYROLYSIS OF CORE INSULATION **(M-C&F-05)**

POSSIBLE SYMPTOMS:

- TEMPERATURE OF FLUX SHIELD ABOVE NORMAL **(S-C&F-07)** (by *normal*, it means temperatures measured on the same shield at previous occasions under similar load conditions)
- TEMPERATURE OF CORE END ABOVE NORMAL **(S-C&F-06)**
- GENERATOR CORE CONDITION MONITOR—ALARM **(S-C&F-01)**
- Some increase of stator winding temperatures above *normal*

OPERATORS MAY CONSIDER:

1. First thing after ascertaining instrumentation is working correctly to assess the intensity of the problem. Comparing temperatures with previous readings may offer a starting point.
2. Operating with a *hot* flux shield is possible, as long as the temperature does not reach the maximum temperature specified by the manufacturer. Reducing load, or just moving away from a leading power factor condition, may maintain the flux-shield temperature in a region that allows continuous operation of the unit till an appropriate time to fix the problem.
3. If the flux-shield temperature reaches its limit, load ought to be reduced until the temperature returns to the allowable operating region. Otherwise, serious damage to the flux screen may occur, and indirectly, core and winding insulation may be affected.
4. Keeping record of pertinent temperatures and load points.
5. In all cases, evaluating not operating the unit in a leading power factor until the problem is fixed.

TEST AND VISUAL INDICATIONS:

- Pressure testing the flux-shield cooling path
- SCW flow tests
- Visual: discoloration of the flux shield

REFERENCES

1. G. Klempner and I. Kerszenbaum. *Handbook of Large Turbo-Generator Operation and Maintenance*. Piscataway, NJ: IEEE Press, 2008, chapters 2 and 8.
2. A. Gonzalez, M.S. Baldwin, J. Stein, and N.E. Nilsson. *Monitoring and Diagnosis of Turbine-Driven Generators*. Englewood Cliffs, NJ: Prentice Hall, Chapter 6.

M-C&F-13: INTERLAMINAR INSULATION FAILURE

APPLICABILITY:

All generators

MALFUNCTION DESCRIPTION:

This malfunction addresses a failure of the interlaminar insulation that exists between all laminates in the stator core (see Figure 5.15). The interlaminar insulation will undergo a degree of degradation throughout the life of the machine (i.e., normal thermal aging). In tightly packed cores in machines that have not experienced a very deleterious incident, such as overfluxing, and where the vibrations were always in the normal zone, the degradation over the life of the core may be minimal. On the other hand, machines with loose or semi-loose cores, and/or which experienced high vibration over a long time period, or which were subject to severe operational incidents, the interlaminar insulation may show signs of significant degradation.

If a serious operational incident occurs, such as overfluxing, the chances of the core not undergoing a runaway situation (such as leading to melting), may be directly dependent on the condition of the interlaminar insulation prior to the event. This is a prime reason why events that may be of concern by themselves (like operation at high frame–core vibrations), may have extreme influence down the line on the integrity of the core.

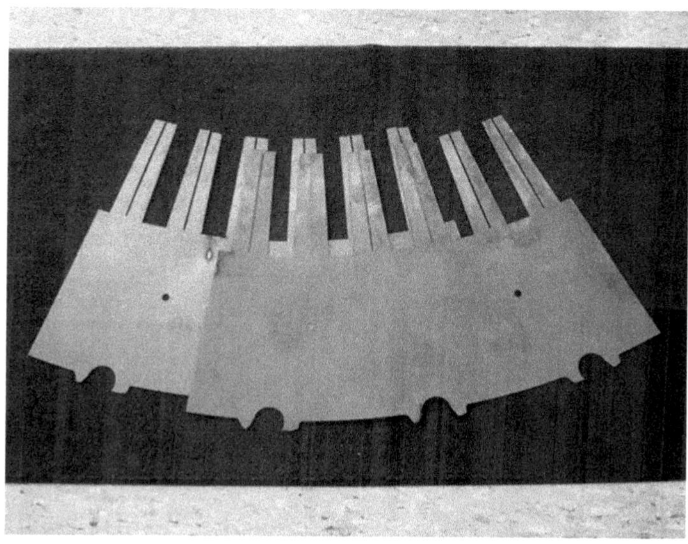

FIGURE 5.15
Interlaminar arcing and insulation breakdown between layers of laminations.

Once the interlaminar insulation fails, it will allow current flow between laminations. This eliminates the designed benefit of laminating the core in first place. The result is an acute increase in core losses, temperature, and further deterioration of the interlaminar insulation.

There are various types of insulation, some organic, some inorganic, or combinations. Some applied to one side of the laminate, some to both. All may have different maximum allowable temperature limits. Thus, when attempting to investigate a problem with the interlaminar insulation, an effort should be made to first learn the pertinent details of the core insulation in the machine in question.

FOR A DISCUSSION OF THE SUBJECT REFER TO:

- BURNING/PYROLYSIS OF CORE INSULATION **(M-C&F-05)**
- LOCAL CORE FAULT **(M-C&F-15)**

M-C&F-14: LEAK OF STATOR COOLING WATER (SCW) FROM A FLUX SHIELD

APPLICABILITY:

All machines with water-cooled flux shields

MALFUNCTION DESCRIPTION:

Many large machines have flux shields installed on both ends of the core. The purpose of these shields or screens is to ameliorate the impact that axial fluxes have on the ends of the core (called *core-end effects*), in particular when the generator operates at leading power factor.

Flux shields, under certain load conditions (leading power factor operation) may become extremely hot if not force-cooled. Thus, a failure of the cooling medium to fully meet its designed requirement can result in serious damage to the shield, and to any insulation that may indirectly be heated by the hot shield.

This problem relates to the case where a SCW leak has developed on the flux shield or from the SCW connections leading to it, resulting in water ingression into the generator.

FOR DISCUSSION OF THIS ISSUE REFER TO:

- BREACH OF FLUX-SHIELD COOLING PATH **(M-C&F-03)**
- HYDROGEN INGRESSION INTO THE STATOR COOLING WATER **(M-S-29)**
- LEAK OF STATOR COOLING WATER (SCW) INTO GENERATOR **(M-S-37)**

M-C&F-15: LOCAL CORE FAULT

APPLICABILITY:

All generators

MALFUNCTION DESCRIPTION:

This malfunction addresses the situation when the stator core is damaged in a small area (see Figure 5.16 for an example of local core damage), or a number of small areas. A damaged core area, for the purpose of this discussion, is not just a mere mechanical indentation, wear, deformation, or other type of anomaly, but it is also resulting in elevated temperatures, compared with the rest of the core. The main concern is the possibility of a local core defect becoming larger to the point of a run-away fault that causes a stator ground or a major core melt.

Core problems are among the most serious in a generator. If they are extensive and/or intensive enough that they require repair, the cost in labor, materials, and lost production may be very high. Thus, core problems must be tackled with the urgency.

PROBABLE CAUSES:

- INTERLAMINAR INSULATION FAILURE (M-C&F-13)
- BLOCKED COOLING VENT IN STATOR CORE (M-C&F-02)

Operation in with leading power factor beyond capability curve (affects only end regions of the core):

- FIELD CURRENT TOO LOW ONLINE (M-E-06)

Stator core fault orginating from loose debris lodging in radial vents and shorting iron packets

FIGURE 5.16
Stator core bore damage due to foreign debris.

Overfluxing of the core during a volts-per-hertz (V/Hz) event (can cause global core damage, but also an original weak spot or spots):

- OVERFLUXING OFFLINE **(M-E-14)**
- OVERFLUXING ONLINE **(M-E-15)**

Foreign metallic object shortening laminations:

- FOREIGN BODY INTERACTION WITH THE STATOR CORE **(M-C&F-07)**
- Damage to core laminations during a maintenance activity (like pulling out or in a rotor)
- A winding short-circuit that resulted in high currents flowing through the core
- Internal short-circuit, for example, between phases via the core's iron

POSSIBLE CONSEQUENCES:

- BURNING-PYROLYSIS OF CORE INSULATION **(M-C&F-04)**
- TEMPERATURE OF A LOCAL CORE AREA ABOVE NORMAL **(M-C&F-20)**
- Damage to the interlaminar core insulation to the extent that the condition will continue to deteriorate after the offending cause has been removed
- Partial meltdown of core material leading to
  - Major core repair
  - Ground faults
  - Stator rewind
  - Damage to the rotor from core material solidifying in the gas gap

POSSIBLE SYMPTOMS:

- GENERATOR CORE-CONDITION MONITOR—ALARM **(S-C&F-01)**
- LOCAL CORE TEMPERATURE ABOVE NORMAL **(S-C&F-05)**
- OPERATION WITH LEADING FACTOR BEYOND ALLOWABLE REGION **(S-E-10)**
- OVERFLUXING (VOLTS per HERTZ) TOO HIGH OFFLINE **(S-E-11)**
- SHAFT CURRENT ABOVE NORMAL **(S-R-04)**
- SHAFT VOLTAGE ABOVE NORMAL **(S-R-05)**

Overvoltage event:

- ANOMALOUS GENERATOR OUTPUT VOLTAGE **(S-E-01)**

OPERATORS MAY CONSIDER:

Localized core defects are difficult to diagnose during operation, unless an RTD is embedded in the vicinity of the core's fault, and/other contributing factors are not present, such as a recent V/Hz event, or the unit was recently or at the present time being operated at a significant leading power factor. Other causes will probably require taking the unit offline for testing and inspection, before they can be identified.

1. Confirming that the readings by the instrumentation (temperature and/or GCM alarm) are not due to instrumentation error.
2. Core burning may be deduced mainly from GCM alarming. Higher-than-*normal* temperatures without GCM information may indicate a problem with the core, global, or local.
3. If a GCM alarm is received, analyzing a hydrogen sample to ascertain the nature of the problem. (If tagged compounds were applied to the stator, analysis of the hydrogen sample may clearly indicate the core as the problem.)
4. Reducing generator loading or remove it from service according to the severity of the indications.
5. Remember that a runaway core problem can be catastrophic to the generator.

TEST AND VISUAL INDICATIONS:

- Analysis of GCM-trapped hydrogen sample
- Visual inspection of core
- EL-CID and/or offline flux tests

REFERENCES

1. G. Klempner and I. Kerszenbaum. *Handbook of Large Turbo-Generator Operation and Maintenance*. Piscataway, NJ: IEEE Press, 2008, chapters 4, 5, and 8.
2. G. Kliman, S.B. Lee, M.R. Shah, M. Lusted, and K. Nair. A New Method for Synchronous Generator Core Quality Evaluation. *IEEE Transactions on Energy Conversion*, 19, 576–582, 2003.

M-C&F-16: LOOSE CORE LAMINATIONS

APPLICABILITY:

All generators

MALFUNCTION DESCRIPTION:

This malfunction addresses the case where the axial compression of the core, applied during manufacturing, is partially or totally lost. This condition results in loose core laminations, particularly at the core ends. Once the core ends become loose, they are susceptible to the push/pull effect of the alternating axial stray fluxes penetrating the core ends.

Looseness of the stator core can be the result of insufficient clamping pressure due to inadequate design, but also due to wrong manufacturing process (e.g., insufficient shaking of the laminations during construction), and looseness introduced during operation because of deficient pressure retention. Some designs are known to be more prone to core looseness than others. Also, core looseness can be the result of continuous operation with high vibrations.

Like any other core problem, excessive looseness of the laminations can develop into a runaway trouble, due to progressive loss of interlaminar insulation. Past certain point, there are no other alternatives than replacing the core, or re-insulating the laminations and restacking the core. Either of those are extremely costly alternatives that included great loss of production. On the other hand, machines are known to operate for many years with cores exhibiting some degree of looseness, without major consequences. *Know your machine* in this case, is key in evaluating the risks for continuing operation in those conditions.

PROBABLE CAUSES:

- Inadequate core axial restraining systems
- Deficient manufacturing process
- Operating the unit with high stator vibrations

POSSIBLE CONSEQUENCES:

- Deterioration of interlaminar insulation
- Reducing the withstand capability of the core against abnormal events such as overfluxing
- Pieces of laminations braking loose and becoming airborne impacting windings and/or core/rotor fans, and magnetic termites (see Figure 5.17)
- Loose laminates shifting and cutting into the wall insulation of the conducting bars

FIGURE 5.17
Loose core end that has succumbed to the push/pull effect of the alternating axial stray fluxes penetrating the core ends. Core-end damage and a stator ground fault was the end result.

- Iron oxide dust deposits
- ARCING BETWEEN CORE AND KEYBARS (M-C&F-01)
- FOREIGN BODY INTERACTING WITH THE CORE (M-C&F-07)
- INTERLAMINAR INSULATION FAILURE (M-C&F-13)
- LOCAL CORE FAULTS (M-C&F-15)

POSSIBLE SYMPTOMS:

- HIGH VIBRATION OF CORE AND/OR FRAME **(M-C&F-03)**
- Elevated sound levels
- Elevated temperatures from core defects produced by loose cores

Ground faults:

- STATOR GROUND ALARM AND/OR TRIP **(S-S-17)**

RF activity from arcing between core and keybars:

- RADIO FREQUENCY (RF) MONITOR HIGH **(S-S-13)**

OPERATORS MAY CONSIDER:

1. The fact is that it is almost impossible to identify a loose core from any other sources of activity inside the generator that may trigger the same *Possible Symptoms* mentioned above. Core looseness will almost exclusively be found during an open generator inspection, and with the rotor outside the bore. Therefore, a core-tightness check ought to be included during open machine outages with rotor out, because these opportunities are not given too often
2. In particular, operators of machines with designs known to be prone to loose cores, or with a history that may indicate a loose core exists or is probable, should see that cores are inspected

FIGURE 5.18
Knife test for loose laminations.

carefully for tightness during those long outages, and the preparations are made for taking repair activities, if required (tightening of the core teeth is doable within a few shifts)

TEST AND VISUAL INDICATIONS:

- Most direct test is the *knife test* (Figure 5.18)
- Loose laminate particles found in the machine
- Shifted laminations
- Laminate iron-oxide deposits
- BCB

REFERENCES

1. G. Klempner and I. Kerszenbaum. *Handbook of Large Turbo-Generator Operation and Maintenance.* Piscataway, NJ: IEEE Press, 2008, chapter 8.
2. EPRI Report #10074411. Repair and Testing Guide for Generator Cores Grounded at the Outside Diameter.
3. S.B. Lee, G.B. Kliman, M.R. Shah, K. Nair, and R.M. Lusted. An Iron Core Probe Based Inter-laminar Core Fault Detection Technique for Generator Stator Cores, *IEEE Transactions on Energy Conversion*, 20, 344–351, 2003.

M-C&F-17: LOOSE GENERATOR FOOTING

APPLICABILITY:

All generators

MALFUNCTION DESCRIPTION:

Looseness of the generator casing and frame footing is an indication that the hold-down bolts may have become loose. This is generally very noticeable by increased noise and vibration of the casing as one walks by the generator. Figure 5.19 shows an example of frame bolting to the foundation.

PROBABLE CAUSES:

- General loosening of the hold-down bolts over time
- Presence of a natural frequency in the stator casing and frame
- Excessive rotor lateral vibration from some type of issue with the rotor, such as many shorted turns
- Misalignment of the stator casing

POSSIBLE CONSEQUENCES:

- Cracked stator casing welds, both internal and external (see Figure 5.20 as an example of cracking of frame weld)
- Hydrogen leaks from the stator casing and frame (possible self-ignition of the hydrogen in H_2-cooled machines)
- Spalling and cracking of the concrete support

FIGURE 5.19
Hold down bolts of the stator casing and frame.

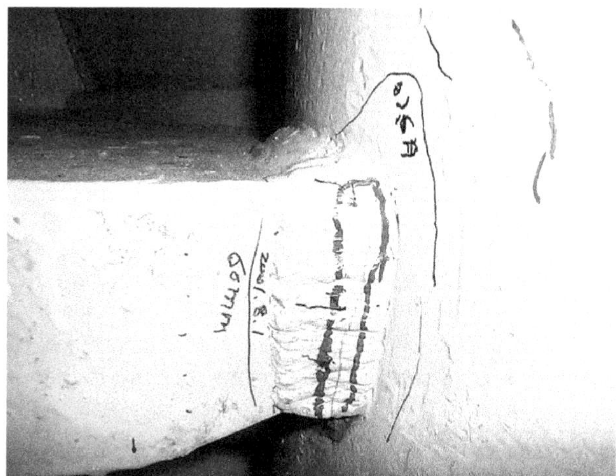

FIGURE 5.20
Major crack in outer stator casing.

POSSIBLE SYMPTOMS:

- HIGH VIBRATION OF CORE AND/OR FRAME **(S-C&F-03)**

OPERATORS MAY CONSIDER:

Maneuvering load and power factor of the generator to minimize the vibration and noise effect, but this should only be considered a temporary measure. It is most likely that the generator will require some investigation and repair/re-alignment, and so on.

TEST AND VISUAL INDICATIONS:

- Visual inspection for frame and casing weld cracking
- Casing pressure test for leaks
- *Bump testing* for presence of natural frequencies

REFERENCE

1. G. Klempner and I. Kerszenbaum. *Handbook of Large Turbo-Generator Operation and Maintenance*. Piscataway, NJ: IEEE Press, 2008, chapter 8.

M-C&F-18: OIL AND WATER LEAKS INSIDE THE GENERATOR

APPLICABILITY:

All generators

MALFUNCTION DESCRIPTION:

This malfunction addresses the situation where a leak of water and/or oil occurs inside the generator.

Water and oil leaks can happen from a number of sources, and their effect on the integrity of the machine depends on the type of leak, the location of the leak, how widespread it is, and what components in the generator are being affected by the leak.

In general, operators of machines with 18Mn-5Cr retaining rings need to be especially careful about how they deal with water leaks inside the generator, due to the affinity of those rings to aqueous stress-corrosion cracking (SCC). However, all leaks in any machine may lead to short- or long-term detrimental effects, and thus should be taken seriously.

The ingression of liquids into generators can happen (and do happen!) also during outages, with the unit offline. The severity of the event is the same, regardless of being on or offline. Actually, during idle conditions, the stress on the retaining rings is highest. Thus, this is the time the retaining rings are more prone to SCC due to water contamination.

PROBABLE CAUSES:

- INGRESSION OF OIL INTO GENERATOR **(M-R-16)**
- LEAK IN THE FLOW-PATH OF STATOR COOLING WATER (SCW) **(M-SCW-09)**
- LEAK OF CONDUCTING-BAR COOLING WATER INTO GENERATOR **(M-S-36)**
- LEAK OF STATOR COOLING WATER (SCW) INTO GENERATOR **(M-S-37)**
- LEAK OF STATOR COOLING WATER (SCW) INTO GENERTOR FROM TERMINALS **(M-S-38)**

- LEAK OF STATOR-COOLING WATER (SCW) FROM A FLUX SHIELD (M-C&F-14)
- PHYSICAL BREACH OF HYDROGEN OR AIR HEAT EXCHANGERS (M-H-15)
- WATER LEAK INTO GENERATOR FROM AIR OR HYDROGEN COOLER (M-H-18)
- Oil leak from bearings/hydrogen seals

POSSIBLE CONSEQUENCES:

- Corrosion of metallic components.
- Contamination of the hydrogen gas inside the casing.
- Ground faults by water creating electric tracking in aged-cracked windings.
- SCC of 18-5 retaining rings and zone rings.
- SCC of 18-18 retaining rings and zone rings if halides are present.
- There are retaining rings in operation made of other materials than 18Mn-5Cr or 18Mn-18Cr. For instance magnetic retaining rings. Operators should know what the RRs in their generators are made of, and to what contaminants/liquids they are sensitive to, and take precautions accordingly.

ELEVATION OF HYDROGEN DEW POINT:

- ELEVATED HYDROGEN DEW POINT (M-H-05)

LOOSENING OF WINDING SUPPORT SYSTEMS BY OIL CONTAMINATION:

- LOOSENING OF STATOR WINDING SLOT WEDGES (M-S-41)

POSSIBLE SYMPTOMS:

- LIQUID-IN-GENERATOR ALARM (S-C&F-04)
- Loss of cooling water (primary and/or secondary systems)
- Loss of oil
- Excessive rates of ingression of hydrogen into primary and/or secondary water
- Indications from the water or lube-oil skids
- Hydrogen dew point/moisture content too high

OPERATORS MAY CONSIDER:

1. After verifying the alarms are *true*, by checking for liquid in the leak detector device, the first thing to do is finding out what the liquid is. This is done by taking a sample and having it analyzed.
2. Estimating the rate of leak. See Figure 5.21 for a typical liquid detector.
3. With the information of 1) and 2) above, assessing the risk for continuing operation.

FIGURE 5.21
Liquid detectors indicate the presence of liquids at the bottom of the generator casing.

4. Bear in mind that some leaks may be tolerated in the short run, but may result in serious problems at the long run; for example, large quantities of oil will loosen the stator winding support systems on the long run, and there is no easy way of reversing the situation. Also, SCW leaks may cause irreversible damage to the wall insulation of the conducting bars by capillary action.
5. From all the above, the most prudent action would be removing the unit from service as soon as possible for repairs for any other than very low leaks, which may be tolerated for a while.
6. It is possible, for large units with several hydrogen heat exchangers (two or four) to isolate the one with the leak and keep running at reduced load until such a time the unit can be taken offline for repairs. This activity may result in thermal unbalance and increased vibrations—depending on the design of the cooling arrangement, thus it must be considered carefully (better in consultation with the OEM).

TEST AND VISUAL INDICATIONS:

- Liquids found inside the generator
- A Thin (or not too thin) layer of oil on the endwindings, bore ID, rotor surface and other areas of the machine
- Excessive quantities of H_2 in the SCW or primary water from the H_2 coolers
- Pressure/vacuum tests.

REFERENCES

1. G. Klempner and I. Kerszenbaum. *Handbook of Large Turbo-Generator Operation and Maintenance*. Piscataway, NJ: IEEE Press, 2008, chapter 4.
2. EPRI Product 3002003589. Risk-Informed Inspection of 18Mn 18 C Generator Retaining Rings, November 2014.
3. EPRI TR-102949. Generator Retaining Ring Moisture Protection Guide, September 1993.
4. EPRI Product 3002006238. Review of Damage to Generator Retaining Rings, December 2015.
5. EPRI TR-104209. Evaluation of Nonmagnetic Generator Retaining Rings, October 1994.
6. G. Stein. *The Development of New Materials for Nonmagnetizable Retaining Rings and Other Applications in the Power Generating Industry*. Steel Forgings, ASTM STP 903, Philadelphia, pp. 237–257, 1986.

M-C&F-19: OVERVOLTAGE

APPLICABILITY:

All generators

MALFUNCTION DESCRIPTION:

This Malfunction addresses the impact that transient overvoltages have on the stator cores.

Transients from grid disturbances occur frequently. Of those, some produce overvoltages that impact the power transformers and generators. These overvoltages are not large enough in magnitude to initiate discharge currents through lightning arrestors and/or MOVs. In smaller generators, they may be absorbed by shunt surge capacitors. In most cases, they are mitigated largely by the main step up transformer impedances. Nevertheless, to some extent, overvoltages reach the main generator on occasion.

Another, less common cause of an overvoltage, can also be operation above the limits of the capability curve, in the lagging power factor range. There are generally alarms for this type of occurrence and the operator would normally be aware that this is occurring. It is usually a man-made type of operating error if it occurs.

PROBABLE CAUSES:

- Significant grid disturbances

POSSIBLE CONSEQUENCES:

- Interlaminar insulation damage, in particular in those areas where the insulation is already weak. The probability of such a damage is very small with the unit online, because the grid controls the voltage, and either the high voltage transient is very short (fraction of

FIGURE 5.22
A number of through bolts are shown. The through bolts are insulated throughout, and also at their ends, with washers made of insulation material.

a second), or if a duration of a few seconds, not more than a few percentage above the maximum allowable continuous voltage.

- INTERLAMINAR INSULATION FAILURE (M-C&F-13).
- Breach of core through bolt insulation. The probability of this type of failure due to online voltage transients is somewhat higher than the probability of interlaminar insulation failure, because even a very short transient can cause a failure of the through bolt insulation. Figure 5.22 shows the end of the through bolts with the core assembled).

POSSIBLE SYMPTOMS:

- None

OPERATORS MAY CONSIDER:

1. If an overvoltage event of significant amplitude and duration (several seconds or more) has been recorded as reaching the generator, scheduling an EL-CID test of the core at the earliest opportunity may be prudent. Core through bolt insulation tests may be scheduled for the next rotor-out inspection

TEST AND VISUAL INDICATIONS:

- Alarms from voltage devices and/or V/Hz protection or V/Hz excitation limiters
- In extreme cases: V/Hz protection trip

REFERENCES

1. G. Klempner and I. Kerszenbaum. *Handbook of Large Turbo-Generator Operation and Maintenance*. Piscataway, NJ: IEEE Press, 2008, chapters 3 and 6.
2. M. Sasic and D. Bertenshaw. *EL-CID Testing of Turbo-Generators at Higher Frequency*. EPRI TGUG Workshop, St. Louis, MO, August 16–17, 2010.

M-C&F-20: TEMPERATURE OF A LOCAL CORE AREA—ABOVE NORMAL

APPLICABILITY:

All generators

MALFUNCTION DESCRIPTION:

This malfunction relates to the situation when the temperature of a section of the core is above *normal*. By *normal*, it means that the temperature of the affected area is at least 10° higher than the rest of the core, or core temperatures previously recorded for this core zone.

Core problems are among the most serious in a generator. If they are extensive and/or intensive enough that they require repair, the cost in labor, materials and lost production may be very high. Thus, core problems must be tackled with the urgency they deserve.

Temperature rise above normal at a localized core area can result from a number of problems. This is covered by the following problem descriptions:

PROBABLE CAUSES:

- BURNING/PYROLYSIS OF CORE INSULATION (M-C&F-05)
- LOCAL CORE FAULT (M-C&F-15)
- Overfluxing by operating beyond capability curve in the lagging power region
- Core vents blocked (see Figure 5.23)
- OVEREXCITATION—ONLINE (M-E-13)
- TEMPERATUE OF LOCAL CORE END ABOVE NORMAL (M-C&F-20)

POSSIBLE CONSEQUENCES:

- BURNING/PYROLYSIS OF CORE INSULATION (M-S-05)

POSSIBLE SYMPTOMS:

- GENERATOR CORE CONDITION MONITOR—ALARM (S-C&F-01)
- TEMPERATURE OF CORE END ABOVE NORMAL (S-C&F-06)
- Above *normal temperature* readings from instruments away from the core ends or that can be affected by high *temperature of the core ends* (such as *hot hydrogen* RTDs)
- Any core temperature measurement above *normal*

FIGURE 5.23
Blocked core vents belonging to a small air-cooled generator.

OPERATORS MAY CONSIDER:

1. If the *core temperatures* are within the *normal* range, keep monitoring them.
2. If the temperature of the core reaches or exceeds the limit values given by the OEM, consider reducing overall output so that temperatures return to within *normal* range. Normal range is usually below the Class B insulation limit of 130°C.
3. If, after reducing load the core temperature remains above *normal*, consider removing the machine from service and troubleshooting it.

Note: If the temperature of the core remains high after returning the unit to service, and even when the load is reduced below rated, it is an indication of a possible loss of interlaminar insulation in the core. This is a very serious situation that should be investigated and repaired.

TEST AND VISUAL INDICATIONS:

- EL-CID and/or offline flux tests may uncover damaged interlaminar insulation.

REFERENCES

1. G. Klempner and I. Kerszenbaum. *Handbook of Large Turbo-Generator Operation and Maintenance*. Piscataway, NJ: IEEE Press, 2008, chapters 5 and 8.
2. M. Sasic and D. Bertenshaw. EL-CID Testing of Turbo-Generators at Higher Frequency. In: *EPRI TGUG Workshop*, St. Louis, MO, August 16–17, 2010.
3. G. Klempner. Experience and Benefit of Using EL-CID for Turbine-Generators. In: *EPRI TGUG Meeting*, Orlando, FL, November 28–30, 1995.

M-C&F-21: TEMPERATURE OF CORE END—ABOVE NORMAL

APPLICABILITY:

All generators

MALFUNCTION DESCRIPTION:

This malfunction addresses the situation where the last packets of lamination at each end of the core (called *core ends*), attain temperatures above *normal*. By *normal*, it means temperatures above those experienced by the same machine, at same or similar locations, on previous occasions under similar load conditions. These temperatures, under certain conditions, can reach values well beyond their maximum limits (as indicated by the OEM), with serious deleterious results to the integrity of the core.

In general, the core ends are subject to additional heating than the rest of the core, because of the so-called *end effects*, which are the generation of extra losses by the axial flux (stray flux) impinging on the core from the endwinding regions. As the axial flux components generated by the endwindings of the stator and rotor reaches the core end, it adds to the radial flux across the gas gap, resulting in higher eddy and hysteresis losses in the last few packets of laminations. The intensity of this superposition is larger when the generator is operated at leading power factors.

Most large machines are designed with features designed to withstand or absorb the extra axial fluxes. Nevertheless, the *capability curve* is always smaller in the leading power factor region.

See Figure 5.24 for an example of end-of-core abnormal heating.

FIGURE 5.24
Shown is a melted tooth at the end of the core, due to excessive core looseness.

PROBABLE CAUSES:

- Overfluxing of the core ends by, most probably, operating beyond capability curve in the leading power region
- FIELD CURRENT TOO LOW ONLINE **(M-E-06)**
- LOCAL CORE FAULT **(M-C&F-15)**
- OPERATION OUTSIDE CAPABILITY CURVE—ONLINE **(M-E-11)** (in the leading power factor range)
- TEMPERATURE OF LOCAL CORE END ABOVE NORMAL **(M-C&F-21)**

POSSIBLE CONSEQUENCES:

- BURNING/PYROLYSIS OF CORE INSULATION **(M-S-05)**

POSSIBLE SYMPTOMS:

- GENERATOR CORE CONDITION MONITOR—ALARM **(S-C&F-01)**
- TEMPERATURE OF CORE END ABOVE NORMAL **(S-C&F-06)**
- Above *normal* temperature readings from instruments away from the core ends or that can be affected by high temperature of the core ends (such as *hot hydrogen* RTDs)

OPERATORS MAY CONSIDER:

1. If the temperatures of the core ends are within the *normal* range, keep monitoring them if the machine is operated in a leading power factor.
2. If the temperature of the core ends reaches or exceeds the limit values given by the OEM, consider moving the load point from leading to lagging power factor, or reducing overall output so that temperatures return to within *normal* range. Normal range is usually below the Class B insulation limit of 130°C.
3. If, after reducing load and moving to a lagging power factor the temperatures of the core ends remain above *normal,* consider removing the machine from service and troubleshooting it.
    a. *Note*: If the temperatures of the core end region remain high after returning the unit to a lagging power factor, and even when the load is reduced below rated, it is an indication of a possible loss of interlaminar insulation in the core-end region. This is a very serious situation that should be investigated and repaired.

TEST AND VISUAL INDICATIONS:

EL-CID and/or offline flux tests may uncover damaged interlaminar insulation.

REFERENCES

1. G. Klempner and I. Kerszenbaum. *Handbook of Large Turbo-Generator Operation and Maintenance*. Piscataway, NJ: IEEE Press, 2008, chapters 5 and 8.
2. EPRI TR 1020275. Generator Stator Core Condition Monitoring by Tracking Shaft Voltage and Grounding Current, February 2010.

M-C&F-22: TEMPERATURE OF FLUX SHIELD ABOVE NORMAL

APPLICABILITY:

All generators with flux shields

MALFUNCTION DESCRIPTION:

Many large machines have flux shields (Figure 5.25) or screens installed on both ends of the core (CEFS). The purpose of these shields or screens is to ameliorate the impact that axial fluxes have on the ends of the core (called *core-end effects*), in particular when the generator operates at leading power factors.

Flux shields under certain load conditions (leading power factor operation) may become extremely hot if not force-cooled. Thus, a failure of the cooling medium to fully meet its designed requirement can result in serious damage to the shield, and to any insulation that may indirectly be heated by the hot shield.

The problem discussed herein addresses the situation where the temperature of the CEFS rises above its *normal* temperature. By *normal*, it

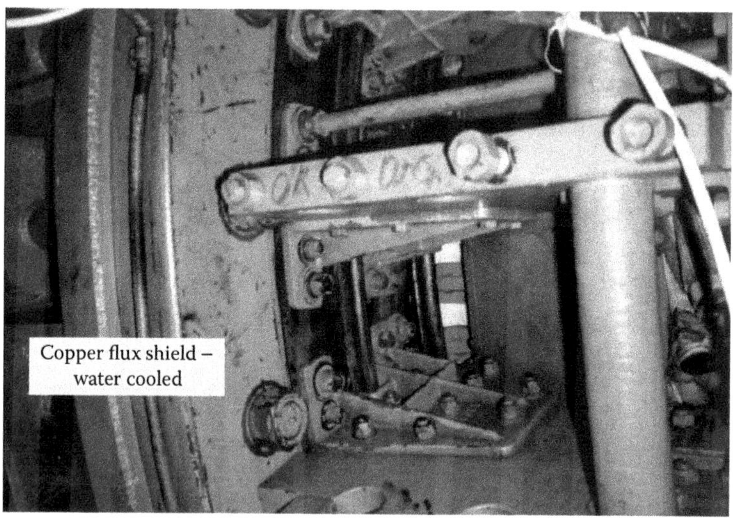

FIGURE 5.25
The figure shows a water-cooled copper shield protecting the end of the cores from overfluxing.

means the temperature measured on previous occasions on the same machine and same of similar location, for a similar load. The difference may be small, or very large, even bringing the temperature of the flux shield above the limits given by the OEM.

PROBABLE CAUSES:

- FIELD CURRENT TOO LOW ONLINE **(M-E-06)**
- HYDROGEN PRESSURE INSIDE GENERATOR BELOW NORMAL **(M-H-09)**

Operation in the leading power factor region beyond the capability of the generator:

- OPERATION OUTSIDE CAPABILITY CURVE—ONLINE **(M-E-11)**

In machines with water-cooled flux screens:

- BREACH OF THE FLUX-SHIELD COOLING PATH **(M-C &F-03)**
- GAS LOCKING IN FLUX-SHIELD COOLING PATH **(M-C &F-08)**
- INSUFFICIENT COOLING OF THE FLUX SHIELD **(M-C &F-12)**
- INSUFFICIENT PERFORMANCE OF AIR OR HYDROGEN HEAT EXCHANGERS **(M-H-12)**

POSSIBLE CONSEQUENCES:

- Severe overheating of the flux shield, leading to deformation and failure of support system, and damage to the flux shield insulation
- Overheating of core and winding insulation on and in the vicinity of the affected shield
- BURNING/PYROLYSIS OF CORE INSULATION **(M-C&F-05)**
- BURNING/PYROLYSIS OF STATOR WINDING INSULATION **(M-S-08)**

POSSIBLE SYMPTOMS:

- TEMPERATURE OF FLUX SHIELD ABOVE NORMAL **(S-C&F-07)**
  - By *normal,* it means temperatures measured on the same shield at previous occasions under similar load conditions
- TEMPERATURE OF CORE END ABOVE NORMAL **(S-C&F-06)**
- Some increase of stator winding temperatures above normal
- LIQUID-IN-GENERATOR ALARM **(S-C&F-04)** (if the flux shield is water-cooled)
- HYDROGEN IS STATOR COOLING WATER (SCW) HIGH **(S-S-04)** (if the flux shield is water cooled)
- Problems with the flow of SCW (if the flux shield is water cooled)

OPERATORS MAY CONSIDER:

1. First, ascertain if the instrumentation is working correctly, and then assess the intensity of the problem. Comparing temperatures with previous readings may offer a starting point.
2. Operating with a *hot* flux shield is possible, as long as the temperature does not reach the maximum temperature specified by the manufacturer. This usually kept to a maximum of the Class B insulation temperature of 130°C. Reducing load or just operating away from a leading power factor condition may maintain the flux-shield temperature in a region that allows continuous operation of the unit till an appropriate time to fix the problem.
3. If the flux-shield temperature reaches its limit, consider reducing load until the temperature returns to the allowable operating region. Otherwise, serious damage to the flux screen may occur, and indirectly, core and winding insulation may be affected.
4. Keeping records of pertinent temperatures and load points.
5. In all cases, evaluate not operating the unit in a leading power factor until the problem is fixed.

TEST AND VISUAL INDICATIONS:

- Pressure testing the flux-shield cooling path
- SCW flow tests
- Megger flux-shield insulation

Note: All of the above are for water-cooled flux shields

- Visual discoloration of the flux shield

REFERENCE

1. G. Klempner and I. Kerszenbaum. *Handbook of Large Turbo-Generator Operation and Maintenance*. Piscataway, NJ: IEEE Press, 2008, chapter 5.

5.3 Stator Malfunctions

M-S-01: BLOCKAGE IN THE FLOW OF STATOR TERMINAL COOLANT

APPLICABILITY:

- Machines with hydrogen-cooled terminals
- Machines with water-cooled terminals

MALFUNCTION DESCRIPTION:

This malfunction addresses the issue of partial or total blockage of flow of coolant to directly cooled stator terminals. The coolant may be water or hydrogen.

Figures 5.26 and 5.27 show a typical arrangement for hydrogen-cooled terminals. Note the *gas inlets* at the top of the HV bushings. These inlets are designed to allow free flow of hydrogen between the hydrogen directly cooling the conductor from the winding to the HV bushings, and the terminal chamber. Figure 5.28 shows a water-cooled connection.

PROBABLE CAUSES:

- In hydrogen-cooled terminals, the source of partial or total blockage may be debris (i.e., leaving a cleaning rag after an outage).
- In hydrogen-cooled terminals, the source of partial or total blockage may be a temporary FME cover installed during maintenance activities, and left there inadvertently.
- Hydrogen gas locking in water-cooled stator terminals.
- GAS LOCKING IN WATER-CARRYING STRANDS OR TUBES (M-S-21)
- In water-cooled terminals, debris such as insulation, broken "O"-ring material that got loose, and so on can be seen.
- Severe SCW leak from hoses connected to water-cooled stator terminals.

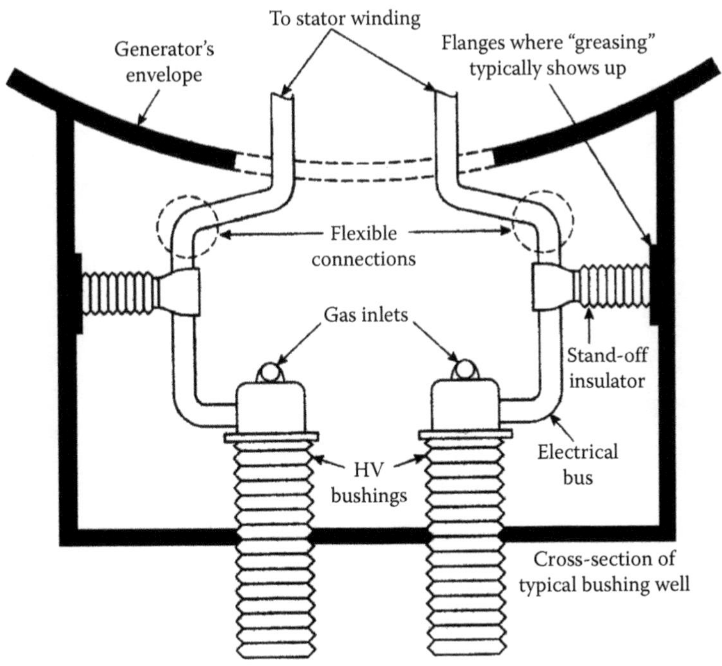

FIGURE 5.26
The figure shows a typical arrangement for hydrogen-cooled terminals.

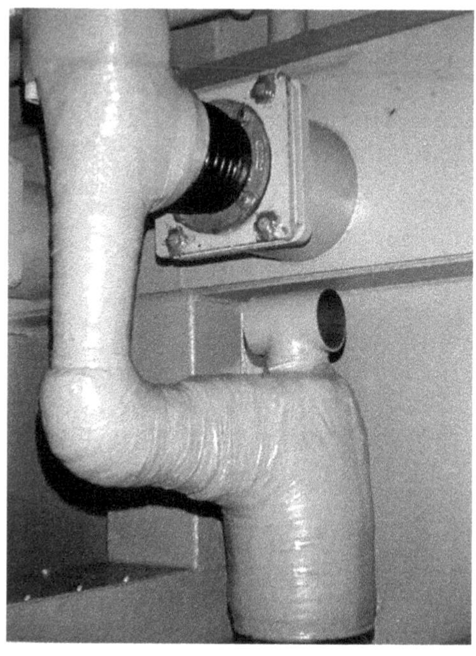

FIGURE 5.27
A real example of a hydrogen-cooled terminal gas inlet for cooling, as well as a stand-off insulator above to support the upper part of the bushing.

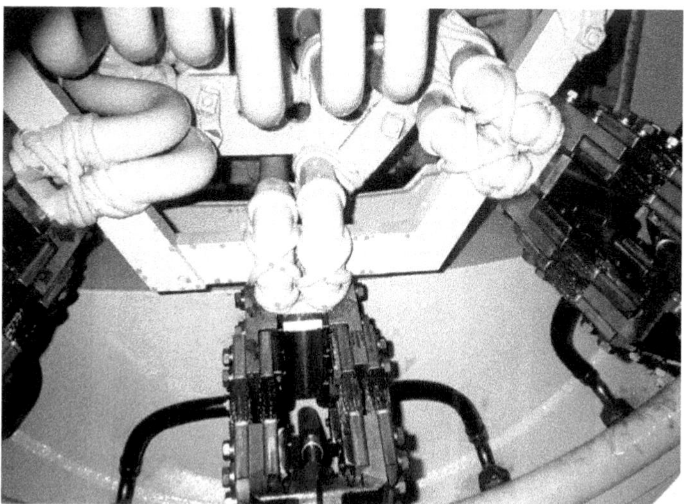

FIGURE 5.28
The water-cooled phase connections to the terminal bushings where the bolted joint is made. The terminals are water-cooled in this machine as shown by the hose connections.

POSSIBLE CONSEQUENCES:

- BURNING/PYROLYSIS OF STATOR WINDING INSULATION **(M-S-08)**
- DAMAGE TO CONDUCTOR BAR INSULATION **(M-S-15)**
- DAMAGE TO HV BUSHINGS **(M-S-17)**
- INSUFFICIENT FLOW OF COOLING MEDIUM IN A STATOR BAR **(M-S-32)**
- TEMPERATURE OF STATOR CONDUCTOR ABOVE NORMAL **(M-S-53)**
- Ground faults on the *terminals* area. See (M-S-26) for stator ground faults

POSSIBLE SYMPTOMS:

- GENERATOR CORE CONDITION MONITOR—ALARM **(S-C&F-01)**
- LIQUID-IN-GENERATOR ALARM **(S-C&F-04)**
- TEMPERATURE OF STATOR TERMINAL ABOVE NORMAL **(S-S-19)**
- TEMPERATURE OF THE GLOBAL OUTLET STATOR COOLING WATER ABOVE NORMAL **(S-SCW-15)**

OPERATORS MAY CONSIDER:

Loss of coolant to the stator terminals of the generator is a very serious problem as it can result in a major failure in a very short time. This will depend on the power output level and degree of cooling loss. If such an event is recognized,

1. Consider reducing load immediately to return the temperatures within normal range.
2. Consider removing the unit from operation as soon as practically possible.
3. Take into consideration that temperatures at the affected area will be much higher than those indicated by *global* RTDs or TCs (meaning, not measuring directly the temperature of the coolant at the terminals, but indirectly in other parts of the machine).

TEST AND VISUAL INDICATORS:

- Gas analysis from hydrogen samples captured by the GCM
- Visual inspection of the terminals (insulation, vent openings for hydrogen-cooled terminals, piping integrity for water-cooled terminals, etc.)
- Megger tests

REFERENCES

1. G. Klempner and I. Kerszenbaum. *Handbook of Large Turbo-Generator Operation and Maintenance*. Piscataway, NJ: IEEE Press, 2008, chapters 2, 3, 5, and 10.
2. EPRI TR 1004704. Guidelines for Detecting and Removing Flow Restrictions of Water-Cooled Stator Windings, July 2002.

M-S-02: BLOCKAGE OF COOLING FLOW IN A STATOR BAR

APPLICABILITY:

- Directly hydrogen-cooled stator windings
- Directly water-cooled stator windings

MALFUNCTION DESCRIPTION:

This malfunction addresses the issue of a partial or total loss of flow in a single bar (Figure 5.29). The symptoms, consequences, and testing are similar to that of a group of coils with the same predicament. For a relevant discussion refer to

- BLOCKAGE OF PHASE-COOLANT FLOW (M-S-04)
- COPPER-OXIDE BUILDUP IN WATER-CARRYING STRANDS (M-SCW-06)
- INSUFFICIENT FLOW OF COOLANT DUE TO BUCKLED COOLING STRANDS OR TUBES (M-S-31)
- INSUFFICIENT FLOW OF PHASE COOLANT (M-S-33)

FIGURE 5.29
Cross section of a stator bar showing the blockage of SCW due to corrosion build-up in the hollow conductor strands.

PROBABLE CAUSES:

- In hydrogen or water-cooled stator bars the source of partial or total blockage can be debris or temporary FME cover installed during maintenance activities, and left there inadvertently
- GAS LOCKING IN WATER-CARRYING STRANDS OR TUBES **(M-S-21)**

POSSIBLE CONSEQUENCES:

- BURNING/PYROLYSIS OF STATOR WINDING INSULATION **(M-S-08)**
- DAMAGE TO CONDUCTOR BAR INSULATION **(M-S-15)**
- Damage to HV bushings **(M-S-17)**
- INSUFFICIENT FLOW OF COOLING MEDIUM IN A STATOR BAR **(M-S-32)**
- TEMPERATURE OF STATOR CONDUCTOR ABOVE NORMAL **(M-S-53)**
- Ground faults on the *terminals* area. See (M-S-26) for stator ground faults

POSSIBLE SYMPTOMS:

- GENERATOR CORE CONDITION MONITOR—ALARM **(S-C&F-01)**
- LIQUID-IN-GENERATOR ALARM **(S-C&F-04)**
- TEMPERATURE OF STATOR TERMINAL ABOVE NORMAL **(S-S-19)**
- TEMPERATURE OF THE GLOBAL OUTLET STATOR COOLING WATER ABOVE NORMAL **(S-SCW-15)**

OPERATORS MAY CONSIDER:

Loss of coolant to a direct (water or hydrogen) cooled stator bar is a very serious problem as it can result in a major failure in a very short time. This will depend on the power output level and degree of cooling loss. If such an event is recognized,

1. Consider reducing load immediately to return the temperatures within normal range.
2. Consider removing the unit from operation as soon as practically possible.
3. Take into consideration that temperatures at the affected area will be much higher than those indicated by *global* RTDs or TCs (meaning, not measuring directly the temperature of the coolant at the terminals, but indirectly in other parts of the machine).

TEST AND VISUAL INDICATORS:

- Gas analysis from hydrogen samples captured by the GCM
- Megger tests

REFERENCES

1. G. Klempner and I. Kerszenbaum. *Handbook of Large Turbo-Generator Operation and Maintenance*. Piscataway, NJ: IEEE Press, 2008, chapters 5 and 8.
2. EPRI TR3002003591. Early Detection of Flow Restriction in Stator Generator Cooling Water Systems, December 2014.

M-S-03: BLOCKAGE OF PARALLEL-PHASE-CONNECTOR COOLING PATH

APPLICABILITY:

- Directly water-cooled stator windings
- Directly hydrogen-cooled stator windings

MALFUNCTION DESCRIPTION:

This malfunction addresses a partial or total blockage of cooling media (hydrogen or water) in a parallel-phase-connector. In the case of hydrogen-cooled stators, this can render the connector and some coils in the associate phase to be not properly cooled. In the case of a water-cooled stator, an entire parallel circuit of a phase may become affected by a lack of cooling. In either case, a significant blockage will result in the affected winding temperatures increasing beyond their class rise, and perhaps, class insulation value.

This failure is very similar in detail to the one described under blocking of phase-connector cooling path, thus the reader is referred to the below causes.

PROBABLE CAUSES:

- BLOCKAGE OF PHASE-CONNECTOR COOLING PATH (M-S-03)

POSSIBLE CONSEQUENCES:

- BLOCKAGE OF PHASE-CONNECTOR COOLING PATH (M-S-03)

POSSIBLE SYMPTOMS:

- BLOCKAGE OF PHASE-CONNECTOR COOLING PATH (M-S-03)

OPERATORS MAY CONSIDER:

- BLOCKAGE OF PHASE-CONNECTOR COOLING PATH (M-S-03)

TEST AND VISUAL INDICATIONS:

- BLOCKAGE OF PHASE-CONNECTOR COOLING PATH (M-S-03)

REFERENCES

1. G. Klempner and I. Kerszenbaum. *Handbook of Large Turbo-Generator Operation and Maintenance*. Piscataway, NJ: IEEE Press, 2008, chapter 8.
2. R. Svoboda and D.A. Palmer. Behaviour of Copper in Generator Stator Cooling-Water Systems. In: *15th International Conference on the Properties of Water and Steam*, Berlin/DE, Germany, September 8–11, 2008; *Power Plant Chemistry* 11(2009)2, 70–76.
3. M. Pourbaix. *Atlas of Electrochemical Equilibria in Aqueous Solutions*. Brussels, Belgium: Pergamon Press, 1966.

M-S-04: BLOCKAGE OF PHASE-COOLANT FLOW

APPLICABILITY:

- Directly hydrogen-cooled stator windings
- Directly water-cooled stator windings

MALFUNCTION DESCRIPTION:

This malfunction addresses the issue of a partial or total loss of flow of cooling medium to one of the stator phases. The cooling media could be hydrogen (directly hydrogen-cooled stator windings) or water (directly water-cooled stator winding).

It is not uncommon in machines with stators directly cooled by hydrogen, for the cooling medium to flow in circuits made of all coils in one phase, or group of coils forming a parallel section of a phase. The same is not very common in water-cooled stators. However, in both cases, a blockage of the flow of hydrogen or water inside the coils will result in a loss—partial or total—of cooling to that phase or parallel.

Loss of flow can be due to a number of causes (see below), and depending on the type of thermal instrument being read and its location, can manifest itself as higher or *lower* than normal temperatures in that phase. For instance, RTDs embedded in the slots where the coils with the lack of cooling flow are located will show higher temperatures than in the rest of the stator. If the blockage of either water or hydrogen flow inside the bars is partial, RTDs and/or TCs installed at the outlets of the bars will show increased levels of temperature. However, if the blockage is total, these TCs may show cooler temperatures than the rest of the stator, because there is no hot gas or water coming out the bars. Therefore, one must exercise caution by looking carefully at different indications. In general, a coil or a group of coils that suddenly shows low outlet temperatures than the rest of the machine most probably indicate total blockage of the cooling medium (unless, of course, the instrumentation for whatever reason failed).

PROBABLE CAUSES/AREA AFFECTED:

- Unintended obstruction introduced during maintenance (see Figure 5.30 for such a case)/All bars and connectors where flow of hydrogen or water is stopped or reduced
- Obstruction due to dislodging of a large amount of copper oxide in a water-cooled winding/All bars where the flow of water is stopped or reduced
- Obstruction due to debris being thrown into the cooling path in a directly hydrogen-cooled winding/All bars and connectors where the flow of coolant is stopped or reduced
- Gas locking in a water-cooled winding/All bars where the flow of water is stopped or reduced

POSSIBLE CONSEQUENCES:

- For mild blockage conditions (fully blocked for a short time or partially blocked for a long time): premature aging of the affected bar insulation
- For serious blockage conditions:
  - Significant insulation degradation of affected bar insulation
  - Permanent deformation of the strands
  - Melting of conductor
- Ground faults
- Failure of insulation in the slot area (ground faults), in the circumferential busses, leads, and terminal box conductors

FIGURE 5.30
Man-made flow restriction from FME dust plugs left in the stator bar end nozzles after a maintenance activity.

POSSIBLE SYMPTOMS:

Too high or too low temperature readings for bars or group of bars (like a phase or phase parallel):

- GENERATOR CORE CONDITION MONITOR—ALARM **(S-C&F-01)**
- OUTLET TEMPERATURE OF STATOR CONDUCTOR BAR COOLANT ABOVE NORMAL (CASE 1) **(S-S-07)**
- TEMPERATURE OF GLOBAL STATOR COOLING WATER ABOVE NORMAL **(S-SCW-15)**
- TEMPERATURE OF STATOR TERMINAL ABOVE NORMAL **(S-S-19)**

Flowrate and/or dp alarms in the SCW skid:

- STATOR COOLING WATER OUTLET PRESSURE BELOW NORMAL **(S-SCW-13)**
- Low hydrogen pressure in machines with stators directly cooled by hydrogen (flowing in the bars)

OPERATORS MAY CONSIDER:

Significant blockage of the flow of cooling medium in directly cooled machines, for any given time, has the potential to resulting in major damage and downtime. Therefore,

1. Once a serious blockage is recognized, a quick reduction of load should be immediately considered, until the temperatures in the affected area are brought back to within insulation class limits.
2. If the temperature readings are lower than normal and lower than the rest of the winding, this may be the sign of a full-blockage situation, in which case the machine should be completely de-loaded (and perhaps taken offline) as soon as possible (only total or almost total removal of load will restore the temperature of the affected bars or group of bars to within allowable limits).

TEST AND VISUAL INDICATIONS:

- Discolored insulation
- Too high or too low polarization index (PI) readings compared with historical readings for this machine and compared with the unaffected phases
- High leakage current
- Failure of Hipot test
- Megger readings very different than the other two phases

REFERENCES

1. G. Klempner and I. Kerszenbaum. *Handbook of Large Turbo-Generator Operation and Maintenance*. Piscataway, NJ: IEEE Press, 2008, chapter 8.

2. J.D. Aspden and C. Maughan. Maintaining the Flow of Revenue from the Production of Power: Optimization of the Chemistry of Generator Cooling Water Systems—One Utility's Experience Over More Than 20 Years of Operation. In: *EPRI's Seventh International Conference on Cycle Chemistry in Fossil Plants,* Houston, TX, June 3–5, 2003.

M-S-05: BLOCKAGE OF PHASE-CONNECTOR COOLING PATH

APPLICABILITY:

- Generators with directly hydrogen-cooled stator windings
- Generators with directly water-cooled stator windings

MALFUNCTION DESCRIPTION:

This malfunction addresses a partial or total blockage of cooling media (hydrogen or water) in a phase connector. In the case of hydrogen-cooled stators, this can render the connector and some coils in the associate phase to be under-cooled. In the case of a water-cooled stator, the entire phase winding may become under-cooled. In either case, a significant blockage will result in the affected winding temperatures increasing beyond their class rise and/or class insulation value.

PROBABLE CAUSES:

- The most probable cause for a blockage of the relatively large inner cross section of a phase connector is a large fragment of insulation or some other form of debris.
- Mistake during a repair activity (see Figure 5.31).

FIGURE 5.31
An FME plug inadvertently left during a maintenance activity.

POSSIBLE CONSEQUENCES:

- BURNING/PYROLYSIS OF STATOR WINDING INSULATION **(M-S-08)**
- GROUND FAULT OF STATOR WINDING **(M-S-26)**
- OVERHEATING OF PHASE-CONNECTOR **(M-S-43)**

  Weakening of braced joints by high temperature stresses:
  - HIGH RESISTANCE OF PHASE-CONNECTOR JOINT **(M-S-27)**
  - HIGH RESISTANCE OF PHASE-PARALLEL-CONNECTOR JOINT **(M-S-28)**

POSSIBLE SYMPTOMS:

High temperature indication on a group of bars or full phase:

- OUTLET TEMPERATURE OF STATOR CONDUCTOR BAR COOLANT ABOVE NORMAL (CASE 1) **(S-S-07)**
- OUTLET TEMPERATURE OF STATOR CONDUCTOR BAR COOLANT ABOVE NORMAL (CASE 2) **(S-S-08)**
- TEMPERATURE OF PHASE CONNECTOR ABOVE NORMAL **(S-S-18)**

OPERATORS MAY CONSIDER:

1. Like any other physical degradation resulting in temperature rises above normal, the urgency of the matter depends on the degree of deviation from the normal. In a major departure, the machine should be taken offline for repairs ASAP. In marginal cases, the operator can attempt to reach a previously scheduled outage, where the repairs for this degradation can be added to existing scope.
2. If the temperatures are above maximum, consider starting de-loading the unit immediately to bring them back within allowable limits. This will also indicate that the problem appears to be severe enough to require taken the unit down for repairs ASAP.

TEST AND VISUAL INDICATIONS:

- Flow tests.
- Visual inspection.
- May require significant effort to find the reason for the temperature increases and the location of the fault. Consulting a winding and cooling diagram will be very advantageous in finding the blockage.

REFERENCES

1. G. Klempner and I. Kerszenbaum. *Handbook of Large Turbo-Generator Operation and Maintenance*. Piscataway, NJ: IEEE Press, 2008, chapter 8.
2. J.L. Drommi and F. Mesnage. How to Prevent Hollow Conductor Plugging: EDG Solution for Aerated Systems. In: *Iris Rotating Machine Conference*, June 10–13, 2002.

M-S-06: BREACH OF THE STRANDS OR TUBES CARRYING COOLING WATER

APPLICABILITY:

Generators with directly water-cooled stators

MALFUNCTION DESCRIPTION:

This malfunction addresses the issue of a breach of the water carrying hollow strands or tubes. The breach can be small or large, with minimal to serious consequences, such as small amounts of hydrogen leaking into the SCW system, up to large leaks of hydrogen into the SCW system causing total gas locking and possible water ingression into the generator and *wet* bars.

There are a number of mechanisms by which the SCW strands can be compromised. Some may be manufacturing deficiencies; for example, bad brazing of strands at the bar ends allowing cracks to develop in the brazed area. Some may be due to maintenance deficiencies, such as applying too much force to a bar in its overhang or involute section of the endwinding, resulting in cracks. Others may be due to operational events, for example, large grid disturbance or continuous operation with high vibrations that results in cracks or breaks of the bars in the endwinding (in particular the welds between conductors). Finally, others may be due to unrelated degradation of the machine, like magnetic termites created by back-of-core fretting or contamination, causing punctures in the water-carrying strands (Figure 5.32).

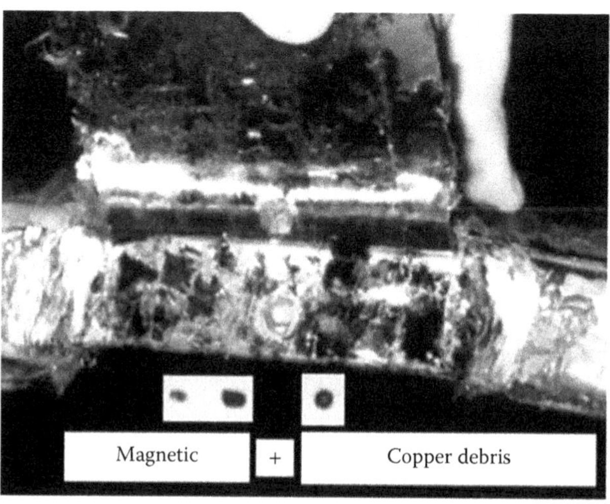

FIGURE 5.32
Breached hollow copper strand. Debris found was both magnetic and copper particles.

Breaches can affect a single bar, a number of bars, or the entire machine, in more than one way. See *Probable Consequences* below.

PROBABLE CAUSES:

- Magnetic termite (also called *worm*) from back-of-core fretting
- Magnetic termite left from drilling, welding, or another maintenance activity inside or on the generator that created small metallic particles that were not cleaned
- Large grid disturbance or short-circuit creating large transient movements of the endwindings
- Inadequate restraining of the endwindings allowing large vibration during operation
- Inadequate restraining of the bars in the slots, allowing large vibration during operation
- Inadequate maintenance activity such as allowing the rotor or rotor skid-plate to press on the endwindings during rotor removal or installation
- Inadequate activity such as over-torquing of the endwindings (leading to cracking of endwinding support hardware or damaging the coils)

POSSIBLE CONSEQUENCES:

Overheating of certain parts of the winding due to a loss of SCW:

- GAS LOCKING IN WATER-CARRYING STRANDS OR TUBES **(M-S-21)**
- INSUFFICIENT FLOW OF COOLING MEDIUM IN A STATOR BAR **(M-S-32)**
- LEAK IN THE FLOW PATH OF THE STATOR COOLING WATER (SCW) **(M-SCW-09)**

Water-damage to wall insulation due to capillary action:

- DAMAGE TO CONDUCTOR WALL INSULATION DUE TO A WATER LEAK **(M-S-16)**
- Rusting of certain components inside the machine due to water ingression
- Stress corrosion cracking of 18Mn-5Cr retaining rings
- Contamination of the stator winding insulation by water and water-carried contaminants
- ELEVATED HYDROGEN DEW POINT **(M-H-05)**

POSSIBLE SYMPTOMS:

- HYDROGEN-IN-STATOR COOLING WATER (SCW) HIGH **(S-S-04)**
- LIQUID-IN-GENERATOR ALARM **(S-C&F-04)**

Temperature changes to a bar or group of stator bars:

- OUTLET TEMPERATURE OF STATOR CONDUCTOR BAR COOLANT ABOVE NORMAL (CASE 1) **(S-S-07)**
- OUTLET TEMPERATURE OF STATOR CONDUCTOR BAR COOLANT ABOVE NORMAL (CASE 2) **(S-S-08)**
- OUTLET TEMPERATURE OF STATOR CONDUCTOR BAR COOLANT BELOW NORMAL **(S-S-09)**

OPERATORS MAY CONSIDER:

A confirmed breach of the strands or tubes carrying SCW has the potential to put the machine integrity in risk. Therefore, unless the leak is very small, operators should consider taking the unit of service for repairs.

TEST AND VISUAL INDICATIONS:

- SCW inside the machine
- Large amounts of hydrogen in the SCW system
- Leaks found by one of the several existing techniques (e.g., bubbles, snooping, bagging, SF_6 injection, helium injection)
- Holes in the insulation left by magnetic termites
- Cracks in the insulation
- Unusual deformation of the bars in the overhangs
- Capacitive mapping indicating wet coils
- Megger tests indicating wet coils

REFERENCES

1. G. Klempner and I. Kerszenbaum. *Handbook of Large Turbo-Generator Operation and Maintenance.* Piscataway, NJ: IEEE Press, 2008, chapters 2 and 5.
2. D. Vermeeren, J.C. Gabriel, and J.L. Dromni: On-Line Flushing of Plugged Hollow Conductors in Stator Bars. In: *EPRI 1997,* Florence, Italy, April 14–16, 1997.

M-S-07: BROKEN/CRACKED STATOR TERMINAL

APPLICABILITY:

All generators

MALFUNCTION DESCRIPTION:

This malfunction applies to the case of a partial or total mechanical failure of a stator terminal. In all cases, if the electrical path is degraded, a *hot* joint is the resulting problem. In terminals directly cooled by water, a mechanical failure that breaches the inner coolant path has the potential for introducing large leaks of water into the generator casing. In all cases,

FIGURE 5.33
A broken strand belonging to a stator lead-to-terminal connection.

a cracked/ruptured/broken terminal has the potential for introducing serious risks to the integrity of the generator. See Figure 5.33 for an example of a broken strand in a terminal connection.

PROBABLE CAUSES:

- Inadequate design
- Substandard manufacturing
- Wrong maintenance
- VIBRATION OF STATOR ENDWINDING—HIGH **(M-S-56)**

POSSIBLE CONSEQUENCES:

- BLOCKAGE IN THE FLOW OF STATOR TERMINAL COOLANT **(M-S-01)**
- BURNING/PYROLYSIS OF STATOR WINDING INSULATION **(M-S-08)**
- DAMAGE TO CONDUCTOR BAR INSULATION **(M-S-15)**
- ELEVATED HYDROGEN DEW POINT **(M-H-05)**
- GAS LOCKING OF WATER-COOLED TERMINALS **(M-S-24)**
- GROUND FAULT ON STATOR WINDING **(M-S-26)**
- HYDROGEN INGRESSION INTO STATOR COOLING WATER **(M-S-29)**
- INSUFFICIENT FLOW OF TERMINAL COOLANT **(M-S-35)**
- LEAK OF STATOR COOLING WATER (SCW) INTO GENERATOR **(M-S-37)**

- LOCALIZED OVERHEATING OF STATOR CONDUCTOR **(M-S-39)**
- TEMPERATURE OF STATOR CONDUCTOR ABOVE NORMAL **(M-S-53)**

POSSIBLE SYMPTOMS:

- RADIO FREQUENCY (RF) MONITOR HIGH **(S-S-13)**

Large increases of hydrogen presence in the SCW system:

- HYDROGEN MAKE-UP RATE ABOVE NORMAL **(S-H-05)**
- HYDROGEN-IN-STATOR COOLING WATER (SCW) HIGH **(S-S-04)**
- LIQUID-IN-GENERATOR ALARM **(S-C&F-04)**
- Temperature instruments located at the terminal chamber (high or low readings)
- GENERATOR CORE CONDITION MONITOR—ALARM **(S-C&F-01)**
- OUTLET TEMPERATURE OF STATOR CONDUCTOR BAR COOLANT ABOVE NORMAL **(S-S-2)**

OPERATORS MAY CONSIDER:

A mechanically faulty stator terminal with potential electric path degradation, with or without loss of coolant represents a very serious problem. If such an event is recognized, consider

1. Reducing load immediately to return the temperatures within normal range.
2. Removing the unit from operation as soon as practically possible.
3. Taking into consideration that temperatures at the affected area will be much higher than those indicated by *global* RTDs or TCs (meaning, not measuring directly the temperature of the coolant at the terminals, but indirectly in other parts of the machine).

TEST AND VISUAL INDICATIONS:

- Gas analysis from hydrogen samples captured by the GCM
- Visual inspection of the terminals (insulation, vent openings for hydrogen-cooled terminals, piping integrity for water-cooled terminals, etc.)
- Megger tests

REFERENCES

1. G. Klempner and I. Kerszenbaum. *Handbook of Large Turbo-Generator Operation and Maintenance*. Piscataway, NJ: IEEE Press, 2008, chapters 2 and 8.
2. J.J. O'connor. "Big Allis" Downed By Stator Vibration, Prime Mover and Electrical Generation. *Power Magazine*, September 1972, pp. 62–64.

M-S-08: BURNING/PYROLYSIS OF STATOR WINDING INSULATION

APPLICABILITY:

All generators

MALFUNCTION DESCRIPTION:

This malfunction would be an indication of serious deterioration of the stator winding insulation due to the presence of a heat source that produces abnormally high temperatures. There are two different processes by which this deterioration can occur: *burning* and *pyrolysis*.

Burning is the physical process of a fuel undergoing combustion, which is a rapid chemical oxidation process creating byproducts and heat. Combustion requires the presence of oxygen, and the subjection of the fuel to a temperature above certain value, specific for each fuel. The fire triangle shows the conditions for the presence of fire (Figure 5.34).

Let us look in the generator for the sources of the three components shown on the fire triangle:

Fuel: All insulation is made either of organic materials, or much more commonly, from a combination of organic and inorganic components (e.g., mica or glass as inorganic material and resins and epoxies as organic materials). The organic materials will burn in the presence of oxygen and high temperatures.

Oxygen: In air-cooled generators, the stator windings are in the presence of air, which is the source of the oxygen required for the initiation and sustenance of burning. In hydrogen-cooled generators, there is no air (i.e., oxygen) present during the normal

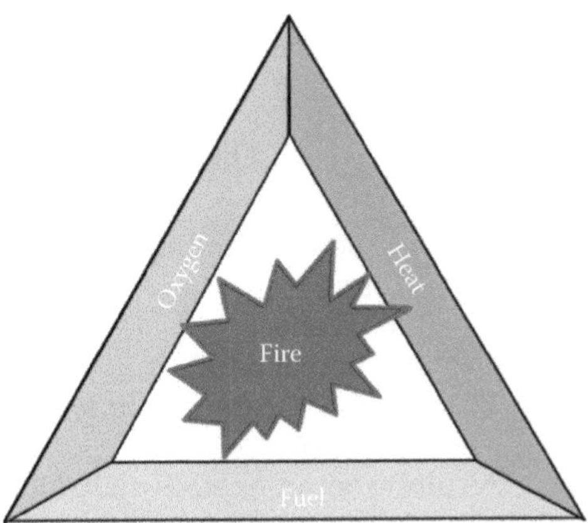

FIGURE 5.34
The so-called *fire triangle* that shows the three elements required for the sustaining a fire.

operation of the unit. Therefore, burning in the presence of fires can only exist if the casing has been severely breached and air is mixed with hydrogen. Depending on the H_2/air ratio, this by itself can produce high-combustion temperatures, lighting up stator insulation, which will then continue to burn.

Heat: The heat required to initiate burning of the insulation may have a number of sources. One indicated above, is an internal explosion of a hydrogen–air mixture. This can only happen if a catastrophic failure of the casing results in large amounts of air mixing with hydrogen inside the machine. Alternatively, it could happen if wrong operator activity results in mixing air with hydrogen inside the machine in ratios that may cause the mixture to self-ignite.

A more likely scenario is the source of heat being an internal electrical failure, such as a phase–phase short-circuit, or inter-turn short-circuit in smaller generators with multiple turns. Also, a serious core failure with melted iron may become a source of very high temperature.

Finally, a major overcurrent event, or loss of cooling can give way to high temperatures. It is hard to imagine these types of events producing winding temperatures required to light up the insulation. Nevertheless, it may be seriously aged in a short time.

Pyrolysis: It is the chemical decomposition of organic materials in the absence of oxygen. Pyrolysis is the mechanism by which stator insulation deteriorates if a large amount of heat is applied in hydrogen-cooled generators.

For the purpose of investigating the development of rapid degradation of stator winding insulation due to abnormal high temperatures, burning, and pyrolysis can be used as a single common factor.

Burning/pyrolysis may cause pyrolization of the insulation resulting in a conductive path capable of creating a short circuit and sustaining the arc (Figure 5.35).

PROBABLE CAUSES/AREA AFFECTED:

- Global loss of cooling/global effect
- Local loss of cooling/section of winding depending on location and nature of loss of cooling
- Significant overload/global effect
- Internal short-circuit/section of winding
- Melting of core or significant core defect; localized portion of winding embedded in the slot
- Operator error resulting in an explosive H_2/O_2 mixture inside the machine
- Large breach of the generator casing resulting in an explosive H_2/O_2 mixture inside and around the machine

FIGURE 5.35
The pyrolysis of a severely overheated stator groundwall.

POSSIBLE CONSEQUENCES:

Depending on type of cause:

- Damage to wall insulation with full or partial loss of insulating capability
- Damage to interturn insulation with full or partial loss of insulating capability
- Deformation of wall insulation (in particular *thermoplastic* insulation, such as with *asphalt* bonding. A consequence of this outcome can be swelling of the wall insulation in the core cooling vent areas and subsequent partial loss of cooling
- Accelerated thermally induced aging of the affected insulation
- Mechanical weakening of the insulation bonding material
- Creation of voids in the insulation bonding material with resultant increase of partial discharge activity (PDA)
- Loosening of the winding in the affected area (*thermoelastic* insulation)
- Increasing the level of hydrogen impurity in the machine
- Ground or interphase faults upon breach of wall insulation
- Shorted-turn in multiturn machines

POSSIBLE SYMPTOMS:

Immediate indications can be obtained through pyrolysis byproducts. They can be detected by the core condition monitor (CCM), also known as GCM:

- GENERATOR CORE CONDITION MONITOR—ALARM **(S-C&F-01)**

Long-term indications can be a significant increase in PDA:
- PARTIAL DISCHARGE ACTIVITY (PDA) HIGH **(S-S-11)**
- RADIO FREQUENCY (RF) MONITOR HIGH **(S-S-13)**

OPERATORS MAY CONSIDER:

1. If GCM alarm received, operators may want to consider checking that the alarm is the result of a *real* pyrolytic activity. They may also check for other indicators of abnormality, such as load condition, air/hydrogen/water cooling system integrity, and ground fault indication.
2. Checking winding RTDs and/or TCs. A global significant increase on winding temperature (*cooled by gas* or *water*) will be captured by those temperature instruments. A localized problem may not be captured; depending how far is the nearest temperature probe.
3. If the GCM alarm has been deemed *true*, then consider reducing load until the situation is cleared, even to the extent of removing the unit from service.
4. Record temperatures versus time of windings and any other abnormal readings and events.

TEST AND VISUAL INDICATIONS:

- Electrical tests such as:
  - Insulation resistance measurement (known as *meggering* or *megger* test)
  - PI
  - High potential testing (*Hipot*)
  - Analysis of a hydrogen sample looking for pyrolysis byproducts
  - And so on
- Visual inspection (directly, borescopic or robotic)

REFERENCE

1. G. Klempner and I. Kerszenbaum. *Handbook of Large Turbo-Generator Operation and Maintenance*. Piscataway, NJ: IEEE Press, 2008, chapters 5 and 8.

M-S-09: COPPER OXIDE BUILDUP IN WATER-CARRYING STRANDS

APPLICABILITY:

Generators with directly water-cooled windings

MALFUNCTION DESCRIPTION:

This malfunction addresses the situation when the flow of SCW is partially or fully blocked by the presence of copper oxides buildup. The chemistry leading to copper oxide buildup in copper strands carrying

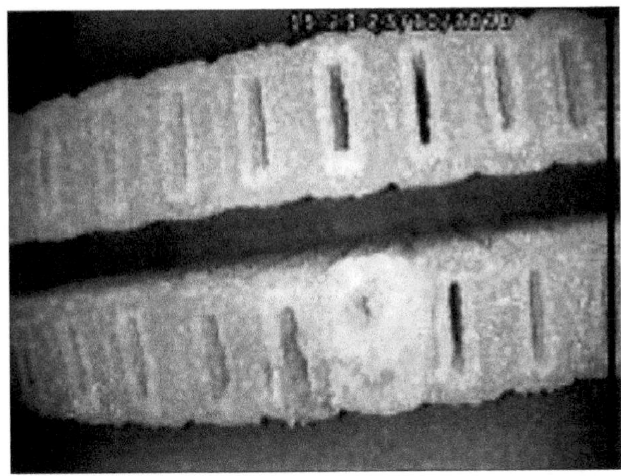

FIGURE 5.36
A stator bar with the hollow copper strands partially blocked by copper oxides.

water is complex. It is covered extensively in the literature, in particular in a number of EPRI reports. One of those reports is shown in the References section. See Figure 5.36 for an example of a stator conductor bar partially blocked by copper oxide buildup.

There are two types of copper oxides: cupric and cuprous. Each one may be found in the stator hollow strands. The References below discuss their differences in the realm of copper corrosion (CuOx) buildup in the machine.

PROBABLE CAUSES:

The most common cause of CuOx buildup is an excursion from the normal chemistry of the SCW; for example, introduction of highly oxygenated water in a low-oxygen SCW system. Another cause for the presence of copper oxides in abnormal quantities is changes in the pH level of the SCW. Oftentimes, the CuOx buildup is the result of less-than-desirable layup practices during outages.

POSSIBLE CONSEQUENCES:

- BURNING/PYROLYSIS OF STATOR WINDING INSULATION **(M-S-08)**
- CONDUCTIVITY OF STATOR COOLING WATER (SCW) ABOVE NORMAL **(M-SCW-04)**
- DAMAGE TO CONDUCTOR BAR INSULATION **(M-S-15)**
- GLOBAL STATOR TEMPERATURE ABOVE NORMAL **(M-S-25)**
- STATOR COOLING WATER (SCW) FILTER/STRAINER RESTRICTION **(M-SCW-13)**
- TEMPERATURE OF STATOR CONDUCTOR ABOVE NORMAL **(M-S-53)**

POSSIBLE SYMPTOMS:

Figure 5.37 shows a strainer with copper oxide deposits.

- CONDUCTIVITY OF STATOR COOLING WATER ABOVE NORMAL **(S-SCW-03)**
- GENERATOR CORE CONDITION MONITOR—ALARM **(S-C&F-01)**
- SLOT TEMPERATURE OF STATOR CONDUCTOR BAR—ABOVE NORMAL (CASE 1) **(S-S-14)**
- STATOR COOLING WATER PRESSURE DROP ACROSS SCW FILTERS/STRAINERS ABOVE NORMAL **(S-SCW-14)**

OPERATORS MAY CONSIDER

1. Like any other problem leading to an increased conductor temperature, controlling the load in such a way that temperatures are maintained in their proper range is paramount.
2. If the presence of CuOx is confirmed as being the source of the problem, online chemical cleaning can be considered as an alternative to removing the unit from the grid for a many-day outage. However, this option is only productive when the blockage of the strands is partial, so that the flow of the SCW with the added chemicals can be effective in removing the buildup. By the way,

FIGURE 5.37
A strainer from a large generator's SCW system. The reddish color is from cuprous oxides released from the copper winding.

the requirement that the chemical agents must be able to flow through partially blocked strands also applies to offline chemical cleaning

3. Cuox buildup ought to be taken very seriously, because once a strand is fully blocked, chemical cleaning attempts become futile. CuOx buildup can lead under certain conditions to the need for rewinding the stator well ahead of its expected end of life. Hence, attempts should also be made to uncover the root cause for the onset of copper oxide buildup

TEST AND VISUAL INDICATIONS:

CuOx deposits on the strainer (see Figure 5.37) can indicate an undesirable rate in the creating of cupper oxides.

REFERENCES

1. G. Klempner and I. Kerszenbaum. *Handbook of Large Turbo-Generator Operation and Maintenance.* Piscataway, NJ: IEEE Press, 2008, chapter 8.
2. EPRI TR 1015669. Turbine Generator Auxiliary Systems Maintenance Guides—Volume 4—Generator Stator Cooling System.
3. GE TIL-1312-2. Stator Water Cooling System YST-1 Strainer Clogging, January 2002.
4. GE TIL-1226-3. Generator Y-Strainer Inspection of Replacement, June 1997.

M-S-10: CRACKED CONDUCTOR STRANDS

APPLICABILITY:

All generators

MALFUNCTION DESCRIPTION:

This malfunction addresses the issue of conductor strands becoming cracked or otherwise loosing electrical continuity (i.e., open circuit) (see Figure 5.38).

Turbine-driven generator stators have conductors made of a number of parallel copper strands. In larger units, the strands in the conductor bar are all brazed at both ends of the bar, forming a single *turn*. In smaller units, there may be two or more turns in a single bar (generally called a multiturn coil). In both those cases, the main reason why a single turn is made from a number of strands instead of a single copper bar is to minimize eddy currents, and therefore, losses in the copper.

Eddy currents depend largely on the cross section of the copper. By dividing the cross section of the turn into smaller insulated strands, the total eddy current induced in the bar is much smaller, and thus so are the generated I^2R losses.

However, copper strands connected in parallel at the ends of the bar provide for the generation of circulating currents (because the different strands in the bar *see* different flux densities, and therefore induce

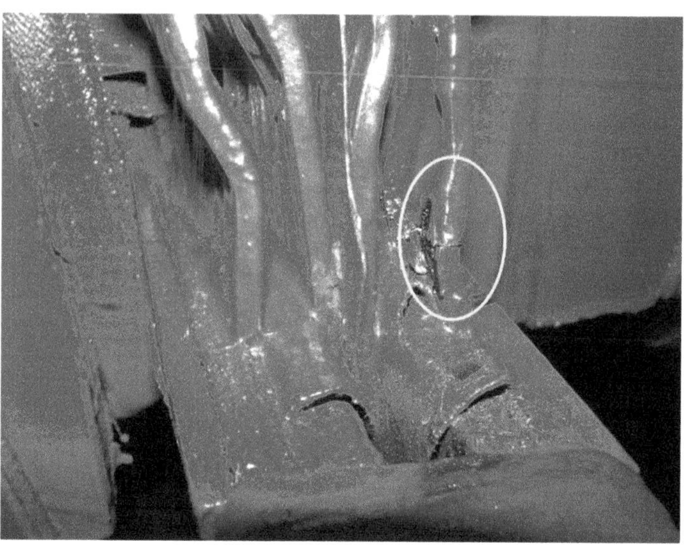

FIGURE 5.38
A broken strand. Many cracked/broken strands can lead to the opening of a connection under load, with catastrophic results to the integrity of the machine.

along each strand a slightly different voltage). To control these circulating currents, the strands are all moved around the bar (i.e., transposed), so that each strand occupies each position at least once. This is called Roebel trans-positioning of the strands (see Figures 5.39a and 5.39b), and all large machines employ this technique in their windings to minimize the voltage difference and circulating currents in the strands. The most optimum transposition arrangement is generally a 540° transposition [1].

In most designs in existence to date, stators with directly water-cooled bars have a number of their strands—if not all—hollow, to allow flow of the cooling water. These hollow strands are prone to cracking if subjected to a number of mechanical and thermal stresses. Cracked or otherwise fractured strands will increase the current load on the neighboring parallel strands resulting in overheating of the bar, as well as hydrogen and water leaks.

Damage to bar insulation due to capillary action in directly water-cooled bars is also possible.

PROBABLE CAUSES:

The most probable causes for fractured or broken strands are as follows:

- High vibration of stator endwindings
- Malsynchronization

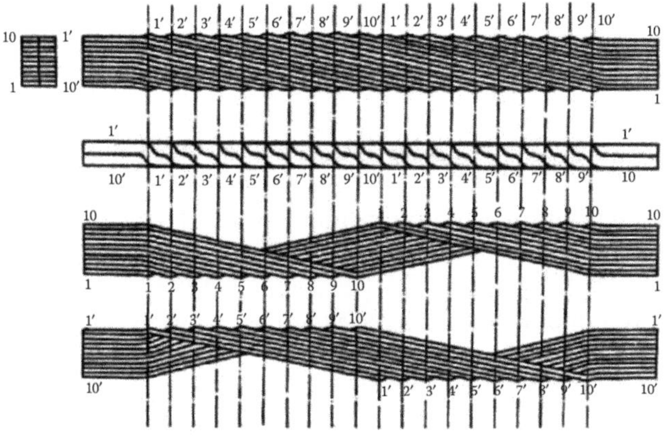

(a)

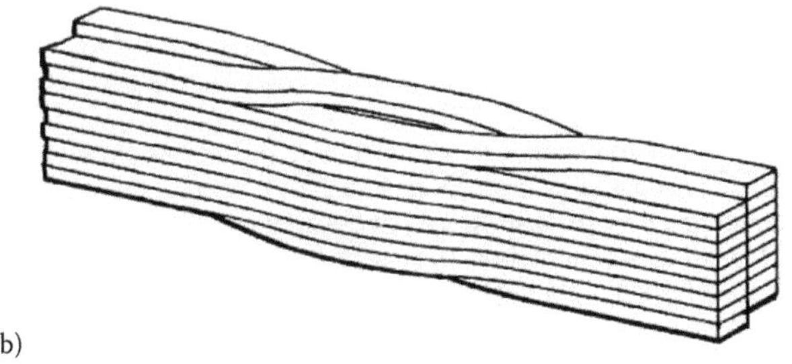

(b)

FIGURE 5.39
Roebel-transposed stator bar strands.

- Maintenance error
- Wrong brazing and/or other manufacturing errors
- Bad design of the connections
- Loose winding leading to high vibration in the slot
- Large grid-induced transient (for instance, due to a close fault)

POSSIBLE CONSEQUENCES:

Further discussion of this and related issues can be found in the following malfunction descriptions:

- BREACH OF THE STRANDS OR TUBES CARRYING COOLING WATER **(M-S-06)**
- BURNING/PYROLYSIS OF STATOR WINDING INSULATION **(M-S-08)**
- DAMAGE TO CONDUCTOR BAR INSULATION **(M-S-15)**

- LEAK OF CONDUCTING-BAR COOLING WATER INTO GENERATOR **(M-S-36)**
- LOCALIZED OVERHEATING OF STATOR CONDUCTOR **(M-S-39)**

POSSIBLE SYMPTOMS:

In machines with RF monitors, increase in RF activity could be the result of cracked strands:

- RADIO FREQUENCY (RF) MONITOR HIGH **(S-S-13)**

Increased temperature (*gas-cooled/water-cooled*) of one or a number of bars:

- OUTLET TEMPERATURE OF STATOR CONDUCTOR BAR COOLANT ABOVE NORMAL (CASE 1) **(S-S-07)**
- OUTLET TEMPERATURE OF STATOR CONDUCTOR BAR COOLANT ABOVE NORMAL (CASE 2) **(S-S-08)**
- OUTLET TEMPERATURE OF STATOR CONDUCTOR BAR COOLANT BELOW NORMAL **(S-S-09)**

Hydrogen-into-SCW leak and/or SCW leak into generator:

- HYDROGEN MAKE-UP RATE ABOVE NORMAL **(S-H-05)**
- HYDROGEN-IN-STATOR COOLING WATER (SCW) HIGH **(S-S-04)**
- LIQUID-IN-GENERATOR ALARM **(S-C&F-04)**

OPERATORS MAY CONSIDER:

Once cracks are diagnosed (very difficult to achieve, as it most probably requires opening sections of the winding, unless on the uninsulated series and parallel connections of some type of hydrogen-cooled machines), it is appropriate to plan for repairs as soon as possible. Cracked strands can possibly result in localized overheating due to strand-to-strand short circuits, *wet bars*, ground faults, and open circuits, if left unattended.

TESTING AND VISUAL INDICATIONS:

- Increased RF activity online.
- Series resistance readings may yield enough of a difference with other bars, but only if carried out on a bar per bar basis and, the bars would have to be disconnected at the ends and electrically *floating* for an accurate comparison to be made.
- Visual indications such as insulation discoloration-darkening may indicate cracked strands.
- Unsatisfactory megger, PI and Hipot tests of the affected bars, if the cracked strands resulted in significant overheating for a long time, and/or a wet bar.
- Capacitance mapping (may uncover wet bars due to a fracture/ broken strand).

REFERENCES

1. G. Klempner and I. Kerszenbaum. *Handbook of Large Turbo-Generator Operation and Maintenance*. Piscataway, NJ: IEEE Press, 2008, chapter 2.
2. EPRI Technical Report 1021774. Generator Stator Endwinding Vibration Guide: Tutorial, August 2011.
3. G. Stone et al. *Electrical Insulation for Rotating Machines*. Piscataway, NJ: IEEE, 2004.

M-S-11: CURRENT OVERLOAD OF STATOR WINDING

APPLICABILITY:

All generators

MALFUNCTION DESCRIPTION:

This malfunction relates to an overload situation. This is not about a transient (such as a shorted-circuit external to the generator or a momentary power swing induced by the grid). This malfunction relates to a steady-state overload, that is, operation with currents above *maximum current* allowable, as indicated by the machine rating and shown on the capability and V-curves for the machine. It is most likely to occur due to operator error. Figure 5.36 shows the limit for stator loading on the capability curve.

It is important to take into consideration that if the hydrogen pressure (in hydrogen-cooled machines) is below rated, the capability of the stator winding to withstand current without damage is also reduced. In some cases, OEMs show a number of capability curves for various pressure values on the same graph, as shown on Figure 5.40.

REFER TO:

- PHASE CURRENT HIGH (M-S-45)

M-S-12: CURRENT UNBALANCE BETWEEN PHASES

APPLICABILITY:

All generators

MALFUNCTION DESCRIPTION:

This malfunction addresses the case where the current magnitudes between the phases are not balanced. In fact, all generators will experience certain degree of unbalance, mainly due to natural load unbalances in the grid they feed into. Machines are designed to withstand a certain degree of current unbalance, both steady state (I_2) and transient (I_2^2t). The standards do not establish a limit on current unbalance, but on negative-sequence currents, which are a direct consequence of phase current unbalances.

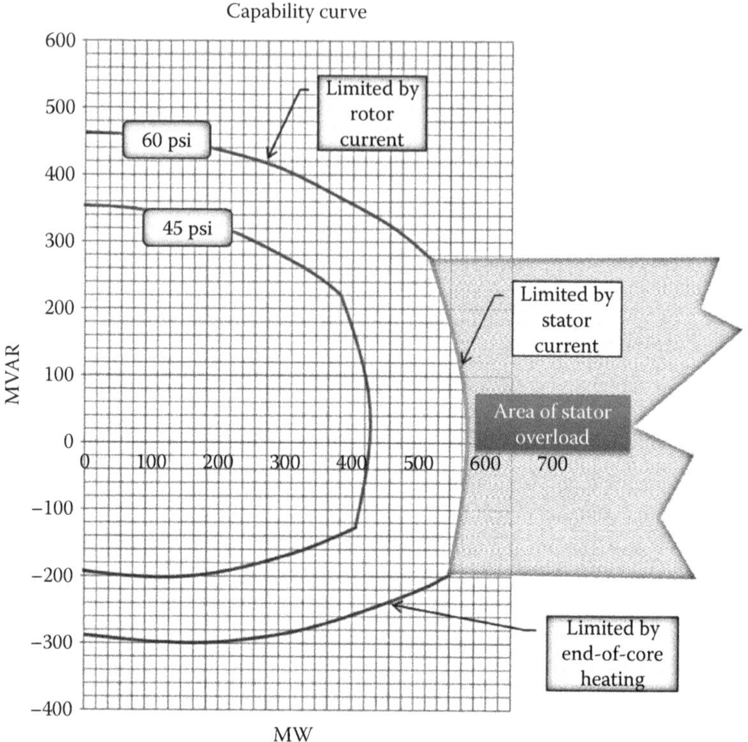

FIGURE 5.40
Various limiting components on the capability curve, including the overload region of the stator winding.

However, negative-sequence current protection exists that is (ought to be) set so that the machine is fully protected against any negative-sequence current event that encroaches on the limits set in the standards.

Some operators do monitor phase current unbalance, and thus do not exclusively depend on negative-sequence current protective devices. This may be a better requirement in units that may operate well below rated values, but it may not be such an effective tool for units operating base load.

There are a few reasons why phase current unbalance exists, some internal to the machine, and some external.

PROBABLE CAUSES:

- *Internal*: Opening of a phase-parallel-connector. In this case, the current in the affected phase will flow on part of the winding.
- *External*: Grid unbalance that can be the result of steady-state unbalance or from short-lived events (here *short-lived* means up to several minutes).

POSSIBLE CONSEQUENCES:

- NEGATIVE-SEQUENCE CURRENTS **(M-S-42)**

See Figure 5.41 for an example of a rotor damaged by high negative-sequence currents.

POSSIBLE SYMPTOMS:

- NEGATIVE-SEQUENCE CURRENTS ALARM AND/OR TRIP **(S-S-06)**
- PHASE CURRENT UNBALANCE **(S-S-12)**

OPERATORS MAY CONSIDER:

Refer to the information in:

- NEGATIVE-SEQUENCE CURRENTS **(M-S-42)**

TEST AND VISUAL INDICATORS:

- For small unbalances, a reduction in load may not eliminate the balance but will reduce the level of resulting negative-sequence current.

For large unbalances, see

- NEGATIVE-SEQUENCE CURRENTS **(M-S-42)**

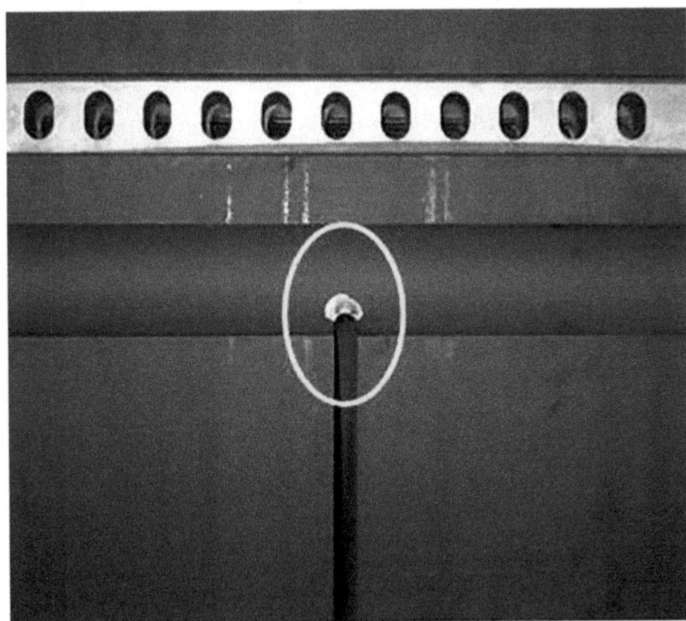

FIGURE 5.41
A hardened corner of a circumferential cross-cut on a rotor, due to negative-sequence currents.

REFERENCES

1. G. Klempner and I. Kerszenbaum. *Handbook of Large Turbo-Generator Operation and Maintenance*. Piscataway, NJ: IEEE Press, 2008, chapter 6.
2. IEEE Std 67. Guide for Operation and Maintenance of Turbine-Driven Generators, Section 7.7.

M-S-13: CURRENT UNBALANCE BEWEEN PHASE PARALLELS

APPLICABILITY:

All generators

MALFUNCTION DESCRIPTION:

In many generators, each phase is made of a combination of parallel and/or series–parallel circuits. If a joint of one of the parallel circuits fails, or the electric circuit is degraded by some other reason (e.g., cracked or broken conducting strands), the current distribution between the parallel circuits may cease from being balanced. In fact, if one of the parallels goes *open circuit* then all of the phase current will go through the unaffected parallel, but the generator protection only sees the phase current after the parallels and will not detect this condition. This type of problem results in temperature differentials between the parallel circuits comprising a phase presenting an opportunity to diagnose this condition. See Figure 5.42 for a winding with two parallel circuits per phase. Some machines may have two, three, or four parallel circuits per phase.

The current flowing in each parallel winding is not measured directly; thus, the operator depends on the temperature probes (be it embedded in slot, or measuring the bar-outlet cooling medium) measuring conducting bar temperature for an indication of such a problem being present.

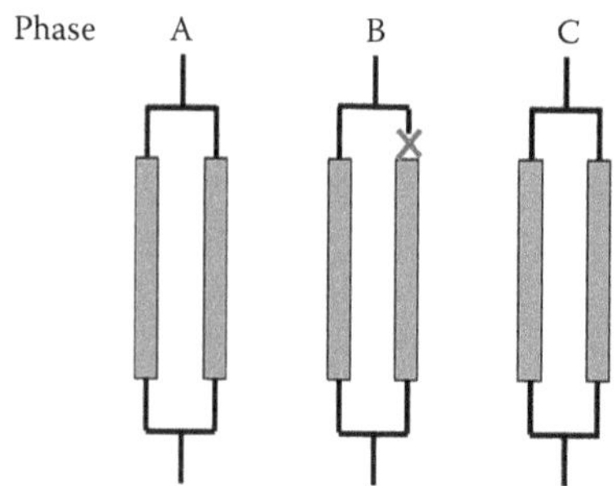

FIGURE 5.42
A winding with two parallel circuits per phase. Phase B has an open parallel circuit.

PROBABLE CAUSES:

- Cracked brazing due to vibrations resulting in less conducting area
- Cracked brazing due to substandard brazing operation
- Cracked joint due to severe grid transient or short-circuit *near* the machine
- Loosening of bolted joint due to relaxation of the bolts
- Bolted joint corrosion buildup resulting in weak electrical contact

POSSIBLE CONSEQUENCES:

- BURNING/PYROLYSIS OF STATOR WINDING INSULATION **(M-S-08)**
- DAMAGE TO CONDUCTOR BAR INSULATION **(M-S-15)**
- NEGATIVE-SEQUENCE CURRENTS **(M-S-42)**

POSSIBLE SYMPTOMS:

- OUTLET TEMPERATURE OF STATOR CONDUCTOR BAR COOLANT ABOVE NORMAL (CASE 1) **(S-S-07)**
- RADIO FREQUENCY (RF) MONITOR HIGH **(S-S-13)**
- TEMPERATURE OF PHASE CONNECTOR ABOVE NORMAL **(S-S-18)**

OPERATORS MAY CONSIDER:

1. It is rather difficult to diagnose a parallel circuit problem. However, any internally caused current unbalance, once recognized, should be addressed with the utmost urgency, because it has the potential to create significant damage.
2. If only temperature indications are present, then depending on the severity of this indications, and in particular how they change in time (operators should trend the affected temperature measurements over time), the operators can consider scheduling a repair at a convenient time.
3. An externally induced unbalance would be present in all generators synchronized to the same bus. An internally induced unbalance will affect only that generator.

TEST AND VISUAL INDICATORS:

- Visual inspection of affected area may show signs of discoloration of the insulation.
- Series resistance measurements can uncover an electrically degraded parallel circuit.

REFERENCE

1. G. Klempner and I. Kerszenbaum. *Handbook of Large Turbo-Generator Operation and Maintenance*. Piscataway, NJ: IEEE Press, 2008, chapter 8.

M-S-14: DAMAGE FROM MAGNETIC TERMITES

APPLICABILITY:

All generators

MALFUNCTION DESCRIPTION:

A *magnetic termite* is the name given to a small, loose piece of magnetic material that becomes lodged in the endwindings or in the slot, between coil and slot (Figures 5.43 and 5.44). Over time, this magnetic material may *drill* into the wall insulation of the coil due to the magnetic forces induced in it by the alternating fluxes generated by the high currents flowing in the conductors. The driving forces are at twice *fundamental* frequency. Thus, in 60 Hz machines, all else being equal, the *drilling* effect will be faster than in 50 Hz machines. Once the wall insulation is tunneled, the magnetic termite will drill into the copper conductor.

Magnetic termites can cause significant faults, as indicated below.

Although the term magnetic termites is applied mainly to ferromagnetic metals, it has been demonstrated that a small metallic particle which is not magnetic, such as aluminum, is also mechanically excited by the interaction between the external flux and the eddy currents induced in the metal. This force is much smaller than the one developed with magnetic particles. Also, aluminum or copper (the main type of nonmagnetic particles may exist inside a generator), are softer than magnetic (steel) particles, and hence, combined with weaker forces, less prone to cause noteworthy damage to the insulation.

FIGURE 5.43
Hole drilled into a stator endwinding due to magnetic material debris from back of core burning.

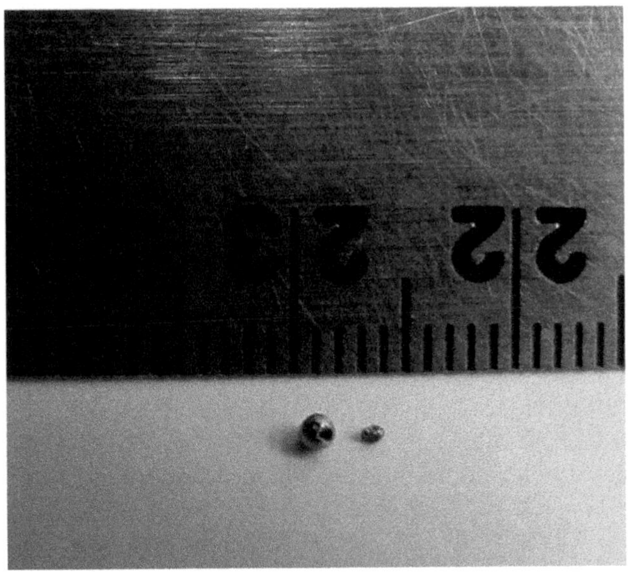

FIGURE 5.44
Magnetic material debris from back of core burning. Chemical analysis showed this to be 3%–4% silicon steel.

PROBABLE CAUSES:

- Debris left from maintenance activities such as drilling and welding inside, above or in the near vicinity of an open generator.
- Debris introduced inadvertently by someone walking into the machine either through the bore (rotor out) or through other opening (such as a removed cooler). It is easy to get a small piece of metal lodged in the sole of a shoe and then dropping it inside the machine. This is one reason why shoe covers or other precautions must be taken seriously during maintenance activities inside, on close or an open generator.
- Debris from *BCB* activity or fretting between core and keybars.
- Any other possible way that a foreign small ferromagnetic fragment can find its way inside the machine during an open-machine outage.

POSSIBLE CONSEQUENCES:

Ground faults. These can occur in any type of generator windings.

- GROUND FAULT OF STATOR WINDING **(M-S-26)**

Broken conducting strands. These can occur in any type of generators.

- BREACH OF THE STRANDS OR TUBES CARRYING COOLING WATER **(M-S-06)**
- DAMAGE TO CONDUCTOR BAR INSULATION **(M-S-15)**

- LEAK OF CONDUCTING-BAR COOLING WATER INTO GENERATOR **(M-S-36)**
- LOCALIZED OVERHEATING OF STATOR CONDUCTOR **(M-S-39)**

POSSIBLE SYMPTOMS:

There are no direct indicators that a magnetic termite is *drilling* a bar. However, there may be a number of symptoms due to a fault created by such a magnetic termite.

- HYDROGEN MAKE-UP RATE ABOVE NORMAL **(S-H-05)**
- HYDROGEN-IN-STATOR COOLING WATER SYSTEM (SCW) HIGH **(S-S-04)**
- LIQUID-IN-GENERATOR ALARM **(S-C&F-04)**
- STATOR GROUND ALARM AND/OR TRIP **(S-S-17)**

Temperature deviations (normally increase) in a bar:

- OUTLET TEMPERATURE OF STATOR CONDUCTOR BAR COOLANT ABOVE NORMAL (CASE 1) **(S-S-07)**
- OUTLET TEMPERATURE OF STATOR CONDUCTOR BAR COOLANT ABOVE NORMAL (CASE 2) **(S-S-08)**
- OUTLET TEMPERATURE OF STATOR CONDUCTOR BAR COOLANT BELOW NORMAL **(S-S-09)**

OPERATORS MAY CONSIDER:

Once a fault alarm or trip has been initiated, consider initiating troubleshooting activities to find the source (see Section *Testing and Visual Indications*). In many cases, if the action by the magnetic termite results in water leaking out from the conductor bar (in water inner-cooled stators), it will be with a leak rate requiring taking the unit offline for repairs sooner than later. Keep in mind, that a *leak* on a water-carrying strand results in hydrogen ingressing the SCW system, and only small amounts of water seeping out the strand, mainly leading to *wet insulation*. Thus, the primary symptom is hydrogen in the SCW in larger-than-normal quantities.

TESTING AND VISUAL INDICATIONS:

- Careful visual inspection may uncover magnetic termites (also called magnetic worms) drilling in the wall insulation.
- If leaks are present in directly water-cooled stators, finding the leak and carefully investigating its root cause may uncover a magnetic termite (Figure 5.44).
- BCB activity may be a precursor for magnetic termites.
- Recent drilling or welding activities inside, above, or very close to an open generator may raise concerns about FME breaches leading to the presence of magnetic particles in the generator.

REFERENCE

1. G. Klempner and I. Kerszenbaum. *Handbook of Large Turbo-Generator Operation and Maintenance*. Piscataway, NJ: IEEE Press, 2008, chapter 8.

M-S-15: DAMAGE TO CONDUCTOR BAR INSULATION

APPLICABILITY:

All generators

MALFUNCTION DESCRIPTION:

This malfunction relates to stator winding insulation that is directly damaged during operation, to be distinguished from another malfunction description [DEFECTIVE STATOR WINDING INSULATION (M-S-18)], that results from factory or maintenance inadequacies. There can be cases straddling the region between a maintenance-caused defective or damaged insulation; thus, some of the issues discussed can be found in both malfunction descriptions.

In smaller multiturn turbine-driven generators, the coil insulation is composed of *strand insulation, turn insulation,* and *groundwall insulation.* In most, if not all, large turbogenerators the only insulation is the *strand insulation* and the *groundwall insulation.*

Strand insulation may fail causing strand-to-strand short-circuits. These faults can be accompanied by circulating currents which are not detected directly by dedicated protective functions, but will be detected in short time by the ground fault protection, after they develop into ground faults by damaging the wall insulation. Both strand-to-strand and wall insulation failures can be accompanied by localized burning/pyrolysis of the insulation, whose released byproducts can be detected by a GCM also known as CCM.

In the case of small turbogenerators with *multiturn* coils (see Figure 5.45), interturn faults behave as a strand-to-strand fault, but the degradation of the coil is much faster.

PROBABLE CAUSES/AREA AFFECTED:

Overload:

- OPERATION OUTSIDE CAPABILITY CURVE—ONLINE (M-E-11)
- DAMAGE FROM MAGNETIC TERMITES (M-S-14) (Also called *magnetic worms*)/localized damage anywhere in the winding—Particularly serious in directly-water-cooled machines
- Mechanical abrasion/mainly in the slot area, but also in endwindings, phase connectors, and terminal box
- A large (often sudden) deformation due to *near* short-circuits or power swings/endwindings

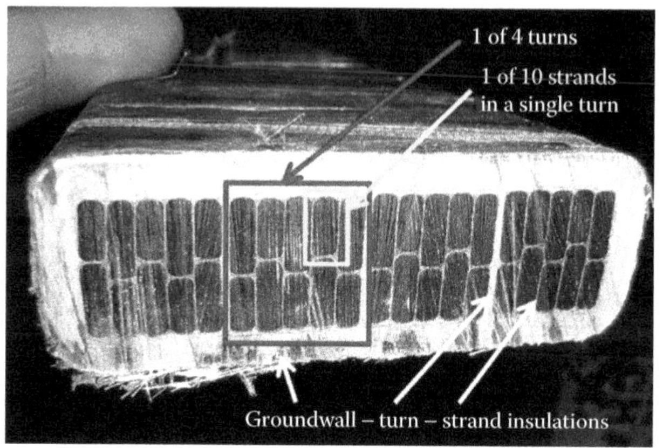

FIGURE 5.45
Multiturn coil cross section showing groundwall strand and turn insulations.

- Electric tracking/overhangs next to the core, and between bars at the tie or blocking area
- Bad joints in phase-connectors or terminal leads causing high temperatures

PDA/slot region:

- RADIO FREQUENCY (RF) ACTIVITY (M-S-46)
  Corona activity/endwindings and phase connector (circumferential busses)
- VIBRATION OF STATOR CONDUCTOR BAR IN SLOT (M-S-55)
  Mainly slot area, however it could affect also the endwindings
- LOOSENING OF STATOR WINDING SLOT WEDGES (M-S-41)
  Loose endwindings/mainly endwindings, but also phase connectors and terminal box
- GLOBAL TEMPERATURE OF CORE ABOVE NORMAL (M-C&F-09)
- LOCAL CORE FAULT (M-C&F-15)
- TEMPERATURE OF A LOCAL CORE AREA ABOVE NORMAL (M-C&F-20)
- TEMPERATURE OF CORE END ABOVE NORMAL (M-C&F-21)

POSSIBLE CONSEQUENCES:

- Interturn short-circuits in machines with two or more turns
- Increased ground leakage currents
- Ground faults: GROUND FAULT OF STATOR WINDING (M-S-26)

POSSIBLE SYMPTOMS:

There is no direct indication of an *incipient* fault due to damaged insulation that has not been fully breached, with exception of byproducts from overheating of the insulation, in those cases that this is a precursor to the fault. Once breached, most indicators will be the activation of turn or ground fault protection:

- STATOR GROUND ALARM AND/OR TRIP **(S-S-17)**
- Interturn fault trip for machines with interturn protection
- GENERATOR CORE CONDITION MONITOR—ALARM **(S-C&F-01)**

OPERATORS MAY CONSIDER:

1. Interturn faults will trip the unit immediately (in machines with interturn protection), and troubleshooting activities are concerned with repairing the affected area.
2. Ground faults can result, depending on the protection scheme, in alarms or alarms and trips. For those schemes that only alarm, there are different philosophies about when to take the unit off-load for repairs. In any case, operating a unit with one ground fault may not be immediately detrimental to a generator with high-impedance neutral grounding (as most large units are), but can result in another ground developing (and that may happen quickly!!), with potentially catastrophic results to the integrity of the machine.

TEST AND VISUAL INDICATIONS:

- IR (also called *megger*) test
- Hipot test
- PDA
- Partial discharge probe test
- DC leakage current test
- Corona deposits on endwindings and/or phase connectors
- Abrasion of the semiconducting coating (needs bars removed for visual access)
- High quantities of ozone next to air-cooled machines
- Visually recognizable deformation of endwindings, phase-connectors or leads
- Cracks, nicks, girths, swelling, bulging, and other localized abnormalities on the wall insulation

REFERENCES

1. G. Klempner and I. Kerszenbaum. *Handbook of Large Turbo-Generator Operation and Maintenance*. Piscataway, NJ: IEEE Press, 2008, chapters 4, 6, and 8.
2. G. Stone et al. *Electrical Insulation for Rotating Machines*. Piscataway, NJ: IEEE, 2004.
3. M.B. Srinivas. Multifactor Aging of HV Generator Stator Insulation Including Mechanical Vibrations. *IEEE Transactions on Dielectrics and Electrical Insulation*, 27(5), 1009–1021.

M-S-16: DAMAGE TO CONDUCTOR WALL INSULATION DUE TO A WATER LEAK

APPLICABILITY:

Generators with water-cooled stator windings

MALFUNCTION DESCRIPTION:

This malfunction relates to a situation when a pinhole water leak or crack in a hollow strand results in water transported by capillary conduction out of the strand and along the ground wall insulation and between strands.

PROBABLE CAUSES:

This is a very rare occurrence except in the case of bar-end water-box brazing, using braze with phosphorus in it. Basically, the phosphorus in the braze combines with the SCW to create phosphoric acid. The water generally has to be stagnant, giving it time to combine with the braze. This then eats through to the insulation and despite the hydrogen gas being at a higher pressure than the braze it will leak back into the insulation by capillary action.

The other ways in which water may leak back into the insulation is from a cracked strand or some other type of breach of a strand that allows water to leak out. High endwinding vibration or bars bouncing in the slot from looseness can aggravate such a problem.

POSSIBLE CONSEQUENCES:

The stator water that would leak into the stator bar groundwall insulation will start off as demineralized and will be a good insulator. But once it leaks it will become contaminated with whatever is present. If there is conductive material present, one could get a ground fault or bar to bar fault.

- GROUND FAULT OF STATOR WINDING (M-S-26)
- LEAK OF STATOR COOLING WATER (SCW) INTO GENERATOR (M-S-37)

POSSIBLE SYMPTOMS:

A likely symptom that this is occurring would be the triggering of a stator leak monitoring system alarm. This is because the hydrogen will likely be leaking into the SCW at the same time the water is leaking out. Another likely symptom from a crevice corrosion-initiated leak at the stator bar-clip that leads to SCW infiltrating the groundwall insulation and gradually degrading it toward the core is a stator ground protection alarm or trip.

- HYDROGEN-IN-STATOR COOLING WATER (SCW) HIGH **(S-S-04)**
- LIQUID-IN-GENERATOR ALARM **(S-C&F-04)**

OPERATORS MAY CONSIDER

1. If a leak monitoring system is present and a leak monitoring alarm is initiated, then the unit should be taken offline for investigation as soon as possible. Same for stator ground fault alarm.
2. Checking the liquid in generator device to determine how much liquid is leaking into the generator. A sample can also be taken to determine what liquid is being collected.

TEST AND VISUAL INDICATIONS:

Pressure and vacuum testing of the stator winding once the unit is offline and opened up are the most common tests for this type of issue. Another is capacitance mapping of the stator bars if the problem is the *crevice corrosion* issue from phosphorus in the braze material. Locating the point of strand cracking or pinhole may be a long and tedious process.

REFERENCES

1. G. Klempner and I. Kerszenbaum. *Handbook of Large Turbo-Generator Operation and Maintenance*. Piscataway, NJ: IEEE Press, 2008, chapter 8.
2. P.J. Tavner, B.G. Gaydon, and D.M. Ward. Monitoring Generators and Large Motors. *Electric Power Applications, IEEE Proceedings B*, 33(3), 169–180.
3. J.A. Worden and J.M. Mundulas, Understanding, Diagnosing, and Repairing Leaks in Water-Cooled Generator Stator Windings, *GE Power Systems*, GER-3751A, August 2001.

M-S-17: DAMAGE TO HV BUSHINGS

APPLICABILITY:

All generators

MALFUNCTION DESCRIPTION:

This malfunction relates to situation when a high voltage terminal bushing is damaged. The damage may be to the copper conductor, ceramic insulator, or the taped insulation.

Figure 5.46a and b shows one high voltage bushing cracked and degraded allowing movement between its parts, and another undergoing electrostatic discharge. Figure 5.47 shows the process of replacing HV bushings.

PROBABLE CAUSES:

Damage can be due to mechanical, electrical, or thermal causes.

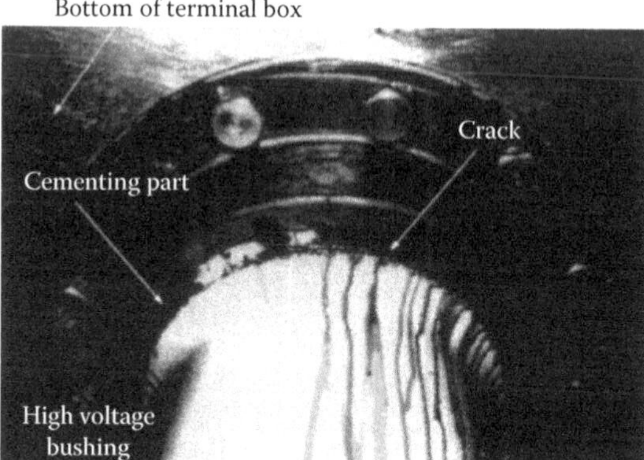

(a)

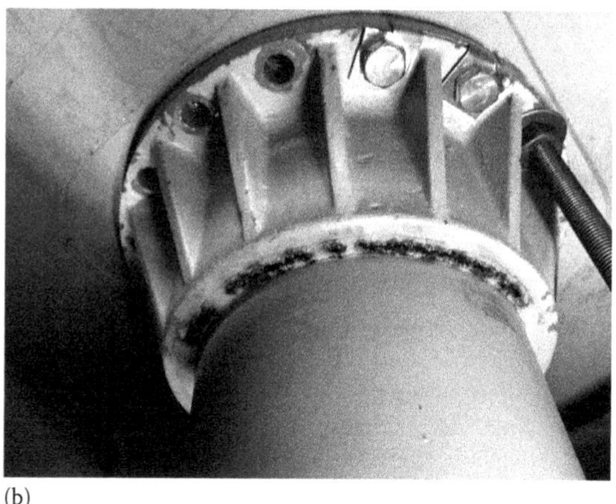

(b)

FIGURE 5.46
(a) HV bushings, allowing leak of the viscous liquid placed inside the terminal box to avoid hydrogen leaking out from the machine. (b) MV bushing on a large generator undergoing electrostatic discharge due to degradation of the electrostatic grounding circuit.

Mechanical vibrations or loosening over time are the most likely causes. Thermal damage is possible when the stator is overloaded or loss of cooling is experienced. Electrical damage can occur if the insulation becomes contaminated or wet and a ground fault occurs.

- LOOSE GENERATOR FOOTING (M-C&F-17)

POSSIBLE CONSEQUENCES:

- GROUND FAULT OF STATOR WINDING (M-S-26)

POSSIBLE SYMPTOMS:

- High vibrations and audible noise from the stator may be present, indicating a problem inside the machine.

FIGURE 5.47
New HV bushing being installed on a hydrogen-cooled generator.

- Typically, these types of problems are discovered in some instances by someone looking at the bushings during routine inspection walks, and only when the bushings are visible to the naked eye with the unit in operation, which is not too often the case.
- HIGH VIBRATION OF CORE AND/OR FRAME **(S-C&F-03)**
- LIQUID-IN-GENERATOR ALARM **(S-C&F-04)**

OPERATORS MAY CONSIDER

It is not likely the operators will have direct information to act on. There will likely be only circumstantial evidence to consider such as a *liquid-in-generator* alarm indicating a breach somewhere in the stator winding circuit in the case of water-cooled bushings, or high audible noise due to vibration of the stator. In either case the operators should notify plant engineering to investigate further.

TEST AND VISUAL INDICATIONS:

Inspection of the bottom of the generator should be made to determine if there is any leakage of oil or hydrogen from around the bushing exit from the stator. Hydrogen leakage from the stator can have serious consequences and is a safety risk. The inspector should be a knowledgeable individual who understands the possibility of hydrogen self-ignition from a pressure vessel.

REFERENCES

1. G. Klempner and I. Kerszenbaum. *Handbook of Large Turbo-Generator Operation and Maintenance*. Piscataway, NJ: IEEE Press, 2008, chapter 8.
2. E.R. Perry and R. Torres. *Generator Bushing Maintenance: Designs & Renovation*, Power Engineering (online), 11/01/2006.

M-S-18: DEFECTIVE STATOR WINDING INSULATION

APPLICABILITY:

All generators

MALFUNCTION DESCRIPTION:

This malfunction addresses defective stator winding insulation. By *defective*, it means that due to poor design, manufacturing or maintenance, the insulation has developed weaknesses that are not characteristic of a well-designed, manufactured, and maintained winding of similar construction, rating, and age. This section does not address the issue of a bar insulation failing due to operation-related matters. The problem of a faulty winding due to operational-related issues is entered as a different problem: DAMAGE TO CONDUCTOR BAR INSULATION (M-S-15).

Defects due to poor design and/or manufacturing can be such as

- Insufficient voltage capability (volts/mil) of the groundwall insulation
- Imperfect cohesion of the mica and glass with the bonding epoxies or resins
- Substandard application of semiconductor coating
- Substandard application of grading paint
- Too many voids left inside the bar
- Wrong bar dimensions vis-à-vis the slot dimensions
- Bars too close to each other in the overhang portion of the windings, and so on

Defects due to poor maintenance can be such as

- Lack of adequate support in the slot (loose wedges and/or slot packing)
- Loose endwindings (inadequate retightening of the support system)
- Wrongly assembled or defective oil seals allowing large amount of oils contaminating the machine
- Lack of adequate FME
- Carelessness during an open-machine inspection or maintenance activity, and so forth

Defects originated at the OEM's factory may or may not show themselves early in the life of the machine. For example, a bar with substandard application of semiconducting material for an air-cooled machine application may exhibit signs of serious degradation after a short time in operation, while a bar that has not been bonded adequately may become delaminated over a number of years. Same can be said about inadequate or simply wrong maintenance activities.

PROBABLE CAUSES/AREA AFFECTED:

- Insufficient wall insulation voltage-withstand capability/mainly in the slot area
- Imperfect bar bonding resulting in dilapidation/slot and endwinding
- Imperfect Roebel twisting/slot and endwinding
- Wrong application of grading coat/endwindings
- Too many voids in bar/slot and endwindings
- Bars too close to each other in overhangs/endwindings
- Loose bars in slot/mainly slot area, but can be shown itself as problems in the endwindings
- Loose endwindings/mainly in the endwindings, but also at the ends of the slot area
- Significant oil contamination/global looseness
- Inadequate endwindings support/endwindings and also end portion of slot area
- Carelessly stepping or otherwise physically deforming the endwindings/endwindings
- Lack of proper FME/Most probably endwindings, but also slot area

POSSIBLE CONSEQUENCES:

- GROUND FAULT OF STATOR WINDING (M-S-26)
- Ground fault in the slot area
- Phase–phase fault in the overhang area
- Ground or phase–phase faults in the terminal box or circumferential busses (phase connectors) areas
- Quick erosion of the wall insulation in air-cooled machines
- Too much leakage current and tracking in the area of the overhand close to the core

POSSIBLE SYMPTOMS:

There is no direct operational indication about such windings defects. See Section *Test and Visual Indications.*

OPERATORS MAY CONSIDER:

These types of defects are found online by PDA, or during offline testing and inspections. Typically, this is not a problem requiring immediate shutdown or load reduction. They may be considered issues to be captured in the life cycle management plan of the unit.

If found online,

1. Consider planning for offline full testing and investigation of the possible causes of elevated *PDA* analysis, and consider making contingency plans for repair or rewind (repairing voids can only be done by stripping the wall insulation and re-insulating. If the problem is global, this is a very time-consuming operation and it may be more economical to rewind with new bars, when the loss of production is factored in.)
2. If found offline: Same disposition as in item 1) above

TEST AND VISUAL INDICATIONS:

- EMI
- RF
- High levels of PDA
- High trending of PDA
- Corona deposits on endwindings
- High levels of ozone next to air-cooled generators
- Unsatisfactory PI (mainly too high)
- Unsatisfactory Hipot test
- Unsatisfactory Tan-Delta test
- Unsatisfactory Tip-Up test

REFERENCE

1. G. Klempner and I. Kerszenbaum. *Handbook of Large Turbo-Generator Operation and Maintenance*. Piscataway, NJ: IEEE Press, 2008, chapters 2, 5, 8, and 11.

M-S-19: FAILURE OF RADIAL PHASE CONNECTOR

APPLICABILITY:

All generators

MALFUNCTION DESCRIPTION:

This malfunction relates to the situation when a phase connection is degraded to the extent that it completely fails resulting in an open circuit, causing a very large flashover inside the generator.

Figure 5.48 shows a failed radial lead connection.

PROBABLE CAUSES:

High endwinding vibration due to looseness or a near resonant frequency condition are the mostly likely causes of such a condition. Fatigue of the connection over time would eventually cause it to fracture and become open-circuited.

FIGURE 5.48
An open radial connection. The connection failed because of high vibrations due to natural frequencies, excited by twice the fundamental frequency.

POSSIBLE CONSEQUENCES:

A connection that becomes open-circuited often can create an arc between the two sides of the connection at the fracture point. Serious thermal damage could occur at the time of the final break but it is more likely that it will go unnoticed except for temperatures in the stator winding becoming abnormal. A good example of this case is a situation in which a winding with two parallel circuits, has one of the parallels on one phase becoming open-circuited due to such a failure. One parallel will carry no phase current (the open parallel) while the other undamaged parallel will carry all of the phase current for that phase. Then it is possible that the undamaged parallel will eventually overheat and sustain consequential thermal damage from overloading.

- OVERHEATING OF PHASE CONNECTOR (M-S-43)
- TEMPERATURE OF A PARALLEL CIRCUIT ABOVE NORMAL (M-S-51)
- TEMPERATURE OF STATOR CONDUCTOR ABOVE NORMAL (M-S-53)

POSSIBLE SYMPTOMS:

Stator winding slot TCs or RTD measured temperature will go down to the local temperature of the stator core and winding of the other conductor of same or another phase in the stator slot, in the parallel with the open connection, while the temperature of the parallel carrying full phase current will rise dramatically.

In hydrogen-cooled generators, the hose outlet temperature on the parallel with the open connection will attain a value similar to the hydrogen's temperature close to the position of the probe, and the probes on the overloaded parallel will rise according to the excessive current flowing in the undamaged parallel.

Measuring EMI or RF should show significant increase in activity, due to the arcing across the fractured strands.

- BYPRODUCTS FROM STATOR INSULATION PYROLYSIS **(S-S-01)**
- CURRENT IN ONE OR TWO PHASES ABOVE NORMAL **(S-S-02)**
- ELECTROMAGNETIC INTERFERENCE (EMI) ACTIVITY HIGH **(S-S-03)**
- OUTLET TEMPERATURE OF STATOR CONDUCTOR BAR COOLANT ABOVE NORMAL **(CASE 1) (S-S-07)**
- OUTLET TEMPERATURE OF STATOR CONDUCTOR BAR COOLANT ABOVE NORMAL **(CASE 2) (S-S-08)**
- RADIO FREQUENCY (RF) MONITOR HIGH **(S-S-13)**
- SLOT TEMPERATURE OF STATOR CONDUCTOR BAR—ABOVE NORMAL (CASE 1) **(S-S-14)**
- SLOT TEMPERATURE OF STATOR CONDUCTOR BAR—ABOVE NORMAL (CASE 2) **(S-S-15)**
- TEMPERATURE OF PHASE CONNECTOR ABOVE NORMAL **(S-S-18)**

OPERATORS MAY CONSIDER

This type of degradation can quickly become unstable to the point of a winding failure. The generator should be taken offline and opened up for inspection of the parallel connections.

TEST AND VISUAL INDICATIONS:

Testing of the stator winding series resistance on each phase may indicate a significant difference, helping to diagnose, as a minimum, that there is a stator winding problem. Confirmation would require opening the generator at the connection end of the stator to inspect the winding connections.

REFERENCE

1. G. Klempner and I. Kerszenbaum. *Handbook of Large Turbo-Generator Operation and Maintenance*. Piscataway, NJ: IEEE Press, 2008, chapter 8.

M-S-20: FAILURE OF SERIES CONNECTION

APPLICABILITY:

All generators

MALFUNCTION DESCRIPTION:

This malfunction relates to the situation when a series connection is degraded to the extent that it completely fails resulting in an open circuit, causing a very large flashover inside the generator.

Figure 5.49 shows series connections on a hydrogen inner-cooled winding.

PROBABLE CAUSES:

High endwinding vibration due to looseness or a near resonant frequency condition are the mostly likely causes of such a condition. Fatigue of the connection over time would eventually cause it to fracture and become open-circuit.

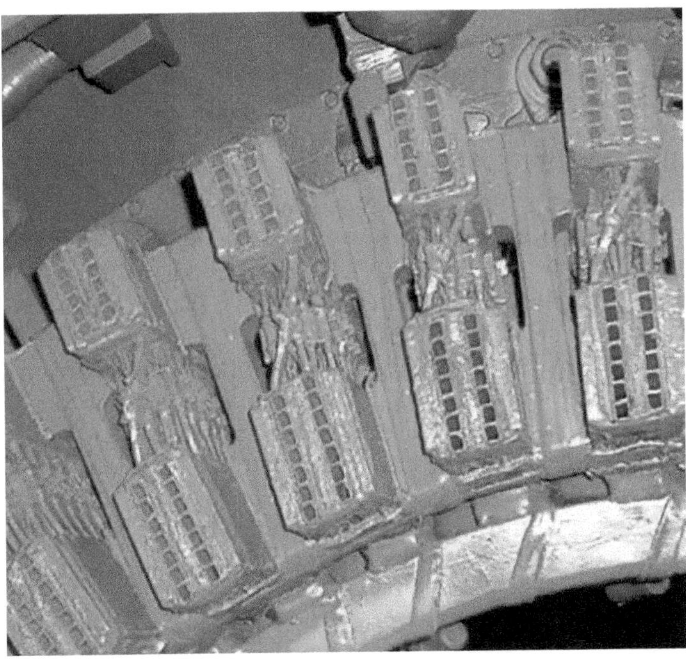

FIGURE 5.49
Series connections of a stator winding directly cooled by hydrogen.

In large generators of certain design and OEM, series connections can be bolted and not brazed [3]. Although very infrequent, such a bolted connection may also fail.

POSSIBLE CONSEQUENCES:

A connection that becomes open-circuited can often create an arc between the two sides of the connection at the fracture point. Serious thermal damage could occur at the time of the final break, but it is more likely that it will go unnoticed except for temperatures in the stator winding becoming abnormal. A good example of this case is a situation in which a winding with two parallel circuits, has one of the parallels on one phase becoming open-circuited due to such a failure. One parallel will carry no phase current (the open parallel) while the other undamaged parallel will carry all of the phase current for that phase. Then it is possible that the undamaged parallel will eventually overheat and sustain consequential thermal damage from overloading.

- CURRENT UNBALANCE BETWEEN PHASES **(M-S-12)**
- CURRENT UNBALANCE BETWEEN PHASE-PARALLELS **(M-S-13)**
- TEMPERATURE OF A PARALLEL CIRCUIT ABOVE NORMAL **(M-S-51)**
- TEMPERATURE OF STATOR CONDUCTOR ABOVE NORMAL **(M-S-53)**

POSSIBLE SYMPTOMS:

Stator winding slot TCs or RTD measured temperature will go down to the local temperature of the stator core and winding of the other conductor of same or another phase in the stator slot, in the parallel with the open connection, while the temperature of the parallel carrying full phase current will rise dramatically.

In hydrogen-cooled generators, the hose outlet temperature on the parallel with the open connection will attain a value similar to the hydrogen's temperature close to the position of the probe, and the probes on the overloaded parallel will rise according to the excessive current flowing in the undamaged parallel.

Measuring EMI or RF should show significant increase in activity, due to the arcing across the fractured strands.

- BYPRODUCTS FROM STATOR INSULATION PYROLYSIS **(S-S-01)**
- CURRENT IN ONE OR TWO PHASES ABOVE NORMAL **(S-S-02)**
- ELECTROMAGNETIC INTERFERENCE (EMI) ACTIVITY HIGH **(S-S-03)**

- OUTLET TEMPERATURE OF STATOR CONDUCTOR BAR COOLANT ABOVE NORMAL (CASE 1) **(S-S-07)**
- OUTLET TEMPERATURE OF STATOR CONDUCTOR BAR COOLANT ABOVE NORMAL (CASE 2) **(S-S-08)**
- RADIO FREQUENCY (RF) MONITOR HIGH **(S-S-13)**
- SLOT TEMPERATURE OF STATOR CONDUCTOR BAR—ABOVE NORMAL (CASE 1) **(S-S-14)**
- SLOT TEMPERATURE OF STATOR CONDUCTOR BAR—ABOVE NORMAL (CASE 2) **(S-S-15)**
- TEMPERATURE OF PHASE CONNECTOR ABOVE NORMAL **(S-S-18)**

OPERATORS MAY CONSIDER

This type of malfunction can quickly become unstable to the point of a winding failure. The generator should be taken offline and opened up for inspection of all the series connections at both ends of the stator.

TEST AND VISUAL INDICATIONS:

Testing of the stator winding series resistance on each phase may indicate a significant difference to diagnose, as a minimum, that there is a stator winding problem. Confirmation would require opening the generator at both ends of the stator to inspect the winding connections.

REFERENCES

1. G. Klempner and I. Kerszenbaum. *Handbook of Large Turbo-Generator Operation and Maintenance*. Piscataway, NJ: IEEE Press, 2008, chapter 8.
2. EPRI Technical Report 1021774. Generator Stator Endwinding Vibration Guide: Tutorial, August 2011.
3. US Patent 7400072 B2. Bolted Spherical Series and Phase Connector for Stator Coils of Electrical Generators.

M-S-21: GAS LOCKING IN WATER-CARRYING STRANDS OR TUBES

APPLICABILITY:

Generators with directly water-cooled stator windings

MALFUNCTION DESCRIPTION:

This malfunction addresses the issue of a partial or total loss of cooling to a stator bar or group of bars in a water inner-cooled stator winding, when enough hydrogen leaks into the SCW system to create a flow restriction. The consequences are a partial or total loss of cooling to a bar or group of bars.

Gas locking may manifest itself as higher or lower than normal temperatures in a bar or group of bars. For instance, RTDs embedded in the slots where the coils with insufficient cooling are located will show higher temperatures than in the rest of the stator. If the water flow inside the bars is partial, TCs installed at the outlets of the bars will show increased levels of temperature. However, if there is a major gas lock such that a certain bar or group of bars does not have any SCW flow at all, the outlet TCs may show cooler temperatures than the rest of the stator, because there is no water coming out the bars, while RTDs or TCs embedded in the slot will show higher temperatures than on the rest of the winding. Therefore, one must exercise caution by looking carefully at the different indications. In general, a coil or a group of coils that suddenly show low outlet temperatures than the rest of the machine most probably indicate a significant slowdown of the flow of cooling water (unless the instrumentation, for whatever reason, has failed).

PROBABLE CAUSES/AREA AFFECTED:

- HYDROGEN INGRESSION INTO STATOR COOLING WATER (M-S-29)

POSSIBLE CONSEQUENCES:

- For mild loss of cooling: premature aging of the affected bar insulation
- For serious loss of cooling:
  - Significant degradation of the affected bar insulation
  - Permanent deformation of the strands
  - Strand-to-strand short circuits
  - Melting of conductor copper
- Ground faults
- Failure of insulation in the slot area (ground faults), in the circumferential busses, leads, and terminal box conductors

POSSIBLE SYMPTOMS:

Temperature readings for a single stator bar or group of bars (i.e., a phase or phase parallel) is too high. In this case, the slot TC or RTD would read the temperature of the stator bar directly through the ground-wall insulation and be higher than normal; and the hose–outlet thermocouple measuring the outlet water temperature would be higher than the rest for a partial restriction, or read the temperature of the hydrogen close to the TC, which would be lower than the temperature of the outlet water in the rest of the winding.

- Temperature readings for a single stator bar or group of bars (i.e., a phase or phase parallel) is too low. In this case, the hose–outlet thermocouple would tend to read the temperature of the cooling gas in the local area, generally the cold gas temperature, because there would be total flow blockage.
- Too low temperature readings for bars or group of bars (like a phase or phase parallel).

- OUTLET TEMPERATURE OF STATOR CONDUCTOR BAR COOLANT ABOVE NORMAL (CASE 1) **(S-S-07)**
- OUTLET TEMPERATURE OF STATOR CONDUCTOR BAR COOLANT ABOVE NORMAL (CASE 2) **(S-S-08)**

Flow alarms in the SCW skid:

- BLOCKAGE IN THE FLOW OF STATOR COOLING WATER (SCW) **(S-SCW-01)**
- BLOCKAGE OF STATOR COOLING WATER (SCW) HEAT EXCHANGER **(S-SCW-02)**
- DIFFERENTIAL PRESSURE ACROSS SCW HEAT EXCHANGER ABOVE NORMAL **(S-SCW-06)**
- STATOR COOLING WATER FLOW BELOW NORMAL **(S-SCW-12)**
- GENERATOR CORE CONDITION MONITOR—ALARM **(S-C&F-01)**
- HYDROGEN-IN-STATOR COOLING WATER (SCW) HIGH **(S-S-04)**
- HYDROGEN MAKE-UP RATE ABOVE NORMAL **(S-H-05)**
- LIQUID-IN-GENERATOR ALARM **(S-C&F-04)**

OPERATORS MAY CONSIDER:

Significant loss in the flow of cooling water in directly cooled machines, for any given length of time, has the potential to resulting in major damage and downtime. Therefore,

1. Once a serious loss of cooling is recognized, consideration should be given to an immediate reduction of load until the temperatures in the affected area are brought back to within insulation class limits
2. If the temperature readings are lower than normal and lower than the rest of the winding, this may be the sign of a complete loss of cooling for a phase or group of coils, or an individual bar, in which case the machine may have to be fully de-loaded (and perhaps taken offline) as soon as possible (only complete or almost complete removal of load will restore the temperature of the affected bars or group of bars to within allowable limits)

TEST AND VISUAL INDICATIONS:

- Discolored insulation
- SCW inside the machine
- Large increase of hydrogen presence in the SCW system
- Too high or too low PI readings compared with historical readings for this machine and compared with the unaffected phases
- High leakage current
- Failure of Hipot test

- Megger readings of the affected phase very different than the other two phases
- Analysis of hydrogen samples

REFERENCE

1. G. Klempner and I. Kerszenbaum. *Handbook of Large Turbo-Generator Operation and Maintenance*. Piscataway, NJ: IEEE Press, 2008, chapter 5.

M-S-22: GAS LOCKING OF PHASE COOLANT

APPLICABILITY:

Generators with directly water-cooled stators

MALFUNCTION DESCRIPTION:

Stators directly cooled by water come in a number of different cooling arrangements. In some machines, global manifolds dispatch and collect the SCW that flows through all the conducting bars. And within these types of manifold arrangements there can be many differing types such as *once-through*, with an inlet manifold (header) at one end and an outlet header at the other end, or the inlet and outlet headers at the same end but a turn-around water connection at the other end (this would indicate *twice-through*) (see Figures 5.50 and 5.51). In others, each phase has its own

FIGURE 5.50
A *once-through* type header arrangement: the inlet header at one end, and outlet header at the other end; each hose servicing two stator bars.

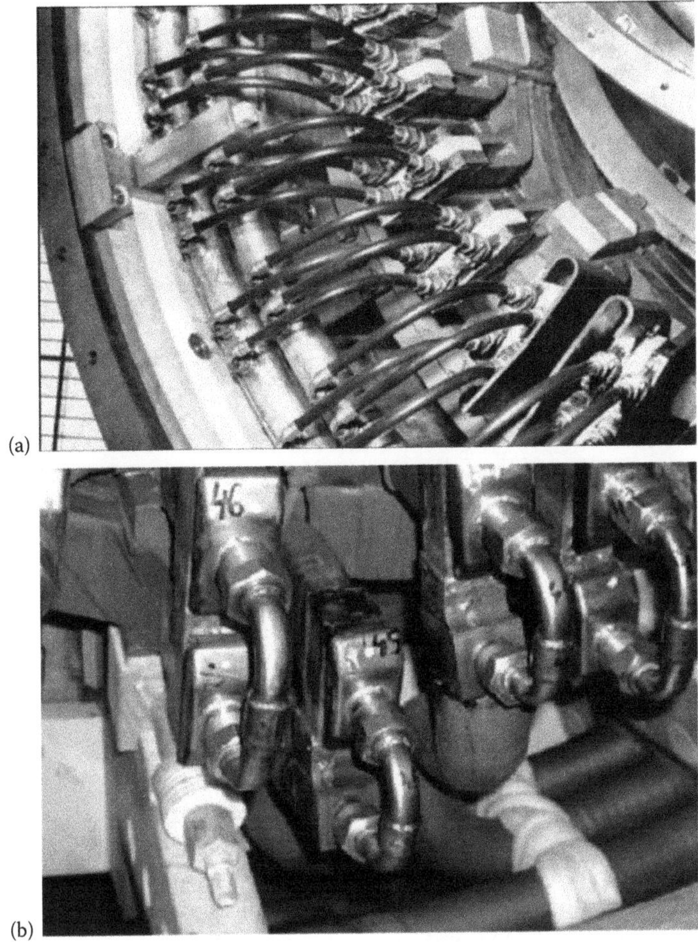

FIGURE 5.51
A *twice-through* type header arrangement (a), with turn-around connection (b) at the other end of the stator.

SCW flow path, usually the stator terminals. In these last ones, a serious ingression of hydrogen into the SCW flow path can cause a partial or total blockage of the flow of SCW by gas locking.

PROBABLE CAUSES:

- Rupture of a phase connector
- Failure of a SCW carrying joint upstream a phase connector
- Any other source of hydrogen leaking into the SCW system

POSSIBLE CONSEQUENCES:

- BLOCKAGE OF PHASE COOLANT FLOW (M-S-04)

POSSIBLE SYMPTOMS:

- BLOCKAGE IN THE FLOW OF STATOR COOLING WATER (SCW) **(S-SCW-01)**
- HYDROGEN INGRESSION INTO STATOR COOLING WATER **(S-S-04)**

OPERATORS MAY CONSIDER:

See description of the following malfunction:

- BLOCKAGE OF PHASE COOLANT FLOW (M-S-04)

TEST AND VISUAL INDICATIONS:

See description of the following malfunctions:

- BLOCKAGE OF PHASE COOLANT FLOW (M-S-04)

M-S-23: GAS LOCKING OF PHASE-PARALLEL COOLANT

APPLICABILITY:

Generators with directly water-cooled stators

MALFUNCTION DESCRIPTION:

Stators directly cooled by water come in a number of different cooling arrangements. In some machines, global manifolds dispatch and collect the SCW that flows through all the conducting bars. In others, each phase has its own SCW flow path. In some, each phase winding has a number of parallel circuits. This problem relates to gas blocking of one those parallel sections of a phase winding.

A serious ingression of hydrogen into the SCW flow path can cause a partial or total blockage of the flow of SCW by gas locking. The area directly affected will be the coils and connections in the path of the gas locking.

For all practical purposes, and with exception of the scope of the affected area, this problem is described in:

- GAS LOCKING OF PHASE COOLANT (M-S-22)

M-S-24: GAS LOCKING OF WATER-COOLED TERMINALS

APPLICABILITY:

Generators with direct water-cooled stators and water-cooled stator terminals

MALFUNCTION DESCRIPTION:

This malfunction relates to a situation where hydrogen penetrates the SCW system, and under the right conditions it partially or completely blocks the flow of SCW to the terminals of the generator. Stator terminals carry all the line current of the generator and therefore, any insufficiency on the flow of the cooling water will cause temperature rises dangerous to the integrity of the insulation.

Gas locking can occur under a number of circumstances; for example, if a significant in-leak of hydrogen to the SCW has happened, or after an outage where the SCW was completely removed from the pipes. In this case, an airlock restriction may exist that needs to be vented. Obviously, some SCW configurations are more prone to gas/air locking than others. Under normal conditions, the SCW pumps ought to produce enough pressure to avoid gas locks; however, if the winding is already somewhat clogged (e.g., by CuOx buildup), then the margin is reduced and the system may become more susceptible to gas/air locking.

PROBABLE CAUSES:

- Some restrictions to the flow of SWC together with hydrogen ingression to the cooling paths
- Hydrogen leak into the SCW due to cracks of the conductor and/or joints due to
    - Inadequate design
    - Substandard manufacturing
    - Wrong maintenance
    - High stator endwinding vibration
    - Connection defects (i.e., *plumbing* leak)

POSSIBLE CONSEQUENCES:

- For mild loss of cooling: premature aging of the affected conductor insulation
- For serious loss of cooling:
    - Significant insulation degradation of affected terminals
    - Melting of copper
    - Damage to HV bushings
    - Ground faults

POSSIBLE SYMPTOMS:

- GENERATOR CORE CONDITION MONITOR—ALARM **(S-C&F-01)**
- HYDROGEN-IN-STATOR COOLING WATER (SCW) HIGH **(S-S-04)**
- LIQUID-IN-GENERATOR ALARM **(S-C&F-04)**

- STATOR COOLING WATER FLOW BELOW NORMAL **(S-SCW-12)**
- STATOR COOLING WATER OUTLET PRESSURE BELOW NORMAL **(S-SCW-13)**
- TEMPERATURE OF GLOBAL OUTLET STATOR COOLING WATER HIGH **(S-SCW-15)**
- TEMPERATURE OF STATOR TERMINAL ABOVE NORMAL **(S-S-19)**

OPERATORS MAY CONSIDER:

1. Once this situation is recognized, consider reducing load until all temperatures return to their normal range.
2. Troubleshooting should be scheduled as soon as practically possible.
3. Record all pertinent data, such as temperatures, SCW flow, hydrogen-in-SCW rates, and so on.

TEST AND VISUAL INDICATIONS:

- Discolored insulation
- SCW inside the generator
- Large presence of hydrogen in the SCW system (indicated by the operation of the stator leak monitoring system)
- Too high or too low PI readings compared with historical values for this machine
- High leakage current
- Failure of Hipot test
- Megger readings very different than on the phases not affected

REFERENCES

1. G. Klempner and I. Kerszenbaum. *Handbook of Large Turbo-Generator Operation and Maintenance*. Piscataway, NJ: IEEE Press, 2008, chapter 11.
2. EPRI Technical Report 105504. Primer on Maintaining the Integrity of Water-Cooled Generator Stator Windings, October 1995.

M-S-25: GLOBAL STATOR TEMPERATURE ABOVE NORMAL

APPLICABILITY:

All generators

MALFUNCTION DESCRIPTION:

This malfunction is concerned with the situation where the entire stator is overheating.

Depending on the type of cooling arrangement and cooling medium, it may be happening for a number of reasons—some of them enumerated below under *Probable Causes.*

An overheating condition has the potential to cause severe damage or degradation to the stator winding and stator core. Thus, it ought to be addressed expediently.

PROBABLE CAUSES:

In air-cooled and indirectly hydrogen-cooled stators, the overheating may be caused by a problem with the heat exchanger; for instance, a degraded flow of primary cooling water, or primary cooling water temperature too high.

In directly hydrogen-cooled stators, a loss of hydrogen pressure within the generator may be the cause.

- HYDROGEN PRESSURE INSIDE GENERATOR BELOW NORMAL **(M-H-09)**

In directly water-cooled stators, the overheating could be caused by a problem with the SCW system, for instance, insufficient flow, or primary water temperature too high.

- TEMPERATURE OF STATOR CONDUCTOR ABOVE NORMAL **(M-S-54)**

In directly water-cooled stators, a partial or total loss of hydrogen pressure will result, primarily, in overheating of the stator core, and to some extent, an increase in the temperature of the stator winding, given that over 90% of the stator copper losses are removed by the SCW. On the other hand, the rotor temperature rise is directly affected by the hydrogen's temperature and pressure.

POSSIBLE CONSEQUENCES:

- BURNING/PYROLYSIS OF STATOR WINDING INSULATION **(M-S-08)**
- DAMAGE TO CONDUCTOR BAR INSULATION **(M-S-15)**
- GLOBAL TEMPERATURE OF CORE ABOVE NORMAL **(M-C&F-09)**

POSSIBLE SYMPTOMS:

All or most RTDs and TCs monitoring stator winding and core show temperatures above *normal.* By *normal,* it means temperatures higher than temperatures previously measured at the load point where the temperatures are being measured now

- HYDROGEN PRESSURE INSIDE THE GENERATOR BELOW NORMAL **(S-H-06)**
- Malfunctions related to the SCW or primary water to the SCW system, such as (M-SCW-01) and (M-SCW-12)
- Symptoms related to lower-than-normal flow of primary water to hydrogen heat exchangers, such as: **(S-H-07)**
- Symptoms related to the flow of SCW, such as: **(S-SCW-08)**, **(S-SCW-11)**, **(S-SCW-13)**, and so on
- GENERATOR CORE CONDITION MONITOR—ALARM **(S-C&F-01)**
- LIQUID-IN-GENERATOR ALARM **(S-C&F-04)**
- Overload alarms, such as: **(S-S-16)** and **(S-E-04)**

OPERATORS MAY CONSIDER:

Global temperature increase of stator winding and/or core has the potential to causing severe damage or long-term degradation of the conductors and interlaminar core insulation.

1. First step is recognition of the type of severity, that is,
   a. Small increase with temperatures still within allowable region?
   b. Large increases with temperatures above insulation class and/or temperature class limits?
   c. Trending quickly higher?
2. Depending on the answers to (1) above, load may or may not have to be reduced, or the unit may have to be taken offline immediately for repairs.

TEST AND VISUAL INDICATIONS:

- Offline electric tests and visual inspection can determine if the abnormal temperatures damaged the insulation of the affected terminals.
- Tests of pressure and control circuits in the SCW skid.
- Test of pressure control circuits of the H_2 skid.
- General inspection of hydrogen and water-cooling systems.
- A very severe case of overheating of the stator core may require an EL-CID and/or full flux test to revalidate the condition of the interlaminar insulation.

REFERENCES

1. G. Klempner and I. Kerszenbaum. *Handbook of Large Turbo-Generator Operation and Maintenance*. Piscataway, NJ: IEEE Press, 2008, chapters 3 and 5.
2. IEEE Std 67-2005 or later. IEEE Guide for Operation and Maintenance of Turbine Generators.

M-S-26: GROUND FAULT OF STATOR WINDING

APPLICABILITY:

All generators

MALFUNCTION DESCRIPTION:

This malfunction addresses the existence of an electric path between any part of the stator winding, including bars or coils, connectors, circumferential busses, leads and bushings, and a grounded part of the machine (usually the stator core or frame). See Figure 5.52 for an example of a ground fault on a stator bar.

It is difficult to distinguish a ground fault inside the generator (i.e., anywhere inside the winding up to and including the bushings), from a ground fault outside the machine, for instance, along the *isolated phase bus* (IPB), or on the generator side of the main step up transformer or other auxiliaries transformers, or on the neutral side of the generator (e.g., on the grounding transformer).

There are number of locations and components where a ground can originate. Also, the range of possible ground currents can be from the small enough to be called *leakage* currents, to the full developed ground

FIGURE 5.52
The figure shows a ground fault on a resin-rich stator bar.

fault, that in most large machines is limited to anything between 5 and 25 amperes, approximately (machines grounded via a grounding transformer or grounding resistor). This range of fault current was selected for providing a compromise between a fault current causing minimal damage to the machine, with enough current for the fault to be detected by the protection instruments.

The threshold value of ground currents that will be noted by the operator depend on the type of detection and/or protection used.

PROBABLE CAUSES:

- CONDUCTIVITY OF STATOR COOLING WATER ABOVE NORMAL (M-SCW-04)
- DAMAGE FROM MAGNETIC TERMITES (M-S-14)
- DAMAGE TO CONDUCTOR BAR INSULATION (M-S-15)
- DEFECTIVE STATOR WINDING INSULATION (M-S-18)
- ELEVATED HYDROGEN DEW POINT (M-H-05)
- Electric tracking along bushings
- Ground on the generator-side-windings of the step-up or auxiliary transformers
- Ground on the isophase bus or terminal cables

POSSIBLE CONSEQUENCES:

- Increased ground leakage currents
- Increased voltage stress across the wall insulation of the other—ungrounded—phases, if the ground is on one phase, away from the machine's neutral point

POSSIBLE SYMPTOMS:

Increase in the normal value of ground current (monitored manually or automatically):

- STATOR GROUND ALARM AND/OR TRIP (S-S-17)

OPERATORS MAY CONSIDER:

Ground faults can result in alarms or alarms and trips, depending on the philosophy of protection at any given power plant or utility. For those that only alarm, there are different approaches about when to take the unit off-load for repairs. In any case, operating a unit with one ground may not be highly detrimental to a generator with high-impedance neutral grounding (as most large units are), but can result in another ground developing because the ungrounded phases will experience higher-than-normal voltages to ground, with catastrophic results to the integrity of the machine.

TEST AND VISUAL INDICATORS:

- IR (also called megger) test
- Hipot test (AC or DC)

- Dc leakage current test
- Visually recognizable deformation of endwindings, phase-connectors or leads
- Cracks, nicks, girths, swelling, bulging, and other localized abnormalities on the wall insulation

REFERENCES

1. G. Klempner and I. Kerszenbaum. *Handbook of Large Turbo-Generator Operation and Maintenance*. Piscataway, NJ: IEEE Press, 2008, chapters 4 and 8.
2. N. Klingerman et al. Understanding Generator Stator Ground Faults and Their Protection Schemes. In: *42nd Annual Western protective Relay Conference*, Spokane, Washington, October 20–22, 2015.
3. R. Ilie. Turbine Generator Rotor and Stator Winding Hipot Testing, *IEEE Electrical Insulation Magazine*, March 2012.

M-S-27: HIGH RESISTANCE OF PHASE-CONNECTOR JOINT

APPLICABILITY:

All generators

MALFUNCTION DESCRIPTION:

This malfunction relates to a deterioration of a bolted or brazed joint of a phase connector. Phase connectors (also called *circumferential ring busses* or *parallel busses*, among other names) connect between the various groups of stator bars that make a parallel circuit and/or a phase. The phase connectors are also the bridge between the phase leads (terminals) and the stator bars of a phase. In most machines, the phase connectors are brazed, but in some cases they may also include bolted connections [5].

Deterioration of a joint belonging to a phase-connector will result, as a minimum, in an increase of temperature. If the joint fails completely, the result is an open circuit in part of the phase winding, or complete loss of one phase.

Depending on the type of winding connection, a bad phase-connector joint may affect a group of coils, a parallel circuit of the phase, a series-parallel circuit, or the entire phase. Depending on the type of winding connection, the symptoms may be different for different machines, and even for different cases within the same machine, depending on the location of the faulty joint.

Figure 5.53 shows a typical terminal lead bolted connection. Damaged braded flexible connectors and/or loose bolts may result in high resistance joint.

PROBABLE CAUSES:

- Cracked brazing due to vibrations resulting in less conducting area
- Cracked brazing due to substandard brazing operation

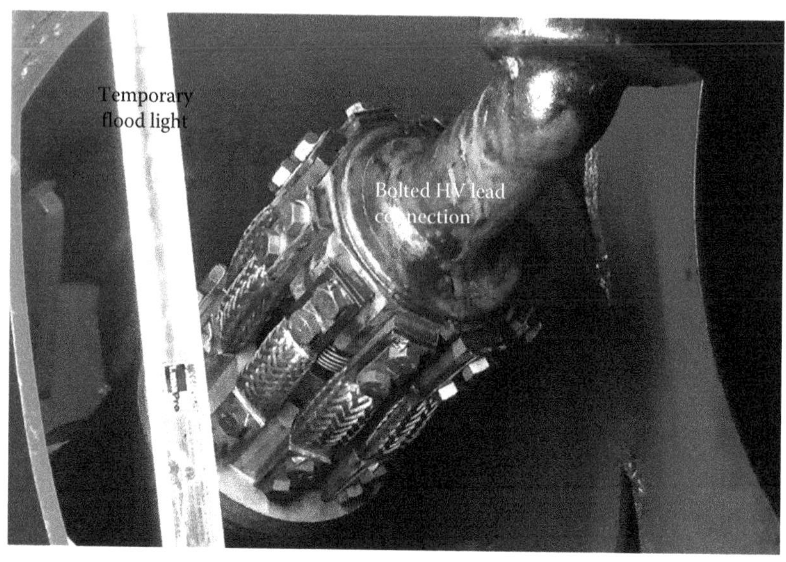

FIGURE 5.53
A bolted connection between a stator winding lead and the HV bushing.

- Cracked joint due to severe grid transient or short-circuit *near* the machine
- Loosening of bolted joint due to relaxation of the bolts
- Bolted joint corrosion buildup resulting in weak electrical contact

POSSIBLE CONSEQUENCES:

- BURNING/PYROLYSIS OF STATOR WINDING INSULATION (M-S-08)
- CURRENT UNBALANCE BETWEEN PHASE PARALLELS (M-S-13)
- CURRENT UNBALANCE BETWEEN PHASES **(M-S-12)**
- DAMAGE TO CONDUCTOR BAR INSULATION **(M-S-15)**
- LOCALIZED OVERHEATING OF STATOR CONDUCTOR (M-S-39)

POSSIBLE SYMPTOMS:

- ELECTROMAGNETIC INTERFERENCE (EMI) ACTIVITY HIGH **(S-S-03)**
- GENERATOR CORE CONDITION MONITOR—ALARM **(S-C&F-01)**
- OUTLET TEMPERATURE OF STATOR CONDUCTOR BAR ABOVE NORMAL RANGE (CASE 1) **(S-S-07)**
- OUTLET TEMPERATURE OF STATOR CONDUCTOR BAR ABOVE NORMAL RANGE (CASE 2) **(S-S-08)**

- RADIO FREQUENCY (RF) MONITOR ALARM **(S-S-13)**
- TEMPERATURE OF PHASE CONNECTOR ABOVE NORMAL **(S-S-18)**

OPERATORS MAY CONSIDER:

1. It is very difficult to diagnose a bad phase-connector joint, from other maladies leading to similar indications. However, any internally caused current unbalance should be addressed with the utmost urgency, because it has the potential to create significant damage, even a catastrophic failure.
2. If only temperature indications are present, then depending on the severity of this indications, and in particular how they change in time (operators ought to trend the affected temperature measurements over time), the operators can schedule a repair at a convenient time.

TEST AND VISUAL INDICATIONS:

- Visual inspection of affected area may show signs of discoloration of the insulation.
- Series resistance measurement can uncover cracked braced joints or loose bolted joint.

REFERENCES

1. G. Klempner and I. Kerszenbaum. *Handbook of Large Turbo-Generator Operation and Maintenance*. Piscataway, NJ: IEEE Press, 2008, chapter 8.
2. IEEE Std 62.2-2004 or later. IEEE Guide for Diagnostic Field Testing of Electric Power Apparatus—Electrical Machinery.
3. EPRI Technical Report 1015336. Nuclear Maintenance Applications Center: Bolted Joint Fundamentals, December 2007.
4. *Electrical Connections for Power Circuits*, Internet Version, USBR, August 2000.
5. US Patent 7400072 B2. Bolted Spherical Series and Phase Connector for Stator Coils of Electrical Generators.

M-S-28: HIGH RESISTANCE OF PHASE-PARALLEL-CONNECTOR JOINT

APPLICABILITY:

All generators

MALFUNCTION DESCRIPTION:

In machines where the stator has parallel connections, one of the parallels may be completely disconnected by the failure of a phase-parallel-connector joint. Alternatively, the joint may not fail completely but exhibit poor electrical contact resulting in a high resistant joint. Figure 5.54 shows a parallel

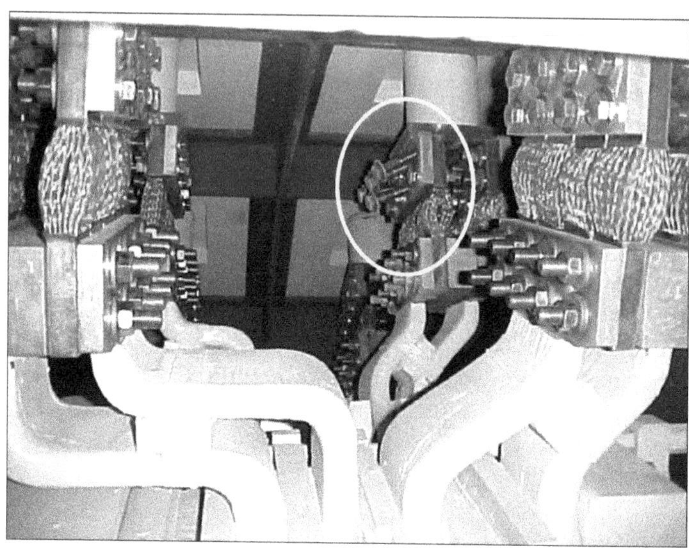

FIGURE 5.54
A parallel circuit of a two-pole hydrogen-cooled generator left partially disconnected by mistake during a maintenance activity.

circuit connection left inadvertently partially disconnected during a maintenance activity.

The effects are similar to those described for the malfunction: HIGH RESISTANCE OF PHASE-CONNECTOR JOINT (M-S-27), with the additional potential of creating large current unbalances between phases.

M-S-29: HYDROGEN INGRESSION INTO STATOR COOLING WATER (SCW)

APPLICABILITY:

Generators with directly water-cooled stators

MALFUNCTION DESCRIPTION:

This malfunction addresses the case of hydrogen ingression into the SCW system. In general, in all directly water-cooled generators there is a small amount of hydrogen that leaks into the SCW system during normal operation. It can occur through not-fully-tight *O-ring* connections, minuscule cracks in brazed connections at the water box area at the bar ends, or through the electrically insulated hoses. These small amounts of hydrogen are vented from the SCW system by a number of different arrangements, depending on the OEM. In most cases, the quantity of vented hydrogen is trended so that any adverse development (a significant increase in H_2 leak rate) can be addressed on time.

Nevertheless, there are cases where the quantity of hydrogen entering the SCW is large enough to constitute a problem that must be rectified in short time. There are a number of significant consequences from an elevated hydrogen leak into the SCW, and these will be enumerated below.

In general, the leak(s) may occur anywhere inside the machine. It can happen in the slot area, or in the phase connectors or circumferential busses, or the terminal box or bushing well.

For a complete discussion of *Probable Causes* and *Possible Consequences* refer to

- BREACH OF THE FLUX-SHIELD COOLING PATH **(M-C&F-03)**
- BREACH OF THE STRANDS OR TUBES CARRYING COOLING WATER **(M-S-06)**
- CRACKED CONDUCTOR STRANDS **(M-S-10)**
- DAMAGE TO CONDUCTOR BAR INSULATION DUE TO A WATER LEAK **(M-S-16)**
- DEGRADED/RUPTURED "O" RINGS OR OTHER SCW PIPING CONNECTIONS **(M-SCW-07)**
- GAS LOCKING IN FLUX-SHIELD COOLING PATH **(M-C&F-08)**
- GAS LOCKING IN WATER-CARRYING STRANDS OR TUBES **(M-S-21)**
- HYDROGEN INGRESSION INTO STATOR COOLING WATER (SCW) FROM FLUX SHIELD **(M-C&F-11)**
- INSUFFICIENT FLOW OF COOLING MEDIUM IN A STATOR BAR **(M-S-32)**
- INSUFFICIENT FLOW OF PHASE COOLANT **(M-S-33)**
- LEAKY SCW HOSES **(M-SCW-11)**
- RUPTURED WATER MANIFOLD (HEADER) **(M-S-47)**
- RUPTURED WATER PARALLEL MANIFOLD (HEADER) **(M-S-48)**
- STATOR COOLING WATER (SCW) LEAK INTO GENERATOR FROM MANIFOLDS **(M-S-50)**
- STATOR COOLING WATER (SCW) LEAK INTO GENERATOR FROM PHASE PARALLEL **(M-S-49)**

POSSIBLE SYMPTOMS:

- HYDROGEN-IN-STATOR COOLING WATER (SCW) HIGH **(S-S-04)**
- HYDROGEN MAKE-UP RATE ABOVE NORMAL **(S-H-05)**
- LIQUID-IN-GENERATOR ALARM **(S-C&F-04)**
- Symptoms related to the malfunctions listed above

OPERATORS MAY CONSIDER:

There is a generally accepted maximum level of hydrogen leakage into SCW which is about 50 ft^3 per day. If hydrogen-into-SCW-leak alarm is in effect, the operators must keep a close watch on the amount of leakage to monitor for an increasing leakage trend. If the limit is reached then shut down for repairs should be considered so that a major failure by hydrogen gas locking, or bar degradation due to water ingression into the insulation due to capillary action do not happen.

TEST AND VISUAL INDICATORS:

Paying attention to an installed *hydrogen-into-SCW-leak monitoring system* would generally be the best defense and early warning of this type of malfunction.

REFERENCES

1. G. Klempner and I. Kerszenbaum. *Handbook of Large Turbo-Generator Operation and Maintenance*. Piscataway, NJ: IEEE Press, 2008, chapters 8 and 11.
2. EPRI Technical Report 3002000420. Turbine Generator Auxiliary System Maintenance Guide Volume 4: Generator Stator Cooling Water System: 2013 Update, December 2013.
3. EPRI Technical Report 111180. Preventing Leakage in Water-Cooled Stator Windings (Phase 2), November 1998.
4. J.A. Worden and J.M. Mundulas. Understanding, Diagnosing, and Repairing Leaks in Water-Cooled Generator Stator Windings, *GE Power Systems*, GER-3751-A, August 2001.

M-S-30: HYDROGEN INGRESSION INTO STATOR COOLING WATER FROM STATOR TERMINALS

APPLICABILITY:

Generators with directly water-cooled stators

REFER TO:

- HYDROGEN INGRESSION INTO STATOR COOLING WATER (SCW) (M-S-29)

M-S-31: INSUFFICIENT FLOW OF COOLANT DUE TO BUCKLED COOLING STRANDS OR TUBES

APPLICABILITY:

- Generators with directly H_2-cooled stator windings
- Generators with directly H_2O-cooled stator windings

MALFUNCTION DESCRIPTION:

This malfunction addresses the issue of a partial or total loss of flow of cooling medium to one or a number of bars, due to a physical distortion of the hollow strands or tubes that carry the cooling medium inside the conductor bar. The cooling medium could be hydrogen (inner-H_2-cooled stators) or water (inner-H_2O-cooled stators). Figure 5.55 shows on the left of the reader, the rectangular tubes that carry hydrogen inside and along the conductor bars to remove the copper losses; on the right of the reader, a typical water-cooled stator bar with hollow strands that carry the cooling water as well as current, interposed with solid strands that only carry current.

The physical deformation and loss of flow can be due to a number of causes (see below), and depending on the type of thermal instrument being read and its location, can manifest itself as higher or lower than normal temperatures in that phase. For instance, RTDs embedded in the slots where the coils with the lack of coolant flow are located will show higher temperatures than in the rest of the stator. If the blockage of either water or hydrogen flow inside the bars is partial, TCs installed at the outlets of the bars will show increased levels of temperature. However, if the blockage is total, these TCs may show cooler temperatures than the rest of the stator, because there is no hot gas or water coming out the bars. Therefore, one must exercise caution by looking carefully at the different indications. In general, a coil or a group of coils that suddenly shows low

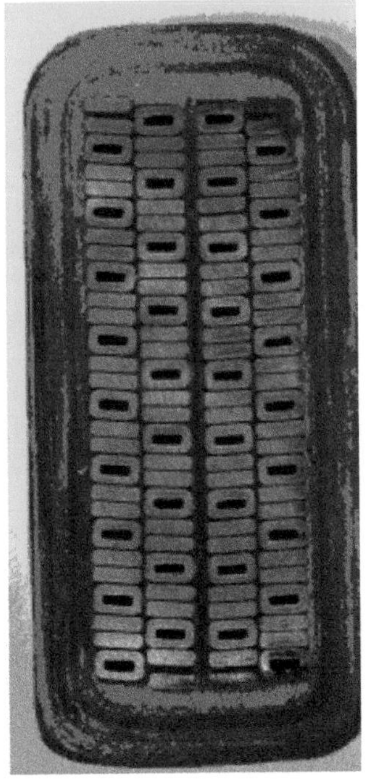

FIGURE 5.55
On the left of the reader it is shown the ends of directly hydrogen-cooled stator winding bars (top and bottom), while at the right of the reader it is shown a directly water-cooled stator winding bar cross section.

outlet temperatures than the rest of the machine most probably indicate a significant blockage of the cooling medium (unless, of course, the instrumentation for whatever reason failed).

PROBABLE CAUSES/AREA AFFECTED:

Deformation of the strands or cooling tubes due to high vibration, maintenance over-torquing of endwinding support hardware, physical event such as a rotor or rotor skid pressing against the endwinding during rotor removal or rotor installation, and so on.

POSSIBLE CONSEQUENCES:

- For mild blockage conditions (fully blocked for a short time or partially blocked for a long time):
  - Premature aging of the affected bar insulation
- For serious blockage conditions (fully blocked for significant time):
  - Significant insulation degradation of affected bar insulation
  - Permanent deformation of the strands
  - Melting of conductor copper
  - Failure of insulation in the slot area (ground faults), in the circumferential busses, leads, and terminal box conductors
- BURNING/PYROLYSIS OF STATOR WINDING INSULATION (M-S-08)
- GROUND FAULT OF STATOR WINDING **(M-S-26)**

POSSIBLE SYMPTOMS:

Too high or too low temperature readings for bars or group of bars (like a phase or phase parallel), such as

- OUTLET TEMPERATURE OF STATOR CONDUCTOR BAR ABOVE NORMAL (CASE 2) **(S-S-07)**
- OUTLET TEMPERATURE OF STATOR CONDUCTOR BAR COOLANT BELOW NORMAL **(S-S-08)**

Flow alarms in the SCW skid, such as

- BLOCKAGE OF STATOR COOLING WATER (SCW) **(S-SCW-01)**
- GENERATOR CORE CONDITION MONITOR—ALARM **(S-C&F-01)**
- OUTLET TEMPERATURE OF STATOR CONDUCTOR BAR ABOVE NORMAL (CASE 1) **(S-S-07)**

OPERATORS MAY CONSIDER:

The type of problem covered in this malfunction description is almost impossible to uncover without physical inspection of the machine. In the rare occasions where deformation of the coils happen to the extent that strands become fully or partially blocked, other conditions would have developed, such as leaks, high vibration, or a physical activity that may lead to such damage. Therefore, it is common for this type of damage to be found as part of an investigation of problems that at first may seem unrelated, and so, operators will only in rare occasions identify the source of a stator problem as damaged strands, before the unit is opened for inspection.

Significant blockage of the flow of cooling medium in directly cooled machines, for any given time, has the potential to resulting in major damage and downtime. Therefore,

1. Once a serious blockage is recognized, a reduction of load should be immediately commenced until the temperatures in the affected area are brought back to within insulation class limits.
2. If the temperature readings are lower than normal and lower than the rest of the winding, this may be the sign of a fully blockage situation, in which case the machine should be fully de-loaded (and perhaps taken offline) as soon as possible (only complete or almost complete removal of load will restore the temperature of the affected bars or group of bars to within allowable limits).

TEST AND VISUAL INDICATORS:

- Discolored insulation
- Too high or too low PI readings compared with historical readings for this machine and compared with the unaffected phases
- High leakage current
- Failure of Hipot test
- Megger readings very different than the other two phases
- Local distortion or damage to the endwindings
- Knowledge of recent damage to the winding by application of force

REFERENCE

1. G. Klempner and I. Kerszenbaum. *Handbook of Large Turbo-Generator Operation and Maintenance*. Piscataway, NJ: IEEE Press, 2008, chapters 8 and 11.

M-S-32: INSUFFICIENT FLOW OF COOLING MEDIUM IN A STATOR BAR

APPLICABILITY:

- Generators with directly H_2-cooled stators
- Generators with directly H_2O-cooled stators

MALFUNCTION DESCRIPTION:

This malfunction addresses the issue of a partial or total loss of flow in a single bar. The symptoms, consequences, and testing are similar than for a group of coils under the same predicament.

For a relevant discussion, refer to the following malfunctions and symptoms.

- BLOCKAGE OF COOLING FLOW IN A STATOR BAR **(M-S-02)**
- BLOCKAGE OF PHASE-COOLANT FLOW **(M-S-04)**
- INSUFFICIENT FLOW OF COOLANT DUE TO BUCKLED COOLING STRANDS OR TUBES **(M-S-31)**
- INSUFFICIENT FLOW OF PHASE COOLANT **(M-S-33)**
- OUTLET TEMPERATURE OF STATOR CONDUCTOR BAR COOLANT ABOVE NORMAL (CASE 1) **(S-S-07)**
- OUTLET TEMPERATURE OF STATOR CONDUCTOR BAR COOLANT ABOVE NORMAL (CASE 2) **(S-S-08)**
- OUTLET TEMPERATURE OF STATOR CONDUCTOR BAR COOLANT BELOW NORMAL **(S-S-09)**

M-S-33: INSUFFICIENT FLOW OF PHASE COOLANT

APPLICABILITY:

- Generators with directly H_2-cooled stator windings
- Generators with directly H_2O-cooled stator windings

MALFUNCTION DESCRIPTION:

This malfunction addresses the issue of a partial or total loss of flow of cooling medium to one of the stator phases. The cooling media could be hydrogen (directly hydrogen-cooled stators) or water (directly water-cooled machines).

It is not uncommon for the cooling medium to flow in circuits made of all coils in one phase, or group of coils forming a parallel section of a phase. In both cases, a partial blockage of the flow of hydrogen or water inside the coils, or a leak of SCW inside the machine will result in a partial loss of cooling to that phase or parallel.

Partial flow can be due to a number of causes (see below), and will manifest itself as higher as or lower than normal temperatures in that phase. For instance, RTDs embedded in the slots where the coils with insufficient cooling are located will show higher temperatures than in the rest of the stator. If the insufficiency in either water or hydrogen flow inside the bars is partial, TCs installed at the outlets of the bars will show increased levels of temperature. However, if there is such a leak of water or gas locking, that certain bars or group of bars does not have any SCW flow at all, these TCs may show cooler temperatures than the rest of the stator, because there is no water coming out the bars. Therefore, one must exercise caution by looking carefully at the different indications. In general, a coil or a group of coils that suddenly shows low outlet temperatures than the rest of the machine most probably indicate a significant loss of the cooling medium (unless, of course, the instrumentation for whatever reason failed).

PROBABLE CAUSES/AREA AFFECTED:

- Unintended obstruction introduced during maintenance/all bars and connectors where flow of hydrogen or water is reduced
- Obstruction due to dislodging of a large amount of copper oxide in a water-cooled winding/all bars where the flow of water is reduced
- Large internal leak of SCW/all bars and connectors where flow of cooling water is negatively affected
- Obstruction due to debris being thrown into the cooling path in a directly hydrogen-cooled winding/all bars and connectors where the flow of coolant is reduced
- Gas locking in a water-cooled winding/all bars where the flow of water is reduced

POSSIBLE CONSEQUENCES:

- For mild loss of cooling: premature aging of the affected bar insulation
- For serious loss of cooling:
    - Significant insulation degradation of affected bar insulation
    - Permanent deformation of the strands
    - Melting of conductor copper
- GROUND FAULT OF STATOR WINDING **(M-S-26)**
- BURNING/PYROLYSIS OF STATOR WINDING INSULATION **(M-S-08)**
- Failure of insulation in the slot area (ground faults), in the circumferential busses, leads, and terminal box conductors

POSSIBLE SYMPTOMS:

Too high or too low temperature readings for bars or group of bars (like a phase or phase parallel):

- BLOCKAGE IN THE FLOW OF STATOR COOLING WATER (SCW) **(S-SCW-01)**
- DIFFERENTIAL PRESSURE ACROSS SCW HEAT EXCHANGER ABOVE NORMAL **(S-SCW-05)**
- OUTLET TEMPERATURE OF STATOR CONDUCTOR BAR COOLANT ABOVE NORMAL (CASE 1) **(S-S-07)**
- OUTLET TEMPERATURE OF STATOR CONDUCTOR BAR COOLANT ABOVE NORMAL (CASE 2) **(S-S-08)**
- OUTLET TEMPERATURE OF STATOR CONDUCTOR BAR COOLANT BELOW NORMAL **(S-S-09)**
- STATOR COOLING WATER FLOW BELOW NORMAL **(S-SCW-12)**

- TEMPERATURE OF GLOBAL OUTLET STATOR COOLING WATER ABOVE NORMAL **(S-SCW-15)**

Flow alarms in the SCW skid:

- GENERATOR CORE CONDITION MONITOR—ALARM **(S-C&F-01)**
- HYDROGEN-IN-STATOR COOLING WATER (SCW) HIGH **(S-S-04)**
- LIQUID-IN-GENERATOR ALARM **(S-C&F-04)**

OPERATORS MAY CONSIDER:

Significant loss of the flow of cooling medium in directly cooled machines, for any given time, has the potential to resulting in major damage and downtime. Therefore,

1. Once a serious loss of cooling is recognized (and instrumentation checked to ascertain it is not an instrument-error situation), a reduction of load should be immediately commenced until the temperatures in the affected area are brought back to within insulation class limits.
2. If the temperature readings are lower than normal and lower than the rest of the winding, this may be the sign of a complete loss of cooling for a phase or group of coils, in which case the machine should be fully de-loaded (and perhaps taken offline) as soon as possible (only complete or almost complete removal of load will restore the temperature of the affected bars or group of bars to within allowable limits).

TEST AND VISUAL INDICATIONS:

- Discolored insulation
- SCW inside the machine
- Large increase of hydrogen presence in the SCW system
- Too high or too low PI readings compared with historical readings for this machine and compared with the unaffected phases
- High leakage current
- Failure of Hipot test
- Megger readings very different than the other two phases

For more information see:

- BLOCKAGE OF PHASE COOLANT FLOW **(M-S-04)**

M-S-34: INSUFFICIENT FLOW OF PHASE-PARALLEL COOLANT

APPLICABILITY:

- Generators with directly H_2-cooled stator windings
- Generators with directly H_2O-cooled stator windings

MALFUNCTION DESCRIPTION:

In machines with two or more parallel windings per phase, partial or total loss of coolant to a parallel section will result in higher temperatures of the affected portion of the phase winding.

For all practical purposes, other than the scope of the affected area, this problem is described in:

- INSUFFICIENT FLOW OF PHASE COOLANT (M-S-33)

M-S-35: INSUFFICIENT FLOW OF TERMINAL COOLANT

APPLICABILITY:

- Generators with H_2-cooled stator winding terminals
- Generators with H_2O-cooled stator winding terminals

REFER TO:

- BLOCKAGE IN THE FLOW OF STATOR TERMINAL COOLANT (M-S-01)

M-S-36: LEAK OF CONDUCTING-BAR COOLING WATER INTO GENERATOR

APPLICABILITY:

Generators with directly water-cooled stator windings

MALFUNCTION DESCRIPTION:

This malfunction addresses the issue of a leak of SCW into the generator.

For a complete discussion refer to

- BREACH OF THE STRANDS OR TUBES CARRYING COOLING WATER (M-S-06)

M-S-37: LEAK OF STATOR COOLING WATER (SCW) INTO GENERATOR

APPLICABILITY:

Generators with directly water-cooled stator windings

MALFUNCTION DESCRIPTION:

This malfunction relates to the leak of SCW inside the generator from any part of the stator winding, including overhangs, conductor bars, phase rings (circumferential ring busses), phase connectors, parallel and series connectors, terminals, water-carrying ***PTFE***[1] hoses, "O"-rings, and so on.

[1] PTFE—Polytetrafluoroethylene also know as Teflon.

There is a long list of causes and effects from this type of trouble, as described below.

PROBABLE CAUSES:

Bouncing bar in slot:

- VIBRATION OF STATOR CONDUCTOR BAR IN SLOT (M-S-55)
- VIBRATION OF STATOR ENDWINDING HIGH (M-S-56)
- Degraded "O"-rings
- Leaky electrically insulated hose(s)
- Cracked joints and/or conductors due to:
  - Inadequate design
  - Substandard manufacturing
  - Wrong maintenance activities

POSSIBLE CONSEQUENCES:

Overheating of certain parts of the winding due to a loss of SCW:

- BURNING/PYROLYSIS OF STATOR WINDING INSULATION (M-S-08)
- GAS LOCKING IN WATER-CARRYING STRANDS OR TUBES (M-S-21)
- INSUFFICIENT FLOW OF COOLING MEDIUM IN A STATOR BAR (M-S-32)
- Water damage to wall insulation due to capillary action
- Rusting of certain components inside the machine due to water ingression

Contamination of the stator winding insulation by water and water-carried contaminants:

- ELEVATED HYDROGEN DEW POINT (M-H-05)

POSSIBLE SYMPTOMS:

- HYDROGEN-IN-STATOR COOLING WATER (SCW) HIGH **(S-S-04)**
- LIQUID-IN-GENERATOR ALARM **(S-C&F-04)**
- TEMPERATURE OF GLOBAL STATOR COOLING WATER ABOVE NORMAL **(S-SCW-15)**

Temperature increase in sections of stator winding, such as

- OUTLET TEMPERATURE OF STATOR CONDUCTOR BAR ABOVE NORMAL (CASE 1) **(S-S-07)**
- OUTLET TEMPERATURE OF STATOR CONDUCTOR BAR ABOVE NORMAL (CASE 2) **(S-S-08)**
- TEMPERATURE OF STATOR TERMINAL ABOVE NORMAL **(S-S-19)**

OPERATORS MAY CONSIDER:

In general, significant leaks of SCW into the generator bring with them a number of undesirable effects to the integrity of the unit. Thus, depending on severity of the leak, the load could be maintained while plans are made for repairs at the earliest convenient opportunity, or the machine might have to undergo a load reduction or taken offline for repairs in a hurry.

TEST AND VISUAL INDICATORS:

- SCW inside the machine
- Large amounts of hydrogen in the SCW system
- Leaks found by one of the several existing techniques (e.g., bubbles, snooping, bagging, SF_6 injection, helium injection)
- Holes in the insulation left by magnetic worms
- Cracks in the insulation
- Partial removal of winding insulation may be required for inspection following leak searches

REFERENCES

1. G. Klempner and I. Kerszenbaum. *Handbook of Large Turbo-Generator Operation and Maintenance.* Piscataway, NJ: IEEE Press, 2008, chapter 4.
2. J. Gonzalez et al. *Monitoring and Diagnosis of Turbine-Driven Generators.* Englewood Cliffs, NJ: Prentice Hall.

M-S-38: LEAK OF STATOR COOLING WATER (SCW) INTO GENERATOR FROM TERMINALS

APPLICABILITY:

Generators with directly water-cooled stator windings

MALFUNCTION DESCRIPTION:

This malfunction relates to a leak of SCW into the generator casing due to a rupture or cracking of the terminal leads and connections in the terminal box (*bushing well*).

Depending of the magnitude of the leak, a number of afflictions may occur, such as coolant-starvation, hydrogen-into-SCW leak, hydrogen-dew point elevation, and so on.

PROBABLE CAUSES:

- Cracked joints and/or conductors due to:
  - Inadequate design
  - Substandard manufacturing
  - Wrong maintenance activities
- VIBRATION OF STATOR ENDWINDING HIGH (M-S-56)

POSSIBLE CONSEQUENCES:

Overheating of certain parts of the winding due to a loss of SCW:

- BURNING/PYROLYSIS OF STATOR WINDING INSULATION (M-S-08)

- GAS LOCKING IN WATER-CARRYING STRANDS OR TUBES **(M-S-2132)**
- INSUFFICIENT FLOW OF COOLING MEDIUM IN A STATOR BAR **(M-S-32)**
- Water damage to wall insulation due to capillary action
- Rusting of certain components inside the machine due to water ingression
- Contamination of the stator winding insulation by water and water-carried contaminants
- ELEVATED HYDROGEN DEW POINT **(M-H-05)**
- MOISTURE IN HYDROGEN HIGH **(M-H-14)**

POSSIBLE SYMPTOMS:

- LIQUID-IN-GENERATOR ALARM **(S-C&F-04)**
- HYDROGEN-IN-STATOR COOLING WATER (SCW) HIGH **(S-S-04)**

Temperature increase of terminals or related conductors, such as

- TEMPERATURE OF STATOR TERMINAL ABOVE NORMAL **(S-S-19)**

OPERATORS MAY CONSIDER:

A confirmed breach of the strands or tubes carrying SCW has the potential to put the machine integrity at risk. Therefore, operators should consider taking the unit offline immediately for inspection and repairs.

TEST AND VISUAL INDICATORS:

- SCW inside the machine
- Large amounts of hydrogen in the SCW system
- Leaks found by one of several existing techniques (e.g., bubbles, snooping, bagging, SF_6 injection, helium injection)
- Holes in the insulation left by magnetic worms
- Cracks in the insulation

REFERENCE

1. G. Klempner and I. Kerszenbaum. *Handbook of Large Turbo-Generator Operation and Maintenance*. Piscataway, NJ: IEEE Press, 2008, chapter 4.

M-S-39: LOCALIZED OVERHEATING OF STATOR CONDUCTOR

APPLICABILITY:

All generators

MALFUNCTION DESCRIPTION:

This malfunction addresses localized overheating of a conductor belonging to the stator winding.

There are a number of mechanisms by which a short length of the conductor may overheat. For instance, breakdown of a number of strands will leave others to carry the current. In this case, the overheating length will be between the closest points where all strands in the bar are brazed (normally the bar knuckles or winding connections to circumferential bases, phase connectors, and terminals). This and other causes for localized conductor heating are included below under *Probable Causes.* Figure 5.56 shows a series connection of an inner water-cooled stator winding, stripped of its insulation.

PROBABLE CAUSES/AREAS AFFECTED:

- Break of a number of strands due to excessive conductor vibration, either in the slot, in the endwindings, or both/overheated area between points where the strands are brazed together (knuckles), and the damaged location.
- Separation of a number of strands due to insufficient brazing/overheated area between points where the strands are brazed together (knuckles), and the damaged location.
- Combination of excessive bar vibration with insufficient brazing/overheated area between points where the strands are brazed together (knuckles), and the damaged location.
- Damage to the insulation between strands. This has the effect to partially nullify the Roebel twist arrangement, with the consequence that circulating currents are created within the affected

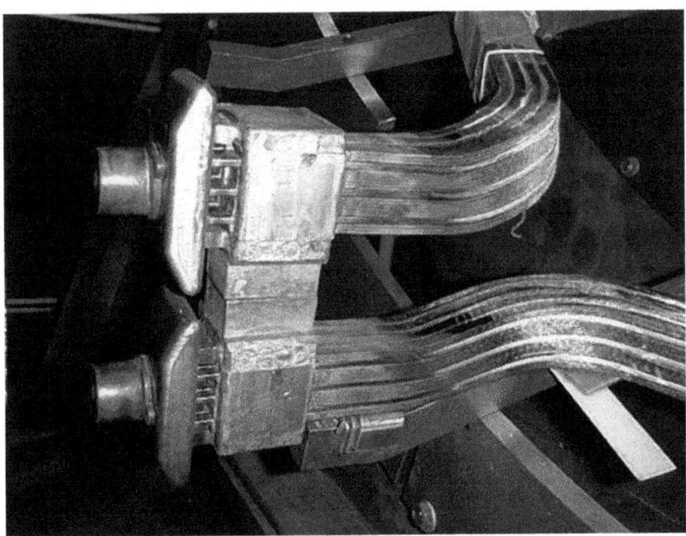

FIGURE 5.56
A series brazed connection at the stator bar end and the *knuckle* portion of the bars where the strands are brazed together.

area. These currents add to the load current carried by the bar, with a resulting increase in temperature./The affected area is between the point where the strands are shorted, to both knuckles at the end of the bar. Common causes for this type of failure is

- ◦ Manufacturing defect
- ◦ Damage during installation of the bar and/or wedging
- ◦ Caused by another failure mechanism such as a local core failure
- ◦ Severe operating stress, such as a nearby shorted-circuit

- BURNING/PYROLYSIS OF STATOR WINDING INSULATION (M-S-08)
- DAMAGE TO CONDUCTOR BAR INSULATION (M-S-15)
- DAMAGE FROM MAGNETIC TERMITES (M-S-14)
- OPERATION OUTSIDE CAPABILITY CURVE—ONLINE (M-E-11)
- SYNCHRONIZATION OUT OF PHASE (M-E-19)

POSSIBLE CONSEQUENCES:

- GROUND FAULT OF STATOR WINDING (M-S-26)
- Phase-to-phase fault on the endwinding area.
- Ground or phase–phase faults in the terminal box or circumferential connections areas.
- If not repaired, progressive damage to the insulation between strands in the vicinity may create a temperature runaway situation, until a ground fault or phase-to-phase fault happens.

POSSIBLE SYMPTOMS:

- Temperature increase of a single bar or group of bars indicated by embedded RTDs or outlet coolant, such as (S-S-07), (S-S-08), (S-S-14), and (S-S-15)
- Temperature increase of a single bar or group of bars indicated by outlet water temperature TCs (directly water-cooled stators), such as (S-S-07) and (S-S-08)
- Temperature increase of a single bar or groups of bars indicated by outlet hydrogen temperature TCs or RTDs (directly hydrogen-cooled stators), such as (S-S-07) and (S-S-08)
- GENERATOR CORE CONDITION MONITOR—ALARM (S-C&F-01)

OPERATORS MAY CONSIDER:

1. If the temperature of the affected bar(s) reaches alarm levels or winding temperature limits, then the machine should be partially de-loaded until the temperatures return to their *normal* values according to the load.
2. This situation (running at lower loads) is clearly not tenable for long periods of time; therefore, the unit should be taken offline for repairs as soon as practicable possible.

Test and Visual Indicators:

- Insulation showing signs of discoloration (trending dark)
- Unsatisfactory megger, PI, and Hipot tests on affected bar(s)

REFERENCE

1. G. Klempner and I. Kerszenbaum. *Handbook of Large Turbo-Generator Operation and Maintenance*. Piscataway, NJ: IEEE Press, 2008, chapter 8.

M-S-40: LOOSENING OF STATOR ENDWINDING BLOCKING

APPLICABILITY:

All generators

MALFUNCTION DESCRIPTION:

This is a malfunction associated with the endwinding support system blocking or packings becoming loose. All stator endwindings have some type of blocking between stator bars and generally ties or wedging systems as well, to assist in keeping the endwinding tight and the vibrations at a minimum. These are all part of the total endwinding support system. The other major function of the support system is to ensure that endwinding vibration is kept away from the once per revolution (50 Hz/60 Hz) and the twice per revolution (100 Hz/120 Hz) natural frequencies of the generator.

This is one of the most serious problems that can occur in the generator as the resulting vibrations can quickly cause a stator winding mechanical failure in a short period of time. Periodic regular inspection and maintenance is required.

Figure 5.57 shows one of several cracked endwinding support brackets, creating a situation that may lead to a loose endwinding. Figure 5.58 shows loose *greasing* blocks.

PROBABLE CAUSES:

- Normal drying of ties and bar insulation can loosen the restrain system of the endwinding.
- Oil ingression to the generator in substantial quantities can act as a lubricant loosening otherwise tightly held components. Over time the oil emulsifies and becomes like grease allowing the conductor bars and blocking to move more freely.
- Loose end-wedges can affect the integrity of the endwindings, and vice versa.
- Wrong installation, deficient design, and poor manufacturing are common causes for slackening of the endwinding.
- Grid events such as nearby short circuits and/or high transient torsional vibrations can also loosen stator windings in both the slot and the endwindings.
- Malsynchronization.

FIGURE 5.57
A number of cracked endwinding support brackets in one generator.

FIGURE 5.58
Stator endwinding ties and blocking greasing from fretting.

POSSIBLE CONSEQUENCES:

- Bar becoming loose in the slot allowing it to bounce around at twice the fundamental frequency (100 Hz for 50 Hz units, or 120 Hz for 60 Hz units), with the following related damage:
    - Cracked insulation
    - Damaged semiconducting coating

  - Erosion of wall insulation, in particular in air-cooled machines, after the semiconducting coating is compromised
  - Broken strands
  - Cracked brazing of the strands in the knuckle area
  - Water/hydrogen leaks in machines with stators directly cooled by water
  - Localized overheating of the conductors
  - Endwinding vibration
  - Ground faults

POSSIBLE SYMPTOMS:

- *Dusting* and/or *dusting* of ties and blockings on the endwindings.
- Series and radial connections vibrating above normal or acceptable levels.
- VIBRATION OF STATOR ENDWINDING HIGH **(S-S-21)**
- PDA measurements can provide the only direct measurement from a slackened bar, other than vibration probes. The loose bar will provide a PDA signature, which is recognizable and different from other types of PDA causes. However, very seldom is PDA measured on a continuous basis. Thus, it is important to take periodic readings and trend them, if PDA probes are installed.
- PARTIAL DISCHARGE ACTIVITY (PDA) HIGH **(S-S-11)**

OPERATORS MAY CONSIDER:

On occasion, excessive endwinding vibration may require limiting the generator's output. Also, excessive endwinding vibrations may lead to leaks in inner-cooled stators, cracked strands and open series, parallel or radial lead connections.

1. Consider reducing load until the accelerometers installed on the endwindings indicate the readings are acceptable to operation, until a planned repair activity is scheduled. However, if the vibration of the endwinding (or any section of it) has changed significantly in a relatively short period, it may indicate a serious condition is developing that may require the unit to be taken offline for troubleshooting and repairs. Figure 5.59 shows an open radial connection most probably due to excessive endwinding vibration (radial lead).
2. A *bump test* of the endwindings with the rotor *in situ* is possible to ascertain the condition of the restraint system related to the existence of undesirable natural frequencies close to the fundamental and twice-fundamental frequency. However, this is beyond operator's response activity, as it required an offline and partially disassembled unit.
3. Wedge survey (also with the unit offline and partially disassembled). If the readings show that some end wedges are loose, and/

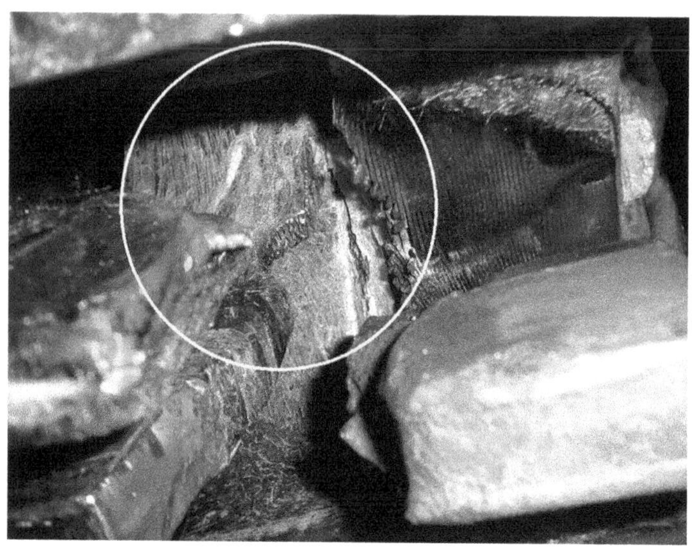

FIGURE 5.59
A radial lead connection that failed during operation in an air-cooled generator.

or three or move contiguous wedges are slack in a given slot, those ought to be retightened.

4. Repairs may require retightening of the endwinding if such a mechanical system is provided, and/or partial or total rewedging of the stator winding. This is normally the case for large machines. Sometimes (e.g., TIL 1966 from GE) advises removing some blocking (in very specific places and circumstances), in order to reduce dusting/greasing and improve the behavior of the endwinding by shifting the natural frequency to a lower value.

TEST AND VISUAL INDICATIONS:

- Tapping the wedges, in particular the end wedges.
- Visual observations of telling signs (such as dust deposits and/or *grease* deposits on the endwindings and also on the end wedges).
- Bump testing of the endwindings during major outages can identify natural frequency issues.

REFERENCES

1. G. Klempner and I. Kerszenbaum. *Handbook of Large Turbo-Generator Operation and Maintenance*. Piscataway, NJ: IEEE Press, 2008, chapter 8.
2. EPRI Technical Report 1021774. Generator Stator Endwinding Vibration Guide: Tutorial, August 2011.
3. R. Sewak, R. Ranjan, and V. Kumar. Experimental Modal Analysis of Stator Overhangs of a Large Turbogenerator. *Energy and Power Engineering*, 3, 221–226, 2011, Scientific Research, published online.

M-S-41: LOOSENING OF STATOR WINDING SLOT WEDGES

APPLICABILITY:

All generators

MALFUNCTION DESCRIPTION:

One of the most common maintenance activities that are carried out during the life of the machine is the verification that a sufficient number of slot wedges are effectively providing strong support to the bars. The method used in checking wedge conditions is the *wedge survey* (Figure 5.61), which can be done manually or automatically. A manual survey of wedge tightness requires the rotor to be out of the stator (see Figure 5.60). Automatic survey can be done with the rotor out, and in some cases, depending on type of wedge, with the rotor *in situ* using crawling devices that move along the airgap.

In generators that underwent the global ***VPI***[2] process during assembly, the wedges are glued to the winding and to the core in a way that very seldom becomes loose. Hence, wedge survey of these machines may be carried very seldom, or not at all, or a very small number of wedges can be tested as a sample of the entire wedging system.

Wedges that are in good operating condition and provide full support to the conductor bar are known as being *tight*. On the other hand, *loose*

FIGURE 5.60
Manually tapping wedges for looseness, in the stator bore.

[2] VPI—Vacuum pressure impregnation.

wedges do not provide enough support to the bar in the slot, with the consequence that the bar vibrates during operation. This vibration can produce a number of onerous effects on the integrity of the winding, as spelled out in the *Probable Effects* section, below.

A typical rule of the thumb is that when the stator has more than 25% of all wedges loose, it is time to re-wedge the entire stator. On the other hand, if there are less than 25% of loose or slack wedges globally, any loose end wedge should be tightened, and any three or more continuous wedges in a slot that are loose should also be retightened.

Historically, wedges were *tapped* with the rotor removed from the stator bore for the purpose of accessing the bore and the rotor itself. This is still the preferred method. In recent years, a number of OEMs are providing a service which allows some wedge systems in some machine designs to be tested with the rotor *in situ*. However, to tighten or replace the wedges, being a few or all the wedges in the stator, the rotor must be out of the bore. Rotors in general are kept continuously in the bore for many years (10 years or more is not uncommon these days). Thus, it is imperative that when a rotor is removed for planned maintenance or other reason, a full wedge survey be made, and the results be trended (unless robotic wedge surveys are carried out with the rotor in place). This activity allows planning ahead for retightening or re-wedging next time the rotor comes out the bore.

PROBABLE CAUSES:

- Normal shrinking of the wedges or the insulation packing under the wedges can happen over time by becoming dryer. The shrunk wedges or packing insulation will allow movement of the bar in the slot.
- Oil ingression to the generator in substantial quantities can get between the wedge and slot allowing more movement of the wedge. Over time, the oil emulsifies and becomes like grease that provides an opportunity for the wedge to move more freely.
- Wedges can also become slack due to cracking or other damage.
- The *sitting* area of the wedge (the dovetail area), can erode under certain circumstances, such as elevated machine vibration. A wedge with eroded edges will move freely. A slot with a number of loose wedges will allow the bar to vibrate, causing further loosening of other wedges in the same slot.
- Wrong installation, deficient design and poor manufacturing are common causes for slackening of the wedges.

POSSIBLE CONSEQUENCES:

- Bar becoming loose in the slot allowing it to bounce around at twice the fundamental frequency (100 Hz for 50 Hz units, or 120 Hz for 60 Hz units), with the following related damage:
    - Cracked insulation
    - Damaged semiconducting coating

- Erosion of wall insulation, in particular in air-cooled machines, after the semiconducting coating is compromised
- Broken strands
- Cracked brazing of the strands in the knuckle area
- Water/hydrogen leaks in machines with stators directly cooled by water
- Localized overheating of the conductors
- Endwinding vibration
- Ground faults

POSSIBLE SYMPTOMS:

- PDA measurements can provide the only direct observation from a slackened bar, anywhere in the machine. Vibration probes installed on wedges can only measure the looseness of that wedge/bar. The loose bar will provide a PDA signature which is recognizable and different from other types of PDA causes. However, very seldom is PDA measured on a continuous basis. Thus, it is important to take periodic readings and trend them, if PDA probes are installed.
- PARTIAL DISCHARGE ACTIVITY (PDA) HIGH **(S-S-11)**

OPERATORS MAY CONSIDER:

Very rarely will loose wedges require imposing an immediate limit on the machine operation. Nevertheless,

1. If loose wedges are suspected, a PDA on line test may be carried out if the probes are installed.
2. During a major generator inspection with the rotor outside the bore, a wedge tightness survey should be carried out and the readings trended and compared with previous readings. If the readings show that some end-wedges are loose, and/or three or more contiguous wedges are slack in a given slot, those ought to be retightened. If the trend shows that more than 25% of wedges will be loose during next major *rotor-out* outage, consider making preparations for retightening or fully re-wedging the stator. If more than 25% wedges are found loose during an outage, strong consideration must be given to fix the situation during the same outage. For a more detailed explanation see References 1 and 2.

TEST AND VISUAL INDICATIONS:

- PDA only with the machine on line
- PDA and PDA probe with the machine offline and rotor outside the bore
- Tapping the wedges
- Visual observations of telling signs (such as wedge material dust deposits and/or *grease* deposits between wedge and slot)

REFERENCES

1. G. Klempner and I. Kerszenbaum. *Handbook of Large Turbo-Generator Operation and Maintenance*. Piscataway, NJ: IEEE Press, 2008, chapters 2, 8, and 11.
2. G. Stone et al. *Electrical Insulation for Rotating Machines*. Piscataway, NJ: IEEE, 2004.

M-S-42: NEGATIVE-SEQUENCE CURRENTS

APPLICABILITY:

All generators

MALFUNCTION DESCRIPTION:

Positive, negative, and *zero sequence* currents are the three set of components, in the form of phasors (rotating vectors), representing a set of three-phase unbalanced currents. This is the basis for an analytical technique allowing electrical engineers to solve problems in three-phase circuits involving unbalanced currents between the three phases. When looking at the effects that unbalanced phase currents have on a generator in operation, the focus is on the negative-sequence currents.

Negative-sequence currents create an airgap (gas gap) magnetic flux rotating at the same speed as the rotor-produced flux, but in the opposite direction. This means that the rotor *sees* a double-frequency magnetic flux sweeping its surface. Thus, in a 60 Hz system, the flux produced by negative-sequence currents sweeps the rotor's surface at 120 Hz. In a 50 Hz system the sweeping frequency is 100 Hz.

The result of having a fast moving flux impacting the nonlaminated rotor is the generation of *skin effect* eddy currents that tend to concentrate on the thin layer of the rotor circumference, closest to the rotor's airgap surface (Figure 5.61). A combination of eddy currents of enough magnitude and duration has the potential for severely damaging rotor components such as retaining rings, wedges, and others (Figure 5.62).

All machines are designed, per applicable standards, to meet a minimum degree of current unbalance, and thus negative-sequence currents, for given times. The problem is when, for a number of causes, these values are exceeded.

PROBABLE CAUSES:

- Large current unbalance due to grid abnormal events
- Short-circuits on the step-up transformer or auxiliary transformer(s)
- Short-circuit between two phases in the terminal box or nonsegregated terminal cables
- Single-pole opening of a generator circuit breaker/switch

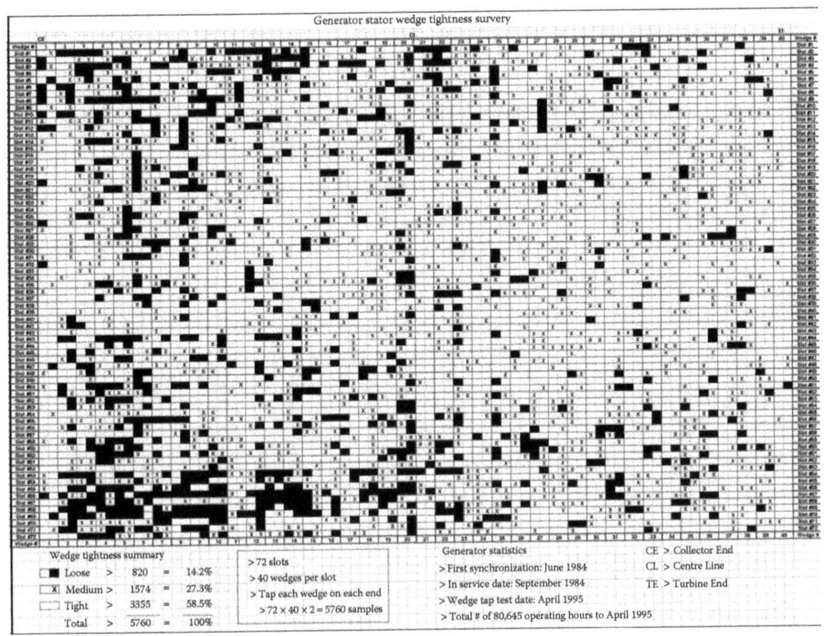

FIGURE 5.61
Typical wedge map after tapping all stator wedges, with the rotor out. Black squares indicate a *loose* half wedge; x'ed squares indicate wedges that are in the process of loosening, that is, *medium*; and blank squares indicate *tight* wedges.

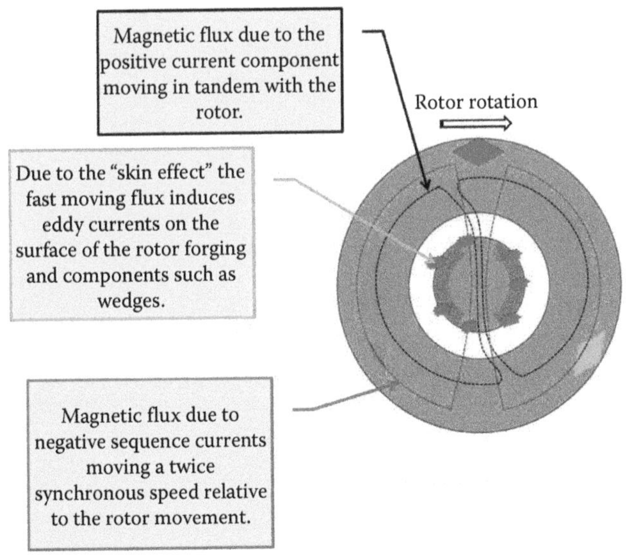

FIGURE 5.62
The negative-sequence current originated magnetic flux in the airgap of the machine creating hot spots on the surface of the rotor.

- Two-phase operation due to loss of one phase of the transmission system
- HIGH RESISTANCE OF PHASE-CONNECTOR JOINT **(M-S-27)**
- Any type of problem resulting in significant unbalance between the currents in the three phases

POSSIBLE CONSEQUENCES:

- Arcing between wedges and slots in the rotor with damage to either one or both
- Arcing between wedges and retaining rings in rotor, resulting in damage to either one or both
- Arcing between retaining ring and forging resulting in damage to either one or both
- Extensive overheating of rotor body, retaining rings, and zone rings (when installed), as well as wedges and other metallic components close to the airgap

POSSIBLE SYMPTOMS:

- NEGATIVE-SEQUENCE CURRENT ALARM AND/OR TRIP **(S-S-06)**
- PHASE CURRENTS UNBALANCE **(S-S-12)**
- TERMINAL VOLTAGES UNBALANCED **(S-S-20)**

OPERATORS MAY CONSIDER:

1. Negative-sequence events have the potential to cause major damage to the machine in a short time. Thus, the immediate response is implemented by a protection device tripping the unit. In those cases where a breach of maximum values of negative-sequence values will trip the machine, the operator's response is to see that the event is properly reviewed by engineers before the machine is re-energized. A large event may require some level of disassembly to inspect for damage to the rotor.
2. If the negative-sequence unbalance only causes an alarm, consider reducing load immediately to minimum until the alarm clears. If the alarm does not clear, consider taking the unit offline. Consider having engineers evaluate the situation before putting the generator back on line again.

TEST AND VISUAL INDICATIONS:

For large events, consider visually inspecting the rotor surface and wedges, retaining and zone rings, and so on for indications of arcing. In some instances, this may require removal of the retaining rings and wedges to look for arcs in otherwise inaccessible areas.

REFERENCES

1. G. Klempner and I. Kerszenbaum. *Handbook of Large Turbo-Generator Operation and Maintenance*. Piscataway, NJ: IEEE Press, 2008, chapters 2, 4, and 5.
2. EPRI TR-1014910. Negative-Sequence Effects on Generator Rotors.

M-S-43: OVERHEATING OF PHASE CONNECTOR

APPLICABILITY:

All generators

MALFUNCTION DESCRIPTION:

This malfunction addresses the issue of a phase connector overheating.

Partial or total loss of flow of cooling medium to one of the stator phase connectors is a high probability cause. The cooling media could be hydrogen (directly hydrogen-cooled stator windings) or water (directly water-cooled stator winding). Partially degraded connection (broken strands or cracked solid connector) may be another.

It is not uncommon in machines with stators directly cooled by water, for the cooling medium to flow in circuits made of all coils in one phase, or group of coils forming a parallel section of a phase. The cooling includes any phase connector that connects the various parallel circuits. The same is not very common in hydrogen-cooled stator windings. However, in both cases, a blockage of the flow of hydrogen or water inside the coils could result in a loss—partial or total—of cooling to that phase or parallel connector.

Loss of flow can be due to a number of causes (see below), and depending on the type of thermal instrument being read and its location, can manifest itself as higher or lower than normal temperatures in that phase. For instance, RTDs embedded in the slots where the coils with the lack of cooling flow are located will show higher temperatures than in the rest of the stator. If the blockage of either water or hydrogen flow inside the bars is partial, RTDs and/or TCs installed at the outlets of the bars will show increased levels of temperature. However, if the blockage is total, these TCs may show cooler temperatures than the rest of the stator, because there is no hot gas or water coming out of the bars. Therefore, one must exercise caution by looking carefully at the different indications. In general, a coil or a group of coils that suddenly shows low outlet temperatures than the rest of the machine most probably indicate total blockage of the cooling medium (unless, and of course, the instrumentation for whatever reason failed).

PROBABLE CAUSES/AREA AFFECTED:

- Unintended obstruction introduced during maintenance/all bars and connectors where flow of hydrogen or water is stopped or reduced

- Obstruction due to dislodging of a large amount of copper oxide in a water-cooled winding/all bars where the flow of water is stopped or reduced
- Obstruction due to debris being thrown into the cooling path in a directly hydrogen-cooled winding/all bars and connectors where the flow of coolant is stopped or reduced
- Gas locking in a water-cooled winding/all bars and connectors where the flow of water is stopped or reduced

POSSIBLE CONSEQUENCES:

- For mild blockage conditions (fully blocked for a short time or partially blocked for a long time): premature aging of the affected bar insulation
- For serious blockage conditions:
  - Significant insulation degradation of affected bar insulation
  - Permanent deformation of the strands
  - Melting of conductor
  - Ground faults
  - Failure of insulation in the slot area (ground faults), in the circumferential busses, leads, and terminal box conductors

POSSIBLE SYMPTOMS:

- Too high or too low temperature readings for bars or group of bars (like a phase or phase parallel)
- TEMPERATURE OF PHASE CONNECTOR ABOVE NORMAL **(S-S-18)**

OPERATORS MAY CONSIDER:

Significant blockage of the flow of cooling medium in directly cooled machines, for any given time, has the potential to resulting in major damage and downtime. Therefore,

1. Once overheating is recognized, a fast reduction of load should be immediately considered, until the temperatures in the affected area are brought back to within insulation class limits.
2. If the temperature readings are lower than normal and lower than the rest of the winding, this may be the sign of a fully blockage situation, in which case the machine should be fully de-loaded (and perhaps taken offline) as soon as possible (only total or almost total removal of load will restore the temperature of the affected bars or group of bars to within allowable limits).

TEST AND VISUAL INDICATIONS:

- Discolored insulation
- Too high or too low PI readings compared with historical readings for this machine and compared with the unaffected phases

- High leakage current
- Failure of Hipot test
- Megger readings very different than the other two phases

REFERENCE

1. G. Klempner and I. Kerszenbaum. *Handbook of Large Turbo-Generator Operation and Maintenance.* Piscataway, NJ: IEEE Press, 2008, chapter 8.

M-S-44: PARTIAL DISCHARGE ACTIVITY (PDA) HIGH

APPLICABILITY:

- All generators
- The generator must be energized such that the stator winding has voltage induced and it is above the level of discharge or corona inception

MALFUNCTION DESCRIPTION:

There are a number of different types of PD that can be monitored inside the generator, including sources of discharge that are not classical *generator PD,* but rather discharges from the ***IPB***[3], collector ring noise, static exciter field forcing spikes, and so on (see Figures 5.63 and 5.64).

FIGURE 5.63
Arcing and welding of wedges to the retaining ring due to large negative-sequence currents.

[3] IPB—Isolated phase bus.

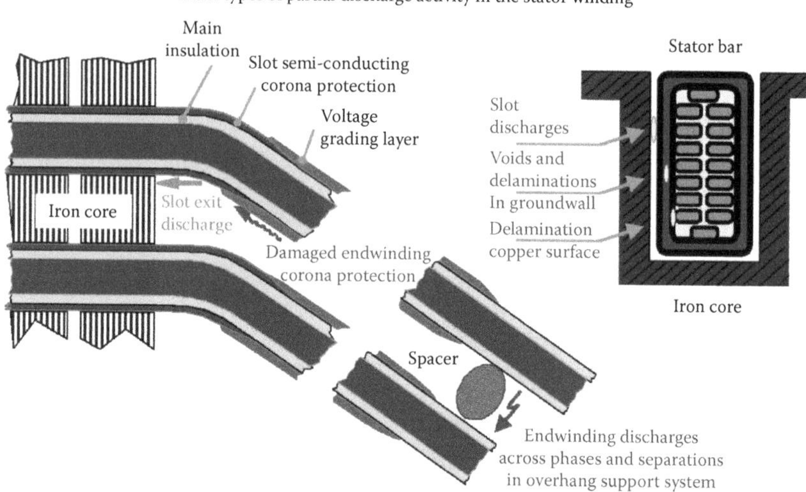

FIGURE 5.64
Various types of partial discharge activity (PDA).

With regard to the stator winding itself, PD can occur in the slot and/or on the endwindings. There are six basic types of PD:

1. *Slot discharge*: positive predominance
2. *Slot discharge*: negative predominance
3. *Slot discharge*: equal positive and negative magnitude and pulses per second
4. *Endwinding*: slot exit discharge
5. *Endwinding*: discharge along bar ends due to damaged endwinding corona protection
6. *Endwinding*: discharge between phases

PROBABLE CAUSES:

Probable causes are PDA activity in the slots, between conductor bar and slot, inside the wall insulation of the conductor bar, or on the stator overhangs, in the form of *corona*.

POSSIBLE CONSEQUENCES:

Generators do not generally fail due to higher levels of PD, but rather from the resulting effects of insulation deterioration associated with higher levels of PD. This can mean mechanical deterioration of the groundwall insulation due to dry out of the resin binders in the groundwall insulation, loosening of the windings in the slot and endwinding, and so on.

POSSIBLE SYMPTOMS:

- PARTIAL DISCHARGE ACTIVITY (PDA) HIGH **(S-S-11)**

There must be some type of EMI and/or PD monitoring system installed to be able to detect PD. Some vendors offer handheld EMI *sniffing* instruments that may detect PD in air-cooled machines, where the EM signal travels from the inside to the outside via the filter opening.

There are a number of devices and methods used to detect and measure PD such as

- Capacitive coupling
- RFCT's (radio frequency current transformers)
- SSC (stator slot coupler)
- EMI (electromagnetic interference)
- EMI handheld *sniffers*

Some devices look at pure PD in the higher frequency range for generator PD and others measure the discharge over a wider or lower frequency range. Therefore, devices such as these may require some type of pattern recognition (see Figure 5.64) to determine the type (location) of the PD in the generator. And some devices mainly look at the frequency spectrum, which can also show electrical discharges that are not necessarily from the generator itself.

OPERATORS MAY CONSIDER

Periodic measurements of PD are required for the appropriate operator action to be taken. Once higher levels of PD are starting to occur, increased PD measurements will be required as a first step. If the level of PD magnitude and pulses per second continue to excessively higher levels, operators of generator exhibiting such symptoms should consider a shutdown inspection and possible stator rewind.

TEST AND VISUAL INDICATIONS:

- Opening of the machine for visual inspection
- Wedge tapping for loose wedges
- On and offline PD testing
- Additional diagnostic testing such as:
    - DC ramp test
    - Power factor
    - Power factor tip-up (not effective on the entire winding, only effective when carried out on a single coil; mainly a factory test)
    - Capacitance measurements per phase

REFERENCES

1. G. Klempner and I. Kerszenbaum. *Handbook of Large Turbo-Generator Operation and Maintenance*. Piscataway, NJ: IEEE Press, 2008, chapters 5 and 11.
2. G.C. Stone. Examples of Premature Stator Winding Failure in Recently Manufactured Motors and Generators. In: *10th Insucon International Conference*, Birmingham, England, 2006, pp. 217–220.

M-S-45: PHASE CURRENT HIGH

APPLICABILITY:

All generators

MALFUNCTION DESCRIPTION:

This malfunction relates to a situation when the current in a single phase of the stator winding, or in multiple phases, is higher than rated. This problem is not about transient conditions, such as during an external short-circuit or a grid transient, but about continuous operation above rated values.

Stator windings are designed to carry no more than a certain (rated) current magnitude at normal conditions (terminal voltage, frequency, hydrogen pressure in hydrogen-cooled generators, and adequate SCW flow in directly water-cooled units), so that their temperatures remain within the temperature rise limits of the insulation class of the stator winding. If the phase currents are allowed to increase beyond their maximum rated value, the result will be temperature levels that may shorten the expected life of the insulation (for an example of heat damage of a bar, see Figure 5.65). In extreme cases, damage to the winding may cause a *forced outage,* such that a major repair (or full rewind) is required on short notice.

It is possible that, due to grid abnormalities, one phase will show higher current values than the other two phases. If the difference is high enough, another problem (negative-sequence current) will come into play. This may happen even if the currents on the three phases are within their normal limits. However, if the current of a machine with current unbalance between phases reaches maximum value, the consequences to that phase from too high temperatures still presents an immediate concern.

PROBABLE CAUSES:

- Grid unbalance due to change in line configuration or other system changes
- Open parallel

POSSIBLE CONSEQUENCES:

- BURNING/PYROLYSIS OF STATOR WINDING INSULATION (M-S-08)
- DAMAGE TO CONDUCTOR BAR INSULATION (M-S-15)
- TEMPERATURE OF STATOR CONDUCTOR ABOVE NORMAL (M-S-53)
- In rare occasions, unbalance between phase currents may indicate a serious generator problem, like an open section of the winding.

POSSIBLE SYMPTOMS:

Ammeter in control room showing current above nominal:

- CURRENT IN ONE OR TWO PHASES ABOVE NORMAL (S-S-02)

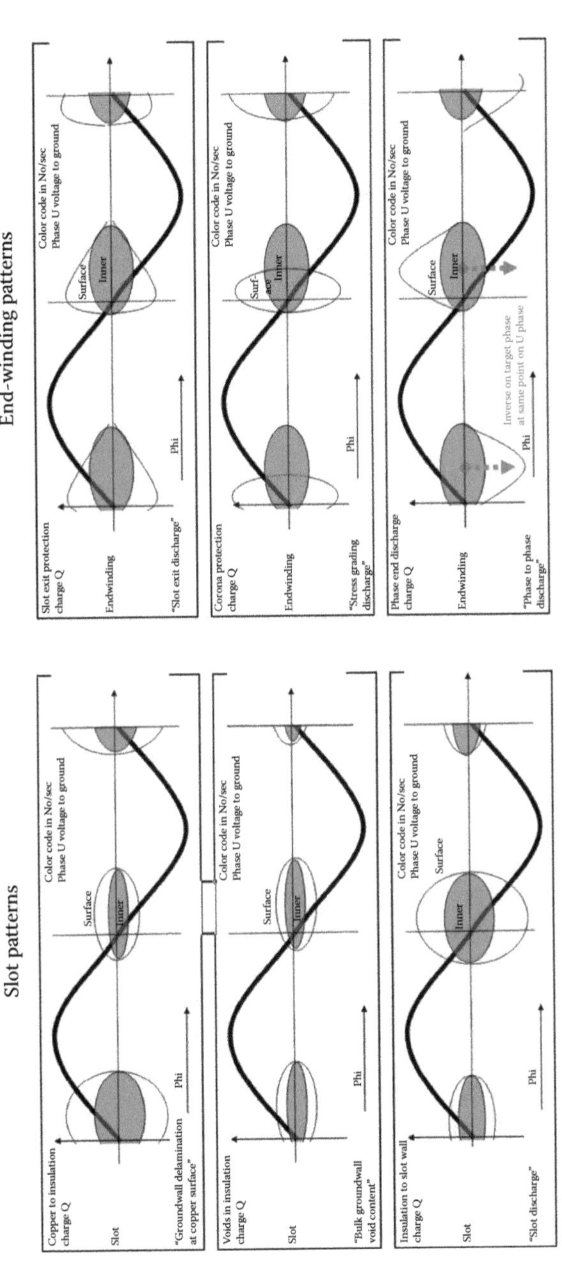

FIGURE 5.65
On the bottom, basic slot PDA patterns are shown. On the top, basic endwinding PDA patterns are shown.

Alarm from high phase (terminal) current:

- STATOR CURRENT ABOVE NORMAL **(S-S-16)**
- Temperature increase in one or more phase coils: **(S-S-07)**, **(S-S-08)**, **(S-S-14)**, and **(S-S-15)**

OPERATORS MAY CONSIDER:

1. Reducing load quickly until currents are returned within allowable limits
2. Monitoring all phase currents to ascertain there is no excessive unbalance, which may be grid-related or from a fault internal to the generator

TEST AND VISUAL INDICATIONS:

If internal fault is suspected, a series of resistance test is recommended.

REFERENCES

1. G. Klempner and I. Kerszenbaum. *Handbook of Large Turbo-Generator Operation and Maintenance*. Piscataway, NJ: IEEE Press, 2008, chapter 4.
2. Y.G. Paithankar and S.R. Bhide. *Fundamentals of Power System Protection*. New Delhi, India: PHI Learning, 2010.

M-S-46: RADIO FREQUENCY (RF) ACTIVITY

APPLICABILITY:

All generators

MALFUNCTION DESCRIPTION:

This malfunction relates to the existence of radio frequency (RF) activity (discharges) in a generator.

RF monitors, also called EMI monitors (*electromagnetic interference*), were the earlier version of PDA monitors. RF monitors are based on the fact that in general, any ionic discharge in the machine may be a sign of a problem or impending problem. RF activity can be produced by partial discharge between slot and conducting bars, PDA in the winding, corona, core–key-bar arcing, arcing across bearings and seals, and so on. RF monitoring can be problematic, because other discharges, such as on the sliprings, may prove to add confusion to the interpretation effort. Actually, there aren't many machines in operation with this type of monitor installed.

Presently, efforts by EPRI and others are focused in assessing this technology as an alternative or complementary to PDA analysis.

The nature of RF monitoring makes it primarily a trending tool, and not an absolute measuring device. Absolute levels of RF activity may provide some insight, if measurements are taken on a number of identical generators.

Figures 5.66 and 5.67 show the typical pick-up location using a split core CT for capturing RF signals. The pick-up is installed on grounded conductors.

FIGURE 5.66
Overheated asphalt bleeding from the groundwall insulation. Asphaltic insulation has not been part of new stator winding insulation for many years.

FIGURE 5.67
The RFCT (High Frequency Current Transformer) is shown installed on a grounded electrical conduit.

PROBABLE CAUSES:

- PDA between slot and conducting bar
- PDA in the winding
- Corona effects at the endwindings
- Electric tracking along conducting bars in the area of gradient potentials (as it leaves the core)
- Electric tracking between two adjacent coils in the overhang zone of the winding
- Arcing across bad joints
- Arcing between core and keybars
- Arcing between shaft and bearings
- Arcing between shaft and seals
- Arcing at the brushgear
- Arcing at the shaft-grounding brushes
- And so on

(See Note 1 in p. 311.)

POSSIBLE CONSEQUENCES:

As stated above, there are a number of possible sources of RF discharges in a machine. Most of those will not present a risk to the integrity of the generator as long as they are not the result of a serious problem. Thus, trending is crucial in using this technique effectively. The following list represents some of the possible serious problems that the RF discharges may allude to.

- CRACKED CONDUCTOR STRANDS **(M-S-10)**
- DAMAGE TO CONDUCTOR BAR INSULATION **(M-S-15)**
- HIGH RESISTANCE OF PHASE-CONNECTOR JOINT **(M-S-27)**
- LOOSENING OF STATOR WINDING SLOT WEDGES **(M-S-41)**
- VIBRATION OF STATOR CONDUCTOR BAR IN SLOT **(M-S-55)**

POSSIBLE SYMPTOMS:

- RADIO FREQUENCY (RF) MONITOR HIGH **(S-S-13)**

OPERATORS MAY CONSIDER:

1. Launching an investigation of possible causes, if the trend of RF activity changes suddenly, or if it trends significantly. RF is a *global* indicator, that is, it is almost impossible to discern between different sources of discharge. Thus, in the case of a significant change, the operator ought to check other instrumentation to look for a second symptom, if any develops.
2. Using a *RF sniffer*, (i.e., an instrument that is handheld and can be moved around the machine) may, in certain cases (e.g., a problem in the brushgear area) help identify the source of the problem.

TEST AND VISUAL INDICATIONS:

- Visual checks for *tracking, corona,* and *BCB*
- Series resistance tests for *bad joints*
- Visual inspection of *bearings, journals,* and *seals* looking for signs of electric pitting
- Loose stator wedges
- Visual inspection of semiconducting coating of stator coils
- Visual inspection for sparking–arcing on the sliprings

Note 1: According to the main proponents of RF or EMI analysis, the technique, when applied to generators, can uncover the following list of problems:

1. Slot discharges resulting from side packing deterioration
2. Slot discharged resulting from stator bar coating deterioration
3. Loose endwindings (broken ties)
4. Loose stator bars (loose wedging)
5. Loose phase rings (circuit rings)
6. Verify maintenance corrected all winding defects
7. Loose flux-shield ground
8. Broken stator bar subconductors (strands)
9. Foreign metal objects on endwindings
10. Shaft oil/hydrogen seal rub
11. Arcing shaft-grounding brush
12. Shaft currents through bearings
13. Verifying shaft-grounding maintenance eliminated bearing currents
14. Contamination on windings (dirt, water, and oil)
15. Contamination in insulation (wet insulation)
16. No contamination present (no maintenance necessary)
17. Arcing in alternator–exciter or on main field sliprings
18. Verifying field ground is not present
19. Defective alternator–exciter diodes
20. Loose brushless exciter components
21. Loose static exciter power circuits
22. Open brushless exciter diode fuses
23. Defective voltage regulator components and/or control settings
24. High DC exciter commutator segment

REFERENCES

1. G. Klempner and I. Kerszenbaum. *Handbook of Large Turbo-Generator Operation and Maintenance*. Piscataway, NJ: IEEE Press, 2008, chapter 5.
2. EPRI Technical Report 1001266. Partial Discharge Testing of Rotating Machine Stator Windings, December 2000.
3. J.E. Timperley. Generator Condition Assessment through EMI Diagnostics. In: *Proceeding of ASME POWER 2008*, Orlando, FL, POWER2008-60166.

M-S-47: RUPTURED WATER MANIFOLD (HEADER)

APPLICABILITY:

Generators with directly water-cooled stator windings with manifolds (headers) or insulated water boxes

MALFUNCTION DESCRIPTION:

This malfunction relates to the rupture (crack, break, or puncture) of the water-carrying manifolds (also called phase headers or headers) or water boxes. These manifolds or water boxes supply and collect the SCW to and from the water-cooled conducting bars (see Figure 5.30 for water box and Figure 5.68 for a header).

Given the large amount of water carried by these headers, any serious breach has the potential for resulting in major hydrogen-into-SCW leaks, as well as large ingression of SCW into the generator. Therefore, the consequences are similar to those described for leaks from conducting bars, but much more intense.

PROBABLE CAUSES:

- VIBRATION OF STATOR ENDWINDING—HIGH (M-S-56)
- Wrong maintenance activities
- Defective welds due to poor quality manufacturing

FIGURE 5.68
The RFCT is shown installed on a grounded electrical conduit.

POSSIBLE CONSEQUENCES:

- GAS LOCKING OF STATOR WATER COOLING (SCW) SYSTEM **(M-SCW-08)**
- HYDROGEN INGRESSION INTO STATOR COOLING WATER **(M-S-29)**
- INSUFFICIENT FLOW OF COOLING MEDIUM IN A STATOR BAR **(M-S-32)**
- INSUFFICIENT FLOW OF PHASE COOLANT **(M-S-33)**
- LEAK OF CONDUCTING-BAR COOLING WATER INTO GENERATOR **(M-S-36)**
- TEMPERATURE OF STATOR CONDUCTOR ABOVE NORMAL **(M-S-53)**

POSSIBLE SYMPTOMS:

- DIFFERENTIAL PRESSURE OF SCW ACROSS STATOR WINDING BELOW NORMAL **(S-SCW-07)**
- HYDROGEN-IN-STATOR COOLING WATER (SCW) HIGH **(S-S-04)**
- LIQUID-IN-GENERATOR ALARM **(S-C&F-04)**
- STATOR COOLING WATER FLOW BELOW NORMAL **(S-SCW-12)**

Abnormal temperatures:

- OUTLET TEMPERATURE OF STATOR CONDUCTOR BAR COOLANT ABOVE NORMAL (CASE 1) **(S-S-07)**
- OUTLET TEMPERATURE OF STATOR CONDUCTOR BAR COOLANT ABOVE NORMAL (CASE 2) **(S-S-08)**
- OUTLET TEMPERATURE OF STATOR CONDUCTOR BAR COOLANT BELOW NORMAL **(S-S-09)**
- TEMPERATURE OF GLOBAL OUTLET STATOR COOLING WATER ABOVE NORMAL **(S-SCW-15)**

OPERATORS MAY CONSIDER:

1. A ruptured SCW manifold or water box is a most serious issue that ought to be repaired immediately, once it is recognized as such.
2. If large amounts of H_2-into-SCW and liquid-in-generator are reported by their respective alarms, operators should consider taking the unit offline as soon as possible for repairs.

TEST AND VISUAL INDICATIONS:

- Visual indications of water leaks inside the machine
- Leak search by one of the used methods (helium, SF_6, CO_2, etc.)
- Cracked/fractured header

REFERENCES

1. G. Klempner and I. Kerszenbaum. *Handbook of Large Turbo-Generator Operation and Maintenance*. Piscataway, NJ: IEEE Press, 2008, chapter 8.
2. A. Gonzales et al. *Monitoring and Diagnosis of Turbine-Driven Generators*. Englewood Cliffs, NJ: Prentice Hall.

M-S-48: RUPTURED WATER PARALLEL MANIFOLD (HEADER)

APPLICABILITY:

Generator with directly water-cooled stator windings with manifolds (headers) or water boxes

MALFUNCTION DESCRIPTION:

This malfunction relates to a failure (crack, break, or puncture) of the water-carrying manifolds (also called header) or water boxes, feeding one phase parallel (i.e., one parallel circuit of a stator phase).

This type of arrangement, where each parallel has its own SCW feed, is not unique to all machines with parallel circuits. Therefore, operators should consult the winding and cooling-path diagram for the affected machine.

Given the large amount of water carried by these headers, any serious breach has the potential of resulting in major hydrogen-into-SCW leaks, as well as large ingression of SCW into the generator. Therefore, consequences are similar to those described for leaks from conducting bars, but much more intense.

Other than the scope of the area affected (a parallel circuit vs. an entire phase), this problem is described in

- RUPTURED WATER MANIFOLD (HEADER) (M-S-47)

M-S-49: STATOR COOLING WATER (SCW) LEAK INTO GENERATOR FROM PHASE- PARALLEL

APPLICABILITY:

Generators with directly water-cooled stator windings

MALFUNCTION DESCRIPTION:

This malfunction describes water leak from a parallel connector into the generator.

For all practical purposes, the description of this problem can be found in

- STATOR COOLING WATER (SCW) LEAK INTO GENERATOR FROM MANIFOLDS (M-S-50)

Additional concern is the insulation in the area around the leak becoming wet between the strands and the wall insulation (so called *wet bar effect*), due to water moving by capillary effect.

M-S-50: STATOR COOLING WATER LEAK INTO GENERATOR FROM MANIFOLDS

APPLICABILITY:

Generators with directly water-cooled stator windings with manifolds (headers) or water boxes

MALFUNCTION DESCRIPTION:

A rupture of one of the water manifolds (or *headers* or *phase headers*) or water boxes is a serious event. The manifolds/water boxes carry all of the cooling water that circulates through the windings. Thus, any crack has the potential for creating a major disruption to the flow of cooling water, in addition to serious contamination of the insulation.

PROBABLE CAUSES:

- Inadequate design
- Substandard manufacturing
- Wrong maintenance activities
- Large grid disturbances
- Failure of a hose-to-manifold joint or a water box sealing gasket
- High machine vibration—in particular the stator endwindings

POSSIBLE CONSEQUENCES:

- Overheating of certain parts of the winding due to a loss of SCW
- GAS LOCKING IN WATER-CARRYING STRANDS OR TUBES (**M-S-21**)
- INSUFFICIENT FLOW OF COOLING MEDIUM IN A STATOR BAR (**M-S-32**)
- Rusting of certain components inside the machine due to water ingression
- Serious concerns with 18Mn-5Cr retaining rings due to stress corrosion cracking
- Contamination of the stator winding insulation by water and water-carried contaminants
- ELEVATED HYDROGEN DEW POINT (**M-H-05**)

POSSIBLE SYMPTOMS:

- DIFFERENTIAL PRESSURE OF SCW ACROSS STATOR WINDING BELOW NORMAL (**S-SCW-07**)

- HYDROGEN-IN-STATOR COOLING WATER (SCW) HIGH **(S-S-04)**
- LIQUID-IN-GENERATOR ALARM **(S-C&F-04)**
- STATOR COOLING WATER FLOW BELOW NORMAL **(S-SCW-12)**

Abnormal temperatures:

- OUTLET TEMPERATURE OF STATOR CONDUCTOR BAR COOLANT ABOVE NORMAL (CASE 1) **(S-S-07)**
- OUTLET TEMPERATURE OF STATOR CONDUCTOR BAR COOLANT ABOVE NORMAL (CASE 2) **(S-S-08)**
- OUTLET TEMPERATURE OF STATOR CONDUCTOR BAR COOLANT BELOW NORMAL **(S-S-09)**
- TEMPERATURE OF GLOBAL OUTLET STATOR COOLING WATER ABOVE NORMAL **(S-SCW-15)**

OPERATORS MAY CONSIDER:

Like any other leak of SCW inside the generator, it should be investigated as soon as practically possible. It is almost impossible differentiating a small leak from a conductor bar from a small leak from a manifold or water box. However, a small leak from a cracked manifold or water box has the potential of becoming a major leak very fast. This type of possibility should be one of the factors weighed by the operators when deciding on when to take the unit offline for repairs.

TEST AND VISUAL INDICATIONS:

- SCW inside the machine
- Large amounts of hydrogen in the SCW system
- Leaks found by one of several existing techniques (e.g., bubbles, snooping, bagging, SF_6 injection, helium injection)
- Signs of rusting and/or corrosion

REFERENCE

1. G. Klempner and I. Kerszenbaum. *Handbook of Large Turbo-Generator Operation and Maintenance*. Piscataway, NJ: IEEE Press, 2008, chapter 8.

M-S-51: TEMPERATURE OF A PARALLEL CIRCUIT ABOVE NORMAL

APPLICABILITY:

All generators with parallel circuits in their stator windings

MALFUNCTION DESCRIPTION:

This malfunction addresses the issue of a group of conductors that together make a parallel circuit of a stator phase winding, exhibiting temperatures above *normal*.

This problem is described in

- TEMPERATURE OF A STATOR CONDUCTOR ABOVE NORMAL (M-S-53)

However, take note of the fact that in this case, the group of affected conducting bars is exactly those that comprise a parallel circuit of a phase. This fact can be of great help in identifying the source of the problem.

M-S-52: TEMPERATURE OF A SINGLE PHASE—ABOVE NORMAL

APPLICABILITY:

All generators

MALFUNCTION DESCRIPTION:

This malfunction addresses the case when all conductors in an individual phase reach temperatures above *normal*. Bear in mind, the *normal* is a function of load; that is, at full load, the *normal* temperature of the affected winding will be higher than at lower loads. So, there are two indications that are important when looking at the temperature of a group of conductors (in this case a phase):

1. Is the temperature of the affected phase higher than the rest of the winding?
2. Is the temperature of the affected phase at or above temperature class limit of the machine in question?

The relevant information about this problem can be found within these malfunctions:

- CURRENT UNBALANCE BETWEEN PHASES (M-S-12)
- TEMPERATURE OF STATOR CONDUCTOR ABOVE NORMAL (M-S-53)

M-S-53: TEMPERATURE OF STATOR CONDUCTOR—ABOVE NORMAL

APPLICABILITY:

All generators

MALFUNCTION DESCRIPTION:

This malfunction addresses the issue of a conductor bar attaining temperatures above normal. Bear in mind, that *normal* is a function of load; that is, at full load, the *normal* temperature of the winding will be higher than at lower loads. So, there are two indications that are important when looking at the temperature of a conductor bar or group of bars:

1. Is the temperature of the affected bar(s) higher than the rest of the winding?
2. Is the temperature of the affected bar(s) at or above the temperature class limit of the machine in question?

In both cases described above the measured temperature is above *normal,* indicating a problem with the unit. Nevertheless, the operator's immediate required actions may be different. To explain this point better, refer to Figures 5.69 and 5.70. Figure 5.69 shows a system that is dynamically measuring conductor bar temperatures in a directly water-cooled stator; *dynamic temperature monitoring* is a method of having temperature limits that change with load changes. One point has reached the alarm setting for the particular load point the unit is operating at the time the readings are taken. Clearly, the affected bar has a problem, but the temperatures are all below the limiting maximum allowed, giving the operator some time to investigate further before reducing load or some other appropriate action. Now, if the problem is left to continue deteriorating, we can see on Figure 5.70, where the temperature of the affected bar is plotted in time, that at certain stage the temperature of the bar reaches and

FIGURE 5.69
A typical stator cooling water manifold feeding a number of conductor bars via TPFE hoses, in a once-through arrangement. A similar manifold is on the other side of the stator.

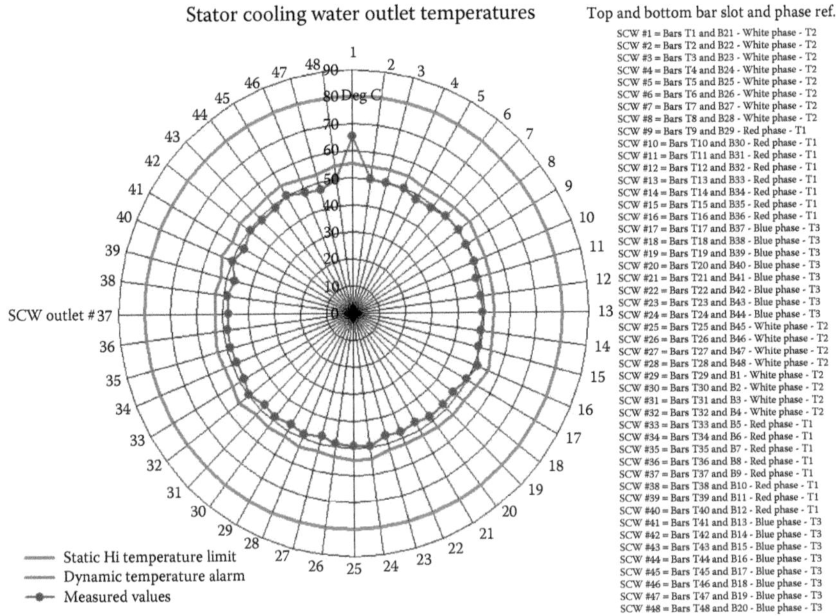

FIGURE 5.70
Stator cooling water hose outlet temperature polar plot, indicating the stator bar in slot 1 is in distress.

goes above the maximum class temperature (80°C in this example). In this case, the operator will obviously have to reduce load immediately to avoid further damaging the bar. Figure 5.71 shows a bar after being severely overheated.

PROBABLE CAUSES:

- BLOCKAGE OF COOLING FLOW IN A STATOR BAR **(M-S-02)**
- BLOCKAGE OF PHASE COOLANT FLOW **(M-S-04)**
- BREACH OF THE STRANDS OR TUBES CARRYING COOLING WATER **(M-S-06)**
- CRACKED CONDUCTOR STRANDS **(M-S-10)**
- GAS LOCKING IN WATER-CARRYING STRANDS OR TUBES **(M-S-21)**
- INSUFFICIENT FLOW OF COOLANT DUE TO BUCKLED COOLING STRANDS OR TUBES **(M-S-31)**
- INSUFFICIENT FLOW OF COOLING MEDIUM IN A STATOR BAR **(M-S-32)**
- INSUFFICIENT FLOW OF PHASE COOLANT **(M-S-33)**
- LOCALIZED OVERHEATING OF STATOR CONDUCTOR **(M-S-39)**

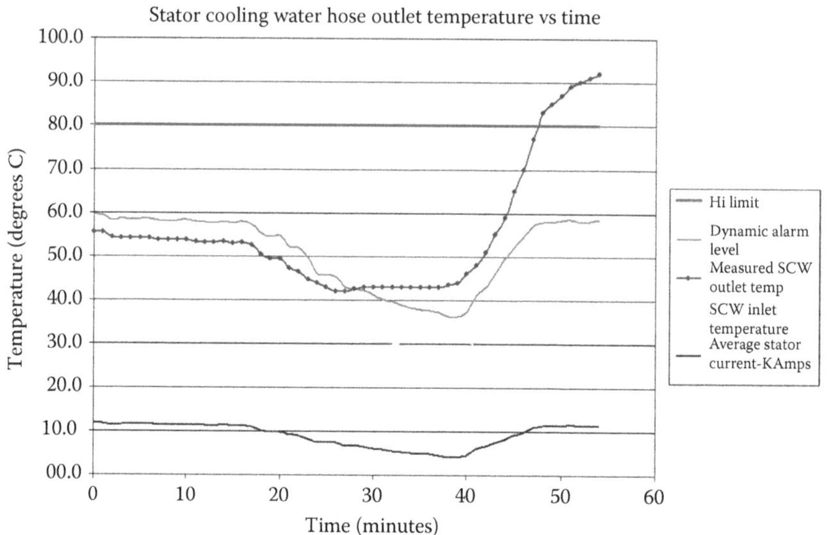

FIGURE 5.71
Stator cooling water hose outlet temperature time plot, indicating the stator bar in slot 1 has reached, and gone over its allowable temperature limit.

POSSIBLE CONSEQUENCES:

- BURNING/PYROLYSIS OF STATOR WINDING INSULATION (M-S-08)
- DAMAGE TO CONDUCTOR BAR INSULATION **(M-S-15)**
- LOCALIZED OVERHEATING OF STATOR CONDUCTOR (M-S-39)

POSSIBLE SYMPTOMS:

High temperature indication on a bar or group of bars via embedded RTDs or bar outlet RTDs/TCs in directly H_2-cooled or H_2O-cooled stators, such as

- OUTLET TEMPERATURE OF STATOR CONDUCTOR BAR COOLANT ABOVE NORMAL (CASE 1) **(S-S-07)**
- OUTLET TEMPERATURE OF STATOR CONDUCTOR BAR COOLANT ABOVE NORMAL (CASE 2) **(S-S-08)**
- SLOT TEMPERATURE OF STATOR CONDUCTOR BAR ABOVE NORMAL RANGE (CASE 1) **(S-S-14)**
- SLOT TEMPERATURE OF STATOR CONDUCTOR BAR ABOVE NORMAL RANGE (CASE 2) **(S-S-15)**

A *liquid-in-generator* alarm and/or *hydrogen-in-stator cooling water* alarm will support the premise that the *hot* bar(s) is (are) real

- HYDROGEN-IN-STATOR COOLING WATER (SCW) HIGH **(S-S-04)**
- LIQUID-IN-GENERATOR ALARM **(S-C&F-04)**

OPERATORS MAY CONSIDER:

If a single instrument shows conductor bar temperature above *normal*, consider first ascertaining that the instrumentation is working properly. If it is, the recommendations below are a good starting point:

1. If a bar or group of bars show an elevated temperature that can be considered not *normal* (what it is meant by *normal* is explained above), but below the temperature class limit of the winding, then consider trending the bar(s)'s temperature and planning for a repairs outage. How soon the outage ought to be done is a function of how much differential temperature exists between the affected bar(s) and the rest of the winding, as well as how it trends.
2. If a bar or group of bars show an elevated temperature that breached the temperature limit for the temperature rise of the machine, consider reducing load immediately until all temperatures are within *normal* range, and then follow with troubleshooting plans.

TESTING AND VISUAL INDICATIONS:

Offline electric tests and visual inspection can determine if the abnormal temperatures damaged the insulation of the affected conductors.

REFERENCE

1. G. Klempner and I. Kerszenbaum. *Handbook of Large Turbo-Generator Operation and Maintenance*. Piscataway, NJ: IEEE Press, 2008, chapters 4, 5, and 6.

M-S-54: TEMPERATURE OF STATOR TERMINAL ABOVE NORMAL

APPLICABILITY:

All generators

MALFUNCTION DESCRIPTION:

This malfunction addresses the case where one or more stator terminals exhibit temperatures higher than normal for the operating load at the time of measuring the temperatures.

There are a number of possible causes for this type of problem. The operator can narrow the search of possible causes by noting how the temperature behaves. All temperature rises will be to some extent dependent on load. However, a temperature rise that came in the form of a step increase and remains constant, can be explained, among other things, as resulting from a permanent partial blockage (in directly hydrogen or water-cooled terminals). On the other end, a temperature that keeps rising when the load is kept constant will indicate a worsening situation, like a bad joint, gas locking of increasing magnitude, or some other cause.

Large temperature increases of the terminals have a potential to causing significant damage and long repair outages.

PROBABLE CAUSES:

- BLOCKAGE IN THE FLOW OF STATOR TERMINAL COOLANT (M-S-01)
- BROKEN-CRACKED STATOR TERMINAL (M-S-07)
- GAS LOCKING IN WATER-COOLED TERMINALS (M-S-24)
- INSUFFICIENT FLOW OF TERMINAL COOLANT (M-S-35)

Low hydrogen pressure in the machine:

- HYDROGEN ESCAPING THE GENERATOR (M-H-08)
- HYDROGEN PRESSURE INSIDE THE GENERATOR BELOW NORMAL (M-H-09)

Malfunction of the SCW system (not enough cooling of the SCW):

- STATOR COOLING WATER FLOW BELOW NORMAL (M-SCW-14)
- COOLER OUTLET TEMPERATURE OF SCW ABOVE NORMAL (M-SCW-05)

Malfunction of the hydrogen heat exchanger (not enough cooling of the hydrogen):

- FAILURE OF HYDROGEN DRYER (M-H-06)

POSSIBLE CONSEQUENCES:

- BURNING/PYROLYSIS OF STATOR WINDING INSULATION (M-S-08)
- TEMPERATURE OF STATOR CONDUCTOR ABOVE NORMAL (M-S-53)

POSSIBLE SYMPTOMS:

- TEMPERATURE OF STATOR TERMINAL ABOVE NORMAL (S-S-19)
- *Global temperature* indicating temperature above normal

- Issues with hydrogen and/or SCW leaks
- GENERATOR CORE CONDITION MONITOR—ALARM (S-C&F-01)

OPERATORS MAY CONSIDER:

1. First step is identification of the type of severity, that is,
   a. Small increase of temperatures while remaining within allowable region?
   b. Large increases with temperatures above insulation class and/or temperature class limits?
   c. Trending higher quickly?
2. Depending on the answers to (1), load may or may not have to be reduced, or the unit may have to be taken offline immediately for repairs.

TEST AND VISUAL INDICATIONS:

- Offline electric tests and visual inspection can determine if the abnormal temperatures damaged the insulation of the affected terminals.
- Tests of pressure and control circuits in the SCW skid.
- Tests of the hydrogen cooling system.

REFERENCES

1. G. Klempner and I. Kerszenbaum. *Handbook of Large Turbo-Generator Operation and Maintenance*. Piscataway, NJ: IEEE Press, 2008, chapters 4, 5, and 6.
2. G. Stone et al. *Electrical Insulation for Rotating Machines*. Piscataway, NJ: IEEE, 2004.

M-S-55: VIBRATION OF STATOR CONDUCTOR BAR IN SLOT

APPLICABILITY:

All generators

MALFUNCTION DESCRIPTION:

This malfunction addresses the case of a conductor bar (or number of bars) vibrating in a slot (or number of slots).

A conductor bar that is vibrating excessively in the slot may result in a number of problems related to the integrity of the stator winding. These are discussed below in this document under *Possible Consequences*. Causes for excessive conductor bar vibration in the slots are enumerated under *Probable Causes*.

PROBABLE CAUSES:

- LOOSENING OF STATOR WINDING SLOT WEDGES (M-S-41)
- TEMPERATURE OF STATOR CONDUCTOR ABOVE NORMAL (M-S-53)
- VIBRATION OF STATOR ENDWINDINGS HIGH (M-S-56)
- The bar is shrinking due to normal drying during years of operation.
- The bar is shrinking due to abnormal drying due to overload operation.
- The bar shrank due to a severe overload event.

POSSIBLE CONSEQUENCES:

- BREACH OF THE STRANDS OR TUBES CARRYING COOLING WATER (M-S-06)
- CRACKED CONDUCTOR STRANDS (M-S-10)
- DAMAGE TO CONDUCTOR BAR INSULATION (M-S-15)
- LOOSENING OF STATOR WINDING SLOT WEDGES (M-S-41)

POSSIBLE SYMPTOMS:

- ELECTROMAGNETIC INTERFERENCE (EMI) ACTIVITY HIGH **(S-S-03)**
- PARTIAL DISCHARGE ACTIVITY (PDA) HIGH **(S-S-11)**
- RADIO FREQUENCY (RF) MONITOR HIGH **(S-S-13)**

OPERATORS MAY CONSIDER:

Refer to:

- LOOSENING OF STATOR WINDING SLOT WEDGES (M-S-41)

TEST AND VISUAL INDICATIONS:

Refer to:

- LOOSENING OF STATOR WINDING SLOT WEDGES (M-S-41)

REFERENCES

1. G. Klempner and I. Kerszenbaum. *Handbook of Large Turbo-Generator Operation and Maintenance*. Piscataway, NJ: IEEE Press, 2008, chapter 2, 5, and 11.
2. G. Stone et al. *Electrical Insulation for Rotating Machines*. Piscataway, NJ: IEEE, 2004.

M-S-56: VIBRATION OF STATOR END-WINDING—HIGH

APPLICABILITY:

All generators

MALFUNCTION DESCRIPTION:

Vibration of the stator endwinding beyond *normal* levels has a number of undesirable consequences, listed below. Endwinding vibration can be the result of a deficient design such as insufficient restraint and/or resonance frequencies too close to once and twice the operating frequencies. Endwinding vibration can also be the result of weakening of the endwinding restraining system due to operational incidents, such as malsynchronization, external close by short-circuits, oil ingression into the generator, overheating of the ties, and so on.

Vibrating endwindings happen in all type of large turbogenerators—and all of them may result in damage to the winding. However, the consequences of high endwinding vibrations in a generator with directly water-cooled windings are much more serious. For instance, water leaks can happen if the brazed connections crack due to the large vibration, resulting in electrical faults inside the machine.

Endwinding vibration due to looseness needs to be corrected as soon as possible. Left by itself, the situation will deteriorate resulting in a shorted life expectancy for the entire winding.

In general, a good endwinding restraining system will have its critical (resonant) frequencies at least 10% below and 15%–20% above 60 Hz and 120 Hz for a 60 Hz generator, as well as 50 Hz and 100 Hz for a 50 Hz machine.

Typical values for normal vibration are different for *conventional* and *cone-type* endwinding support systems. Also, the OEM may give different values for horizontal, vertical, and axial vibrations. The values are also most probably different for 50 and 60 Hz machines. Finally, they most probably will be different for each manufacturer. One can thus see that the most practical guidance is to obtaining pertinent values specific for the machine under scrutiny. For purpose of an example: values of must shut down can be encountered from 2 to 12 mils peak-to-peak displacement. With such a large variation, proper values for your machine should be obtained from the OEM. References 2 and 3 also provide general guidance on desirable vibration limits.

PROBABLE CAUSES:

- Designers miscalculated resonance frequencies of endwinding resulting in close proximity to 1× and 2× running frequency. This type of problem will tend to show early in the life of the machine.

- Deficient restraint due to wrong design and/or manufacturing. It may show up as increased vibrations in the first few years of operation.
- Major grid disturbance in the vicinity of the unit resulting in power swings between the generator and the grid.
- Malsynchronization or synchronizing the generator often with a wide angle.
- Large short-circuits in the vicinity of the generator (e.g., a fault on the step-up or auxiliaries transformers).
- Ingression of large quantities of oil over a short period of time, or small quantities of oil deposited on the endwindings over a long period of time.
- Dry and/or torn endwinding ties, felt coil separators and any other type of restrain components. This is the normal result of thermal aging during operation. This mechanism will show up as increased vibrations after a substantial number of years.
- Overtorquing of endwinding support system resulting in cracked supporting brackets and/or fasteners.
- Undertorquing of endwinding support system resulting in loose support.
- Serious frame vibration resulting in shacking of the entire core and endwindings.
- Serious rotor vibration resulting via magnetic coupling in large frame vibration. In addition, in machines where one or both bearings are end-bracket-mounted, large rotor vibrations will also cause frame and core vibration via the bearing-bracket coupling.
- Uprating of the generator. Current magnitude is a major driving factor on endwinding vibration (forces proportional to the current squared). Thus, a machine that has been uprated will probably result in higher endwinding vibrations. If the vibration levels were already high prior to the uprate, post uprate they may be of such an extent that damage to the endwindings may ensue at some time.
- Excessive number of loose wedges can affect endwinding tightness in a negative way if left unaddressed.
- LOOSENING OF STATOR WINDING SLOT WEDGES (M-S-41)

AREAS THAT MAY BE AFFECTED:

In all machines:

- Endwinding restrain systems
- Endwinding insulation condition

- Endwinding conductor brazing
- Endwinding conductor cracking and breaking
- Phase connectors (circumferential busses) cracking/insulation damage/loss of restrain/brazing damage
- Loosening of slot end wedges
- Loosening of the slot-embedded part of the winding
- Loose connections in the machine's terminal box (bushing well)
- Cracked/damaged stand-off insulators in the terminal box
- Damaged terminal bushings and/or bushing connections

In directly water-cooled windings, in addition to the above:

- Leaks due to cracks in the brazing between water-carrying strands/tubes and the water boxes
- Leaks due to loosening of the PTFE hose connections to the winding knuckles and water manifold

POSSIBLE SYMPTOMS:

The only direct indicators are vibration probes installed on the endwindings, directly on the conductors and/or on the supporting brackets. Many large machines and some medium-rating ones have endwinding vibration pick-ups, and a few use them online to take readings on a continuous basis. Given that in general the population of machines sporting endwinding vibration pick-ups is not large, this type of situation is mostly found after a resulting problem has been detected (such as a leak in a directly water-cooled winding), or during offline testing (*bump* test).

- STATOR CURRENT ABOVE NOMINAL **(S-S-16)**

OPERATORS MAY CONSIDER:

1. For those few machines that have online vibration probes being monitored continuously, and those that take periodic readings with the unit in operation, the most significant indication of a problem is a sudden increase in vibrations, or a sudden upward change in trend, respectively. This most probably signals a serious deterioration. Consideration should be given to have the load immediately reduced until the endwinding vibrations reach levels prior to the sudden change; a planned shutdown and inspection/test should be done ASAP.
2. In the meantime, look for leaks in directly water-cooled machines (they will show most probably as an increase of H_2 in the cooling water.

TEST AND VISUAL INDICATIONS:

- *Bump test*. It requires *hitting* the endwindings with a *calibrated hammer* and reading the amplitude and frequency of the responses through vibration pick-ups permanently or temporarily mounted on the endwindings and/or their support systems. In directly water-cooled machines one must remember that readings with the water removed from the windings will result in resonance frequencies somewhat higher than when the water fills the entire winding. The difference may be of a few Hz. In all cases, one must also take into consideration that a warm winding in a running generator will tend to have its resonant frequencies a little lower than when the coils are cold (such as during an offline test)
- *Greasing* of the ties/blockings/fasteners and other endwinding support components signal movement of the endwindings resulting from looseness. This condition may or may not necessarily result in high endwinding vibration
- Loose/greasing end-wedges may indicate a vibrating endwinding
- Cracked/broken endwinding support components may indicate a highly vibrating endwinding

Figures 5.72 and 5.73 show typical traces obtained from a *bump test* on a four-pole, 60 Hz, directly water-cooled generator. Figure 5.74 shows results from a bump test done on an air-cooled two-pole generator.

FIGURE 5.72
A severely overheated stator bar with asphaltic groundwall insulation.

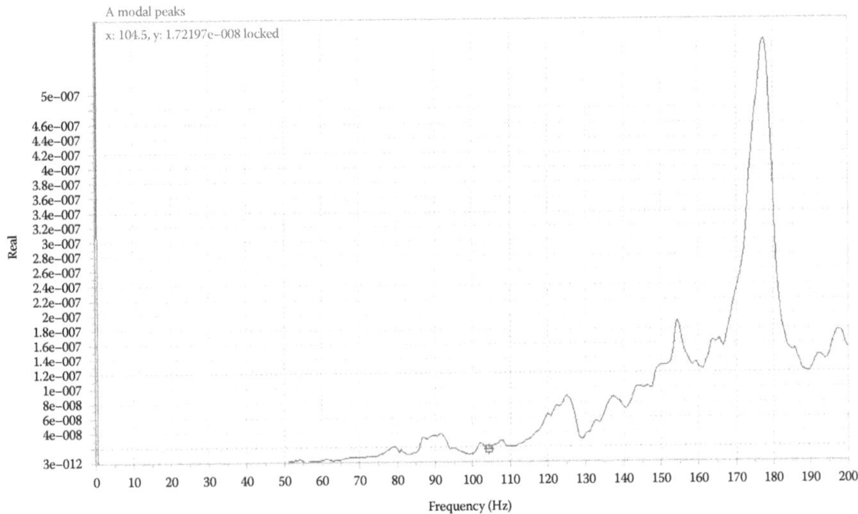

FIGURE 5.73
Results from a bump test on the turbine-side endwinding. The high peak indicates a natural frequency between 170 and 180 Hz—not problematic given the main exciting frequencies are 60 and 120 Hz.

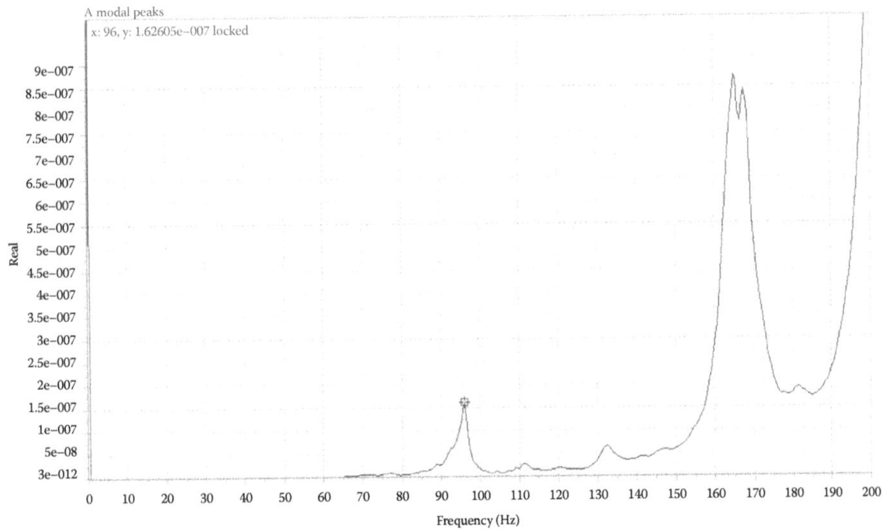

FIGURE 5.74
Results from the same machine as in Figure 5.72, but on the exciter end. Also in this case resonant frequencies are in the 170–180 Hz region.

REFERENCES

1. G. Klempner and I. Kerszenbaum. *Handbook of Large Turbo-Generator Operation and Maintenance*. Piscataway, NJ: IEEE Press, 2008, chapter 5.
2. IEEE Std 1129-2014. IEEE Guide for Online Monitoring of Large Synchronous Generators (10 MVA and Above).
3. IEEE Std 1665-2009. IEEE Guide for the Rewind of Synchronous Generators, 50 Hz and 60 Hz, Rated 1 MVA and Above.
4. EPRI Technical Report 1021774. Generator Stator Endwinding Vibration Guide: Tutorial, August 2011.

M-S-57: VOLTAGE OF GRID—UNBALANCED

APPLICABILITY:

All generators

MALFUNCTION DESCRIPTION:

This malfunction relates to a situation where a grid unbalance between phases creates a voltage unbalance at the terminals of the generator. There is no hard limit on the degree of voltage unbalance at the terminals of a generator in terms of the voltage itself. However, voltage unbalance creates *negative-sequence currents* in the machine. These are expressed and both the continuous I_2 and the transient I_2^2t. These can be very onerous if their magnitude is above certain values. These values are clearly specified in the pertinent standards.

Refer to:

- CURRENT UNBALANCE BETWEEN PHASES (M-S-12)
- NEGATIVE-SEQUENCE CURRENTS (M-S-42)

M-S-58: WINDING (STATOR) OPEN CIRCUIT

APPLICABILITY:

All generators with two or more parallel stator circuits

MALFUNCTION DESCRIPTION:

This malfunction addresses the case of a stator parallel becoming open during operation or left open during a winding repair activity (very low likelihood). This type of failure (during operation) is more often than not the result of higher-than-normal endwinding vibration coupled, in many cases, with a deficient design of the endwinding and/or inadequate brazing of the conductor bars at the series connections and at the radial lead connections.

PROBABLE CAUSES:

- VIBRATION OF STATOR ENDWINDINGS—HIGH **(M-S-56)**
- Oversight during a winding repair activity (very low likelihood)

POSSIBLE CONSEQUENCES:

An open circuit can result in various degrees of damage to the generator. It depends mainly on the type of cooling. For example, water inner-cooled windings will spill large amounts of water inside the machine during an open parallel event. In all cases, and depending on the generator's output, overheating of the other parallel may happen, and negative-sequence currents may exceed the continuous withstand level.

POSSIBLE SYMPTOMS:

- BYPRODUCTS FROM STATOR INSULATION PYROLYSIS **(S-S-01)**
- ELECTROMAGNETIC INTERFERENCE (EMI) ACTIVITY HIGH **(S-S-03)**
- HYDROGEN IS STATOR COOLING WATER (SCW) HIGH **(S-S-04)**
- LIQUID-IN-GENERATOR ALARM **(S-C&F-04)**
- PARTIAL DISCHARGE ACTIVITY (PDA) HIGH **(S-S-11)**
- PHASE CURRENT UNBALANCE **(S-S-12)**
- RADIO FREQUENCY (RF) MONITOR HIGH **(S-S-13)**

OPERATORS MAY CONSIDER:

An open parallel is a serious condition that requires immediate removal from operation and repair. In many cases, the unit will trip automatically due to secondary related events.

TEST AND VISUAL INDICATIONS:

Electrical tests (in particular series resistance) and visual inspection.

5.4 Rotor Malfunctions

M-R-01: ARCING ACROSS BEARINGS AND HYDROGEN SEALS

APPLICABILITY:

- All generators (arcing across bearings and wipers)
- Hydrogen-cooled generators (arcing across bearings, wipers, and hydrogen seals)

MALFUNCTION DESCRIPTION:

This malfunction describes electric arcing activity across bearings, wipers, and hydrogen seals.

Shaft voltages are common to all electric rotating machinery. They come from a number of sources. In large turbine-driven generators, it is not uncommon for shaft voltages to have amplitudes of up to about 150 volts. If left unchecked, these voltages can give rise to currents (called *shaft currents and/or bearing currents*). Shaft currents can flow from the shaft to ground via arcing across the thin oil film in bearings, wipers, and hydrogen seals. In the process, they create electropitting of the bearing, seals, and shaft surfaces, causing, in extreme cases, the failure of the seals and bearings. To avoid this situation, it is imperative to maintain good bearing/seal insulation, as well as shaft-grounding circuits. See Figures 5.75 and 5.76 for examples of bearing degradation due to electric arcing.

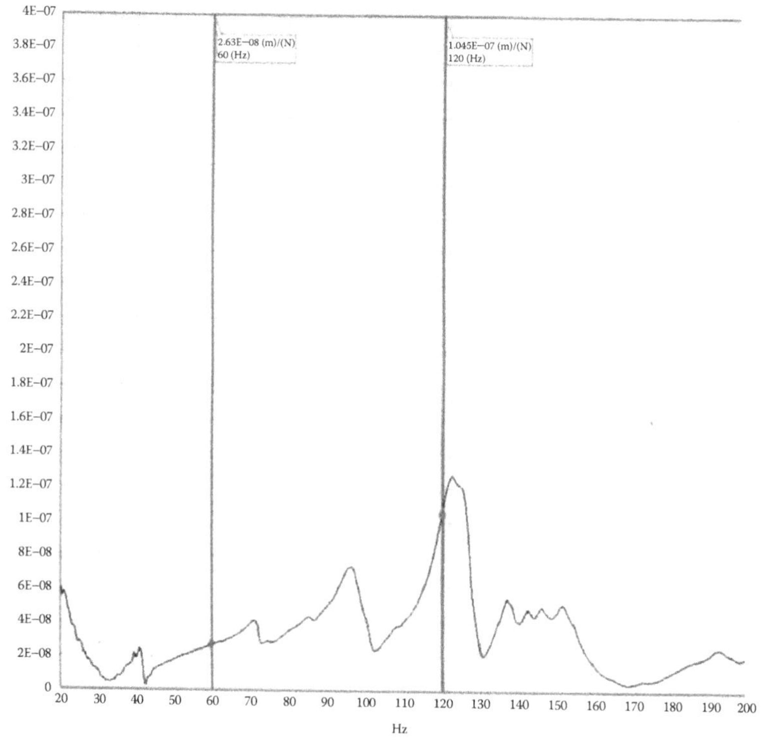

FIGURE 5.75
The result of a bump test on a radial connection of a two-pole generator. It is clear from the test results that there is a natural frequency very close to the main 120 Hz driving forces acting on the winding.

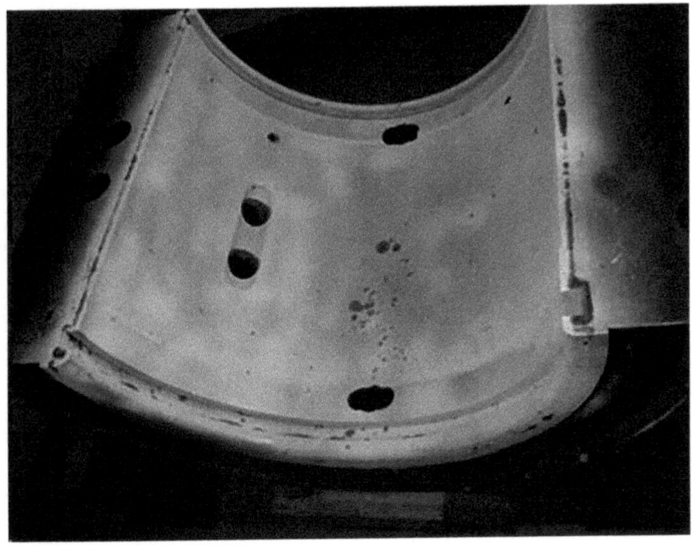

FIGURE 5.76
Pitting/arcing damage on a bearing surface (shown by dye penetrant testing).

PROBABLE CAUSES:

- FAILURE OF BEARING INSULATION **(M-R-09)**
- FAILURE OF H_2 SEAL INSULATION **(M-R-10)**
- INEFECTIVE SHAFT GROUNDING **(M-R-15)**
- SHAFT VOLTAGE—HIGH **(M-R-23)**
- A large stator core fault that shorts out a significant number of laminations, axially. This is not a common cause for bearing/seal arcing, but it has happened. This will show up as stator core-related problems and the reader should also refer to these in this book. Very severe core faults can even short out the entire core of an axially cooled stator core. The result will be full stator bar voltage (i.e., voltage across the full core length, impressed back on the rotor shaft). The result will be severe arcing at the rotor bearing journals (see Figure 5.77).

POSSIBLE CONSEQUENCES:

- Electropitting of bearing/seal/shaft surfaces
- Failure of bearings/seals

POSSIBLE SYMPTOMS:

Alarm from shaft voltage monitoring circuit:

- SHAFT CURRENT ABOVE NORMAL **(S-R-04)**
- SHAFT VOTAGE ABOVE NORMAL **(S-R-05)**

FIGURE 5.77
Severe pitting/arcing damage on an H_2-seal surface.

RF monitoring circuits—increased activity:

- RADIO FREQUENCY (RF) MONITOR—HIGH **(S-S-13)**
- VIBRATION OF ROTOR—HIGH **(S-R-10)**

Degradation of the lube and/or seal oil:

- TEMPERATURE OF SEAL OIL ABOVE NORMAL (S-SOL-10)
- TEMPERATURE OF LUBE OIL AT BEARING OUTLET—HIGH **(S-LOS-06)**

OPERATORS MAY CONSIDER:

1. Shaft voltages are always present in electric machines. As stated above, voltages of up to 150 volts are not uncommon. What is important is to monitor the resulting shaft currents. An increase in shaft currents means that the bearing/seal insulation is deficient, and/or the grounding brushes are not clamping the shaft voltages effectively. If any of these occur, the problem should be addressed ASAP. If left as is, bearing and/or seal failure can eventually happen, and depending on the severity of the situation, can happen quickly.
2. A machine that is experiencing a significant increase in shaft voltage amplitude may undergo a serious stator core failure. Therefore, this type of voltage increase should be investigated diligently.

TEST AND VISUAL INDICATIONS:

- Bearing/seals insulation testing
- Visual inspection of babbitt and journal
- EL-CID and/or flux tests of stator core

REFERENCES

1. G. Klempner and I. Kerszenbaum. *Handbook of Large Turbo-Generator Operation and Maintenance.* Piscataway, NJ: IEEE Press, 2008, chapters 5, 8, 9, and 11.
2. EPRI TR-107137s. Main Generator On-Line Monitoring.

M-R-02: BEARING MISALIGNMENT

APPLICABILITY:

All generators

MALFUNCTION DESCRIPTION:

This malfunction addresses misalignment of bearings.

There are many factors affecting the operation of friction (sleeve and thrust) bearings in large turbine-driven generators. For instance, angular misalignment, linear misalignment, concentricity, journal and/or babbitt deformations, oil viscosity, oil temperature, oil impurities, thermal expansion, and so on. When it comes to misalignment in multishaft units coupled together (in a large nuclear unit up to 11 bearings can be found with 5 or so coupled shafts), misalignment can preload bearings significantly. The end result is increased vibration and increase in bearing temperature.

PROBABLE CAUSES:

Misalignment often occurs during coupling of the generator rotor to the turbine rotors. Careful attention to the airgap measurements and overall catenary of the generator rotor is required.

POSSIBLE CONSEQUENCES:

High vibrations:

- HIGH VIBRATION OF CORE AND/OR FRAME **(M-C&F-10)**
- VIBRATION OF STATOR ENDWINDING—HIGH **(M-S-56)**

High temperature:

- TEMPERATURE OF BEARING ABOVE NORMAL **(M-R-25)**
- Damage to the babbitt (the bearing is *wiped*)

POSSIBLE SYMPTOMS:

- TEMPERATURE OF BEARING ABOVE NORMAL **(S-R-08)**
- VIBRATION OF ROTOR—HIGH **(S-R-10)**

OPERATORS MAY CONSIDER:

The most common cause of a bearing misalignment is incorrect alignment during an outage. Bearing misalignment could also be the result of a shift in the sitting of pedestal-type bearings, due to a major grid disturbance or malsynchronization. Bearing misalignment should be corrected as soon as possible, given the deleterious consequences to the machine by the high vibrations, and the possibility of bearing failure, which may cause a number of additional problems to the rotor.

TEST AND VISUAL INDICATORS:

- Alignment measurements
- Damage to babbitt

REFERENCES

1. G. Klempner and I. Kerszenbaum. *Handbook of Large Turbo-Generator Operation and Maintenance*. Piscataway, NJ: IEEE Press, 2008, chapters 4, 5, and 9.
2. EPRI Technical Report 1026566. Field Guide: Bearing Damage Mechanisms, November 2012.

M-R-03: BLOCKAGE OF ROTOR VENTILATION PATH

APPLICABILITY:

All generators

MALFUNCTION DESCRIPTION:

This malfunction addresses the case when there is a partial or total blockage in the path or paths of the rotor-cooling medium. In the overwhelming majority of cases, this implies hydrogen or air. However, in a few rare cases, rotors of large units are directly cooled by water.

The main predicament in discovering blocked cooling passages in the rotor, is that the temperature of the rotor's conductor is measured indirectly (by dividing V_{field} by I_{field} for rotors with external excitation, or some derived method for self-excited rotors), and thus, localized faults tend to be masked by the global winding temperature. Therefore, most of the problems with blocked vents tend to show as augmented vibrations due to asymmetric heating of the rotor.

PROBABLE CAUSES:

- Shifted insulation and/or blocking (see Figures 5.78 and 5.79)
- Clogged vents by debris (see Figure 5.80) or contamination (air-cooled generators are more prone to contamination, especially

FIGURE 5.78
Severe arcing due to a major stator core fault, shorting out the entire length of the core iron.

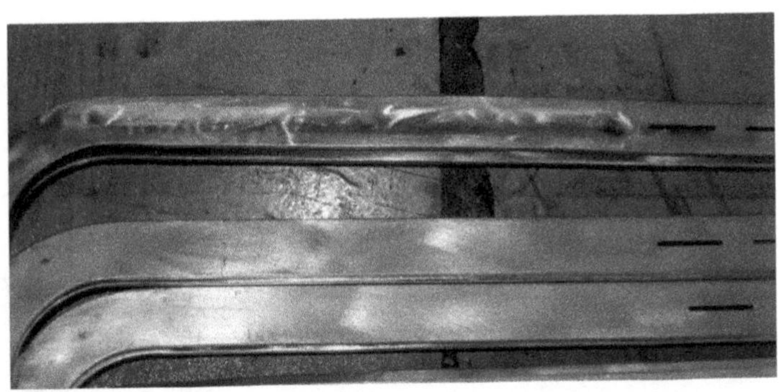

FIGURE 5.79
Collapsed vent paths due to distorted conductors.

those with external excitation, and those that underwent significant oil ingression from the bearings lubricating oil)

- Collapsed vent paths due to distorted conductors (see Figure 5.81)
- Shifted coils and/or insulation blocking radial (and in some cases axial) cooling paths
- Foreign material left during maintenance (see Figure 5.82)

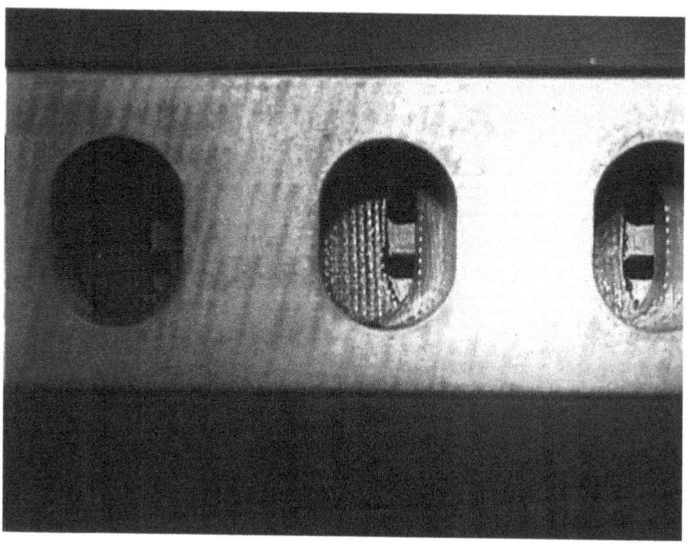

FIGURE 5.80
Shifted insulation in the winding slot almost completely blocking cooling gas from flowing through the vents.

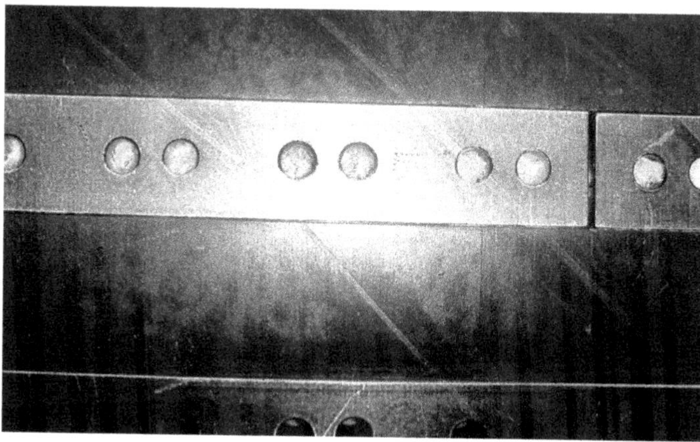

FIGURE 5.81
Clogged cooling vents by debris.

POSSIBLE CONSEQUENCES:

- Loss of expected life of the insulation
- BURNING/PYROLYSIS OF ROTOR INSULATION (M-R-04)
- DEGRADED/DAMAGED FIELD WINDING ISULATION (M-R-07)
- GROUNDED FIELD WINDING (M-R-13)

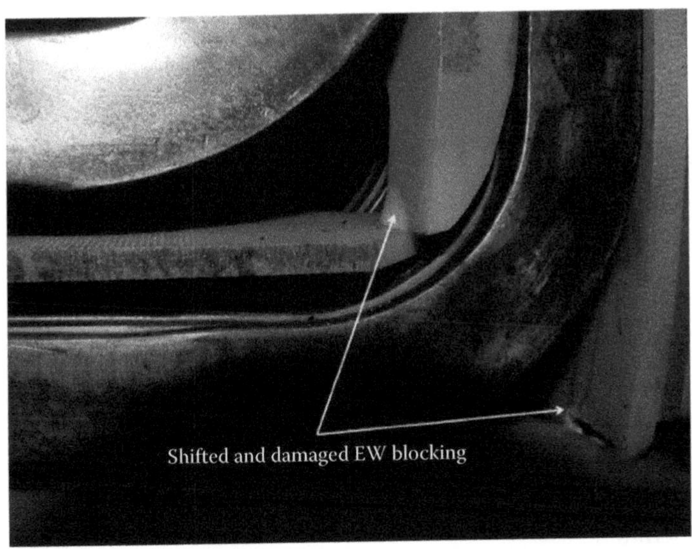

FIGURE 5.82
Shifted and damaged insulation blocking.

- SHORTED TURNS IN FIELD WINDING **(M-R-24)**
- THERMAL ASYMMETRY OF ROTOR **(M-R-28)**
- VIBRATION OF ROTOR—HIGH **(M-R-30)**

POSSIBLE SYMPTOMS:

- GROUNDED FIELD WINDING—ALARM AND/OR TRIP **(S-R-01)**
- TEMPERATURE OF FIELD WINDING ABOVE NORMAL **(S-R-09)**
- VIBRATION OF ROTOR—HIGH **(S-S-10)**

OPERATORS MAY CONSIDER:

1. The degree of urgency for operator response depends on the magnitude of the problem; that is, for temperatures above normal but within limits, there is more time to investigate the source of the problem without having to reduce load, than the case when the temperature limits given by the OEM and/or standards are exceeded
2. Keeping in mind that Arrhenius law means the expected life of the insulation is halved by every 8°C–10°C increase of temperature
3. Keeping in mind the problem may appear to be minor, as measured by the global temperature increase of the rotor winding, but actually, it may be a serious localized fault

4. Temperature rises combined with increase in vibration, erratic MVAR output, a ground-fault or new shorted turns may be signs of a serious problem

TEST AND VISUAL INDICATORS:

- Mainly visual clues, such as discolored insulation, blocked radial vents, and shifted blocking
- THERMAL SENSITIVITY TEST (M-R-29)

REFERENCE

1. G. Klempner and I. Kerszenbaum. *Handbook of Large Turbo-Generator Operation and Maintenance*. Piscataway, NJ: IEEE Press, 2008, chapter 9.

M-R-04: BURNING/PYROLYSIS OF ROTOR INSULATION

APPLICABILITY:

All generators

MALFUNCTION DESCRIPTION:

This malfunction addresses the drastic deterioration of the insulation of the rotor winding, due to the abnormal presence of a source of heat resulting in excessively high temperatures (see Figure 5.83). There are two different processes by which this deterioration can occur: *burning* and *pyrolysis*.

Burning is the physical process of a fuel undergoing combustion, which is a rapid chemical oxidation process creating byproducts and heat.

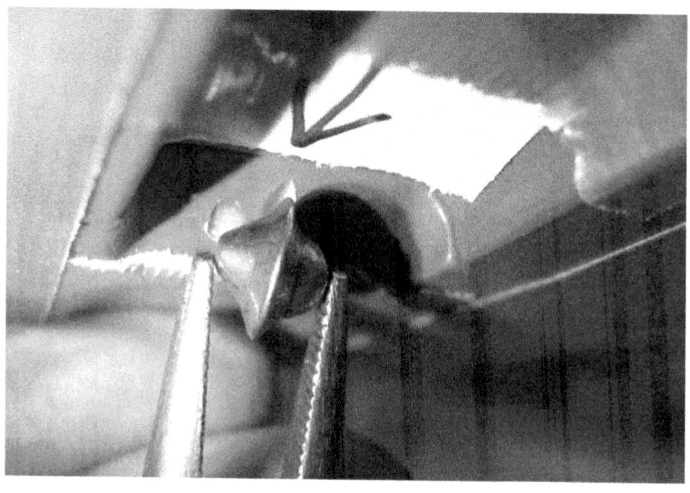

FIGURE 5.83
Foreign material left during maintenance. The FME are BBs that were left in the wedge vent hole during an overhaul.

Combustion requires the presence of oxygen, and the subjection of the fuel to a temperature above certain value, specific for each fuel. The fire triangle shows the conditions for the presence of fire.

Let us look in the generator for the sources of the three components shown on the fire triangle:

Fuel: All insulation is made either of organic materials, or much more commonly, from a combination of organic and inorganic components (e.g., mica or glass as inorganic material and resins and epoxies as organic materials. The organic materials will burn in the presence of oxygen and high temperatures.

Oxygen: In air-cooled generators, the stator windings are in the presence of air, which is the source of the oxygen required for the initiation and sustenance of burning. In H_2-cooled generators, there is no air (and oxygen) present during the normal operation of the unit. Therefore, burning in the presence of fires can only exist if the casing has been severely breached and air is mixed with hydrogen. Depending on the H_2/air ratio, this by itself can produce high-combustion temperatures, lighting up stator insulation, which will then continue to burn.

Heat: The heat required to initiate burning of the insulation may have a number of sources. One indicated above, is an internal explosion of a hydrogen–air mixture. This can only happen if a catastrophic failure of the casing results in large amounts of air mixing with hydrogen inside the machine. Alternatively, it could happen if wrong operator activity results in mixing air with hydrogen inside the machine in ratios that may cause the mixture to self-ignite.

A more likely scenario is heat created by an internal electrical failure, such as a rotor winding short-circuit between turns or to ground.

Finally, a major overcurrent event or loss of cooling can give way to high temperatures. It is hard to imagine these types of events producing winding temperatures required to light up the insulation. Nevertheless, it may be seriously aged in a short time.

Pyrolysis: It is the chemical decomposition of organic materials in the absence of oxygen. Pyrolysis is the mechanism by which stator insulation deteriorates if a large amount of heat is applied in hydrogen-cooled generators and in air-cooled generators without a fire.

For the purpose of investigating the development of rapid degradation of rotor winding insulation due to abnormal high temperatures, burning and pyrolysis can be used as a single common factor.

Burning/pyrolysis may cause pyrolization of the insulation resulting in a conductive path capable of creating short-circuits and maintaining the arc.

PROBABLE CAUSES/AREA AFFECTED:

- Global loss of cooling (see Figure 5.84)
- Local loss of cooling (e.g., blocked cooling passage). This will affect a section of winding depending on location and nature of the loss of cooling
- Significant overload (global effect)
- Internal short-circuit affecting all or a section of the field winding
- Large breach of the generator casing resulting in an explosive H_2/O_2 mixture inside and around the machine

POSSIBLE CONSEQUENCES:

Depending on type of cause:

- Damage to slot insulation with full or partial loss of insulating capability
- Damage to interturn insulation with full or partial loss of insulating capability
- Accelerated thermally induced aging of the affected insulation
- Mechanical weakening of the insulation bonding material
- Increasing the level of hydrogen impurity in the machine
- Ground faults
- Shorted turns

POSSIBLE SYMPTOMS:

- ERRATIC BEHAVIOR OF GENERATOR FIELD TEMPERATURE **(S-E-02)**
- GENERATOR CORE CONDITION MONITOR—ALARM **(S-C&F-01)**

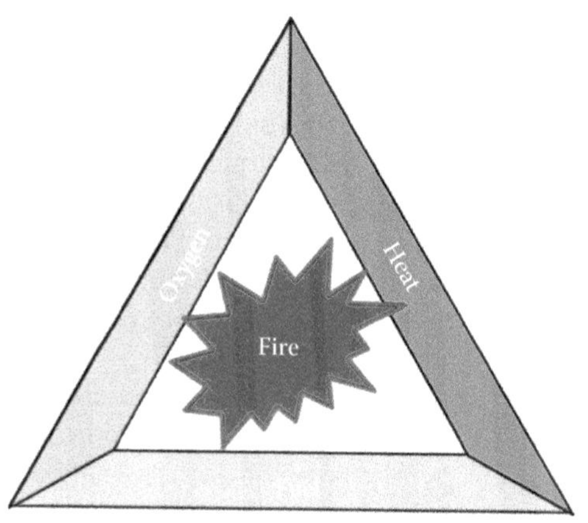

FIGURE 5.84
Fire triangle.

- SHORTED TURNS IN FIELD WINDING **(S-R-07)**
- TEMPERATURE OF FIELD WINDING ABOVE NORMAL **(S-R-09)**
- VIBRATION OF ROTOR—HIGH **(S-R-10)**

OPERATORS MAY CONSIDER:

1. If GCM alarm received, operator should check that the alarm is the result of a *real* pyrolytic activity. Check for other indicators of abnormality, such as load condition, vibration, new shorted turns, *field* ground fault indication, MVAR swings, and so on.
2. Check rotor-winding temperature. A global significant increase on winding temperature will be captured by the temperature measurement. A localized problem may not be detected, depending on the magnitude of the problem.
3. If the GCM alarm has been deemed *true*, then consider reducing load until the situation is cleared, even to the extent of removing the unit from service.
4. Record temperatures versus time of windings and any other abnormal readings and events.

TEST AND VISUAL INDICATIONS:

- Electrical tests such as
  - Insulation resistance measurement (*megger*)
  - PI
  - High potential testing (*Hipot*)
  - And so on
- Visual inspection (directly, borescopic, or robotic)

REFERENCE

1. G. Klempner and I. Kerszenbaum. *Handbook of Large Turbo-Generator Operation and Maintenance*. Piscataway, NJ: IEEE Press, 2008, chapters 1, 4, and 9.

M-R-05: CRACKED SHAFT

APPLICABILITY:

All generators

MALFUNCTION DESCRIPTION:

Having a cracked rotor shaft is considered a major failure and is very rare. The best case is finding it while the crack is still growing and has not gone into a runaway situation. The worst case is when the crack opens up quickly and there is no enough advanced warning to shut the machine down before a catastrophic failure occurs. In either case, it is classed as a major failure because the rotor is no longer operable.

PROBABLE CAUSES:

For a crack to form in a rotor forging, there are a number of possibilities:

- Defects in the rotor bore hole, the bore surface, growing and becoming unstable
- Inclusions in the rotor forging material being overstressed and initiating a crack
- High bending stress in long thin rotors (large length to diameter ratio) that have a forging defect
- Sharp corners with high stress risers at different radii (such as threaded hole for the radial studs—see Figure 5.85)
- Torsional events from the system or the machine itself (see Figure 5.86)
- Uprating beyond the torsional limits of the shaft
- LOSS OF STABILITY **(M-R-17)**
- MECHANICAL FAILURE OF ROTOR COMPONENT **(M-R-18)**

POSSIBLE CONSEQUENCES:

- VIBRATION OF ROTOR—HIGH **(M-R-30)**

Generally, the first consequence of a cracked rotor shaft would be high rotor vibration that would tend to worsen. If the crack is allowed to grow,

FIGURE 5.85
Severely overheated insulation due to a global loss of cooling.

FIGURE 5.86
Cracked shaft initiated at the thread of a radial stud restraint nut.

the worst case is that the shaft would fail completely causing a catastrophic failure of the generator and likely cause collateral damage.

POSSIBLE SYMPTOMS:

- VIBRATION OF ROTOR—HIGH **(S-R-10)**

OPERATORS MAY CONSIDER

There are lateral vibrations on all rotors and limits provided by the manufacturer. These can be monitored by vibration probes of different types. When a rotor vibration alarm is received, operators should follow OEM recommendations and plant operating procedures. However, when rotor vibration becomes excessively high and a secondary alarm is received, the usual course of action is to bring the generator to minimal load and then offline to zero speed for inspection and testing.

TEST AND VISUAL INDICATIONS:

Visual indications are not generally possible unless the cracking is large and in an open place on the shaft where it can be seen.

If the location of the cracking is in the bore of the rotor, or a place where dismantling is required, then the inspection process will become long and expensive. ***NDE***[4] of some type will surely be required to find the location of the crack in the shaft and the extent of it.

REFERENCE

1. G. Klempner and I. Kerszenbaum. *Handbook of Large Turbo-Generator Operation and Maintenance*. Piscataway, NJ: IEEE Press, 2008, chapters 2 and 9.

[4] NDE—Nondestructive examination.

M-R-06: DEFECTIVE FIELD WINDING INSULATION

APPLICABILITY:

All generators

MALFUNCTION DESCRIPTION:

This malfunction addresses defective rotor winding insulation. By *defective*, it means that due to poor design, manufacturing or maintenance, the insulation has developed weaknesses that are not characteristic of a well designed, manufactured and maintained winding of similar construction, rating, and age.

This section does not address the issue of faulty coil insulation. The problem of a faulty winding is entered as different malfunctions:

- DEGRADED/DAMAGED FIELD WINDING INSULATION (M-R-07)

Defects due to poor design and/or manufacturing can be such as

- Insufficient voltage capability (volts/mil) of the wall insulation
- Imperfect bonding, wrong coil dimensions vis-à-vis the slot dimensions
- Coils not properly supported in the overhangs, and so on.

Defects due to poor maintenance can be such as

- Lack of adequate support in the slot (loose wedges and/or slot packing)
- Degraded blocking of the overhangs
- Wrong oil seals allowing large amount of oils contaminating the machine
- Lack of adequate FME
- Carelessness during an open-machine inspection or maintenance activity, and so forth

Defects originated at the OEM's factory may or may not show themselves early in the life of the machine. For example, debris left during assembly of the rotor winding may be the root cause for the development of shorted turns many years later.

PROBABLE CAUSES/AREA AFFECTED:

- Insufficient wall-insulation voltage-withstand capability (affects the slot area)
- Imperfect bonding between interturn insulation and conductor (affects slot and endwinding)
- Bars not properly blocked in overhangs (endwindings)
- Loose bars in slot (mainly slot area, but can show itself as problems in the endwindings)

- Significant oil contamination (global looseness and/or blockage of cooling vents in air-cooled machines with brushgear excitation)
- Lack of proper FME (affected areas are most probably endwindings, but also slot region)

POSSIBLE CONSEQUENCES:

- Ground fault in the slot area
- Coil-to-coil fault in the overhang area
- GROUNDED FIELD WINDING (M-R-13)
- SHORTED TURNS IN FIELD WINDING (M-R-24)

POSSIBLE SYMPTOMS:

There is no direct indication about such windings defects, until they develop into degraded or faulty conditions captured under (M-R-07). See below *Test and Visual Indications.*

OPERATORS MAY CONSIDER:

Typically, this is not a problem requiring immediate shutdown or load reduction.

1. If a situation like this is recognized, plan for repairs or rewinds, as necessary, during the next convenient opportunity and estimate the risks of continued operation as is.

TEST AND VISUAL INDICATIONS:

- Unsatisfactory PI
- Unsatisfactory Hipot test
- Visual inspection

REFERENCES

1. G. Klempner and I. Kerszenbaum. *Handbook of Large Turbo-Generator Operation and Maintenance.* Piscataway, NJ: IEEE Press, 2008, chapter 9.
2. G. Stone et al. *Electrical Insulation for Rotating Machines.* Piscataway, NJ: IEEE, 2004.

M-R-07: DEGRADED/DAMAGED FIELD WINDING INSULATION

APPLICABILITY:

All generators

MALFUNCTION DESCRIPTION:

This malfunction relates to rotor winding insulation that is directly damaged during operation, to be distinguished from another malfunction: DEFECTIVE FIELD WINDING INSULATION (M-R-06), which is a result of factory or maintenance inadequacies. There can be cases which straddle the region between a maintenance-caused defective or damaged insulation, thus some of the issues discussed can be found in both problem descriptions.

Damaged *turn insulation* (also called interturn insulation) may fail causing an interturn short-circuit (called *shorted turns*—see Figure 5.87). Interturn faults might be accompanied by low or high current magnitudes, depending on the resistance of the short. *Wall insulation* may fail causing ground short-circuits which can be detected by the ground-fault protection (see Figure 5.88). Both interturn and wall insulation failures can

FIGURE 5.87
Cracked shaft at the coupling shrink fit keyway. Suspected issue is a torsional event causing the forging to crack at a stress riser.

FIGURE 5.88
Interturn short circuit or rotor winding *shorted turn* due to damaged turn insulation.

be accompanied by localized burning/pyrolysis of the insulation, whose released byproducts can be detected by a GCM also known as CCM.

PROBABLE CAUSES/AFFECTED AREAS:

- Overload (results in global damage)
  - Mechanical abrasion (Mainly in the slot area, but also in end-windings, phase connectors, and terminal box). Mechanical abrasion may be the result of *copper dusting* in the slot area, or abrasion of conductor against glass fiber-based *blocking* in the overhangs
- Large (often sudden) deformations due to uneven forces acting on the winding
- Bad joints causing high temperatures (see Figure 5.89)

POSSIBLE CONSEQUENCES:

- BURNING/PYROLYSIS OF ROTOR INSULATION **(M-R-04)**
- GROUNDED FIELD WINDING **(M-R-13)**
- SHORTED TURNS IN FIELD WINDING **(M-R-24)**
- THERMAL ASYMMETRY OF ROTOR **(M-R-28)**

POSSIBLE SYMPTOMS:

There is no direct indication of an *incipient* fault due to damaged insulation that has not been fully breached, other than trending down of rotor-to-ground insulation. Once breached, most symptoms will be: the activation of *field ground fault protection*, and online monitoring of gas gap (airgap) magnetic flux:

- GROUNDED FIELD WINDING—ALARM AND/OR TRIP **(S-R-01)**
- SHORTED TURNS IN FIELD WINDING **(S-R-07)**

FIGURE 5.89
Slot liner abrasion—severe case.

OPERATORS MAY CONSIDER:

1. Ground faults can result in an alarm, or in alarm and trip. For those that only alarm, there are different philosophies about when to take the unit offload for repairs. In any case, operating a unit with one ground may not be highly detrimental to the generator, but can result in another ground developing, with catastrophic results to the integrity of the machine. See suggestions to the operators in GROUNDED FIELD WINDING (M-R-13).
2. For disposition of shorted turns, see SHORTED TURNS IN FIELD WINDING (M-R-24).

TEST AND VISUAL INDICATIONS:

- Online testing for shorted turns via flux probe (if installed)
- Offline testing for shorted turns via RSO (repetitive surge oscillation), pole balance, volt drop, growler, and so on
- Insulation Resistance (IR, also called megger) test
- Hipot test
- Visually recognizable deformation of endwindings
- Visually recognizable damage to the insulation

REFERENCE

1. G. Klempner and I. Kerszenbaum. *Handbook of Large Turbo-Generator Operation and Maintenance.* Piscataway, NJ: IEEE Press, 2008, chapters 9 and 11.

M-R-08: DYNAMIC ROTOR UNBALANCE (MECHANICAL)

APPLICABILITY:

All generators

MALFUNCTION DESCRIPTION:

This malfunction addresses the case when a rotor experiences dynamic unbalance during operation.

PROBABLE CAUSES:

- Asymmetric weight distribution:
  - Movement of balancing weight(s)
  - Loss of balancing weight/bolt/nut and so on
  - Shifting of retaining ring
  - Shifting of insulation bloc
  - Shifting of any shrunk member
  - Loss of fan blade
  - and so on
- Hydrogen-seal rub

- Oil whirl
- Oil-wiper rubs
- Misalignment
- Rotor cracks
- Foundation/frame resonance
- Foundation/frame looseness

POSSIBLE CONSEQUENCES:

- Metal fatigue
- Bearings/seals/wipers failures
- HIGH VIBRATION OF CORE AND FRAME **(M-C&F-10)**
- VIBRATION OF ROTOR—HIGH **(M-R-30)**
- VIBRATION OF STATOR END-WINDING HIGH **(M-S-56)**

POSSIBLE SYMPTOMS:

- TEMPERATURE OF BEARING ABOVE NORMAL **(S-R-08)**
- TEMPERATURE OF LUBE OIL AT BEARING-OUTLET—HIGH **(S-LOS-06)**
- TEMPERATURE OF HYDROGEN SEAL ABOVE NORMAL **(S-SOS-08)**
- TEMPERATURE OF SEAL OIL ABOVE NORMAL **(S-SOS-10)**
- VIBRATION OF ROTOR—HIGH **(S-R-10)**

OPERATORS MAY CONSIDER:

1. Depending on the level of vibration, operator reaction may be required immediately or on a longer time frame. Keep in mind high vibrations can be very detrimental to the integrity of the unit and can lead to unplanned outages.
2. If cursory troubleshooting efforts do not reveal the cause of the vibrations, a more rigorous effort will be required. A THERMAL SENSITIVITY TEST **(M-R-29)** is an excellent tool for discriminating mechanical from electrically induced vibrations.

TEST AND VISUAL INDICATIONS:

It is much easier finding the source of vibrations while the unit is in operation, than when standing still. If the unit is not spinning, then

- Visual inspection of rotor for loose/shifted/dislocated/lost components
- Nodes for cracks
- Visual inspection of hydrogen seals/bearings/wipers
- Shifted coils
- Shifted slot insulation
- See (M-R-29) for additional information

REFERENCES

1. G. Klempner and I. Kerszenbaum. *Handbook of Large Turbo-Generator Operation and Maintenance*. Piscataway, NJ: IEEE Press, 2008, chapters 2, 5, and 11.
2. EPRI Technical Report 1004951. Optimized Maintenance of Generator Rotors, November 2004.

M-R-09: FAILURE OF BEARING INSULATION

APPLICABILITY:

All generators

MALFUNCTION DESCRIPTION:

Failure of the bearing insulation will cause current to flow from the rotor shaft to the bearing surface. This would be through the oil film, discharging to the bearing babbitt surface. The breakdown of the bearing insulation will cause high current discharge and the shaft voltage to drop.

PROBABLE CAUSES:

Bearing insulation breakdown is a low-probability type of malfunction, but it can happen from the following:

- High mechanical load on the bearing insulation causing it to crack (such as misalignment)
- A system event which causes impacting on the bearing from the rotor shaft, enough to transmit instantaneous forces from the rotor
- BEARING MISALIGNMENT **(M-R-02)**
- DYNAMIC ROTOR UNBALANCE (MECHANICAL) **(M-R-08)**
- RUB BETWEEN BEARING AND JOURNAL **(M-R-20)**
- VIBRATION OF ROTOR—HIGH **(M-R-30)**

POSSIBLE CONSEQUENCES:

- BURNING/PYROLYSIS OF ROTOR INSULATION **(M-R-04)**
- GROUNDED SHAFT (MOMENTARILY) **(M-R-14)**
- RUB BETWEEN BEARING AND JOURNAL **(M-R-20)**
- TEMPERATURE OF BEARING ABOVE NORMAL **(M-R-25)**

POSSIBLE SYMPTOMS:

- SHAFT CURRENT ABOVE NORMAL **(S-R-04)**
- SHAFT VOLTAGE BELOW NORMAL **(S-R-06)**
- TEMPERATURE OF BEARING ABOVE NORMAL **(S-R-08)**

OPERATORS MAY CONSIDER

1. It is unlikely that the operators will notice anything unless there is a shaft-ground alarm present and it is triggered by an actual failure of the bearing insulation. At that point, the operators will have to consider shutting down to avoid severe damage to the bearings and/or rotor shaft.
2. If there is no shaft-ground alarm device, the operators may not notice anything until higher rotor lateral vibration occurs and eventually a failure if nothing is done.

TEST AND VISUAL INDICATIONS:

Megger test of bearing insulation

REFERENCES

1. G. Klempner and I. Kerszenbaum. *Handbook of Large Turbo-Generator Operation and Maintenance*. Piscataway, NJ: IEEE Press, 2008, chapters 9 and 11.
2. EPRI Product ID 3002001758. Turbine-Generator Topics for Plant Engineers: Residual Magnetism, August 2013.
3. P. Nippes, D. David, and A. Peniazev. Monitoring of Shaft Voltages and Grounding Current Brushes. In: *EPRI Motor and Generator Predictive Maintenance & Refurbishment Conference*, Orlando, FL, November 1995.
4. M.J. Costello. Shaft Voltages in Rotating Machinery. *IEEE Transactions on Industry Applications*, 29(2), pp. 419–426, March–April 1993.
5. J.S. Sohre. Are Magnetic Currents Destroying your Machinery? In: *Hydrocarbon Processing*, April 1979, pp. 207–212.

M-R-10: FAILURE OF H_2-SEAL INSULATION

APPLICABILITY:

All hydrogen-cooled generators

MALFUNCTION DESCRIPTION:

Failure of the hydrogen seal insulation will cause current to flow from the rotor shaft to the H_2-seal surface. This would be through the oil film, discharging to the H_2-seal babbitt surface or bronze surface, depending on the type of H_2 seal employed. The breakdown of the insulation will cause the shaft voltage to drop and high shaft current discharge.

PROBABLE CAUSES:

H_2-seal insulation breakdown is a low probability type of malfunction, but it can happen from the following:

- High mechanical load on the H_2-seal insulation causing it to crack (such as misalignment)

- A system event which causes impacting on the H_2 seal from the rotor shaft, enough to transmit instantaneous forces from the rotor
- DYNAMIC ROTOR UNBALANCE (MECHANICAL) **(M-R-08)**
- MISALIGNMENT OF HYDROGEN SEAL **(M-R-19)**
- RUB OF HYDROGEN SEAL **(M-R-21)**
- VIBRATION OF ROTOR—HIGH **(M-R-30)**

POSSIBLE CONSEQUENCES:

- BURNING/PYROLYSIS OF ROTOR INSULATION **(M-R-04)**
- GROUNDED SHAFT (MOMENTARILY) **(M-R-14)**
- RUB OF HYDROGEN SEAL **(M-R-21)**
- Temperature of hydrogen seal above normal

POSSIBLE SYMPTOMS:

- SHAFT CURRENT ABOVE NORMAL **(S-R-04)**
- SHAFT VOLTAGE BELOW NORMAL **(S-R-06)**
- TEMPERATURE OF BEARING ABOVE NORMAL **(S-R-08)**

OPERATORS MAY CONSIDER

1. It is unlikely that the operators will notice anything unless there is a shaft-ground alarm present and it is triggered by an actual failure of the H_2-seal insulation. At that point, the operators will have to consider shutting down to avoid severe damage to seals and/or rotor shaft.
2. Higher H_2-seal temperatures may also be seen and operators should consider reducing load to see if the axial rotor position change from the load change, affects the temperature of the H_2 seal. If it does not make a difference, then it may be time to consider taking the machine offline for investigation of the alarms.
3. If there is no shaft-ground alarm device and no temperature indication, the operators may not notice anything until higher rotor lateral vibration occurs and eventually a failure if nothing is done.

TEST AND VISUAL INDICATIONS:

Megger test of the H_2-seal insulation

REFERENCES

1. G. Klempner and I. Kerszenbaum. *Handbook of Large Turbo-Generator Operation and Maintenance*. Piscataway, NJ: IEEE Press, 2008, chapters 9 and 11.
2. EPRI Product ID 3002001758. Turbine-Generator Topics for Plant Engineers: Residual Magnetism, August 2013.

3. P. Nippes, D. David, and A. Peniazev. Monitoring of Shaft Voltages and Grounding Current Brushes. In: *EPRI Motor and Generator Predictive Maintenance & Refurbishment Conference*, Orlando, FL, November 1995.
4. M.J. Costello. Shaft Voltages in Rotating Machinery. *IEEE Transactions on Industry Applications*, 29(2), pp. 419–426, March–April 1993.
5. J.S. Sohre. Are Magnetic Currents Destroying your Machinery? In: *Hydrocarbon Processing*, April 1979, pp. 207–212.

M-R-11: FAILURE OF HYDROGEN SEALS

APPLICABILITY:

All hydrogen-cooled generators

MALFUNCTION DESCRIPTION:

This malfunction addresses the failure of a hydrogen-seal on a hydrogen-cooled generator.

Hydrogen seals are the means by which the hydrogen in these generators, at operating pressures much higher than atmospheric pressure, is kept from escaping the casing through the gap between the shaft and end brackets. Seals can be of the radial type or of the axial (thrust collar) type. Both types work with the same principle: a layer of oil at pressure higher than the hydrogen's pressure fills the gap between the seal's babbitt and the shaft running surface, keeping the hydrogen from escaping.

There are a number of causes for seal failure (see below under *Probable Causes*). Seal failure can have dramatic negative consequences to the integrity of the generator, as well as to the safety of the people working in the vicinity of the machine.

Carbon seals are more forgiven to oil deficiency and/or oil temperature. In any case, they are also susceptible to damage and loss of hydrogen containment, under certain circumstances.

PROBABLE CAUSES:

- Inadequate cooling of the seal oil (high *inlet-oil* temperature)
- Insufficient flow of seal oil
- Failure of the seal insulation
- Failure of the shaft-grounding circuit
- Shaft-seal rub:
  - Wrong seal dimensions
  - Excessive rotor/frame vibrations
  - Seal misalignment

POSSIBLE CONSEQUENCES:

- Seal failure:
  - Hydrogen escaping the casing:
    - Explosion in brushgear house when attached to the EE of the machine
    - Hydrogen fire
    - Partial loss of cooling by reduced hydrogen pressure
- Shaft damage
- Seal-oil contamination
- Oil ingression inside the casing

POSSIBLE SYMPTOMS:

- DEGRADED HYDROGEN PURITY **(S-H-02)**
- ELEVATED HYDROGEN DEW POINT **(S-H-03)**
- HYDROGEN MAKE-UP RATE ABOVE NORMAL **(S-H-05)**
- HYDROGEN PRESSURE INSIDE GENERATOR BELOW NORMAL **(S-H-06)**
- RADIO FREQUENCY (RF) MONITORS HIGH **(S-S-13)**
- TEMPERATURE OF HYDROGEN SEAL ABOVE NORMAL **(S-SOS-08)**
- TEMPERATURE OF SEAL OIL ABOVE NORMAL **(S-SOS-10)**
- VIBRATION OF ROTOR—HIGH **(S-R-10)**

OPERATORS MAY CONSIDER:

1. A seal may fail catastrophically, or may degrade with time. Depending on the situation, remedial actions may be more or less urgent in nature and drastic in scope.
2. The symptoms stated under *Possible Symptoms* above provided hints about the seriousness and scope of the problem.
3. A proven hydrogen-seal failure should be acted upon immediately by shutting down the unit and repairing it.
4. Special precautions must be taken, such as caution in approaching confined spaces (brushgear housing and other closed areas). For safety from fire and explosive mixtures of H_2 and air, the generator should be purged with CO_2 before working in the area of the damaged seal.
5. The unit may not be able to maintain hydrogen from escaping in turning-gear or even standing still, so may have to be degassed immediately and purged with CO_2

TEST AND VISUAL INDICATIONS:

Damage to seals and/or running surfaces

REFERENCE

1. G. Klempner and I. Kerszenbaum. *Handbook of Large Turbo-Generator Operation and Maintenance.* Piscataway, NJ: IEEE Press, 2008, chapters 2 and 9.

M-R-12: FAILURE OF OIL-WIPER

APPLICABILITY:

Hydrogen-cooled generators

MALFUNCTION DESCRIPTION:

This malfunction addresses the failure of oil wipers.

Oil wipers are used to inhibit H_2 seal and bearing oil from entering into the generator on the hydrogen side of the seals and bearings, and blowing out of the H_2 seal and bearing mounts on the air side of the hydrogen seals and bearings (Figure 5.90).

PROBABLE CAUSES:

- RUB OF OIL WIPER (M-R-22)
- High rotor vibration
- Wiper misalignment
- Shaft misalignment
- Wrong clearances
- Improper wiper installation

POSSIBLE CONSEQUENCES:

- FAILURE OF THE HIDROGEN SEALS (M-R-11)
- INGRESSION OF OIL INTO GENERATOR (M-R-16)
- VIBRATION OF ROTOR—HIGH (M-R-30)
- Damage to the running surface on the shaft

FIGURE 5.90
Overheated rotor winding joint.

POSSIBLE SYMPTOMS:

- LIQUID-IN-GENERATOR ALARM **(S-C&F-04)**
- VIBRATION OF ROTOR—HIGH **(S-R-10)** (Vibrations may be irregular in nature)

OPERATORS MAY CONSIDER:

Once a wiper failure is recognized as such, operators should consider taking the unit down for repairs. Running with a failed wiper can cause significant distress to the generator, with some long-term degradation (from oil ingression).

TEST AND VISUAL INDICATIONS:

Visual inspection of wiper and shaft running surfaces

REFERENCE

1. G. Klempner and I. Kerszenbaum. *Handbook of Large Turbo-Generator Operation and Maintenance*. Piscataway, NJ: IEEE Press, 2008, chapter 7.

M-R-13: GROUNDED FIELD WINDING

APPLICABILITY:

All generators

MALFUNCTION DESCRIPTION:

The rotor field windings (*field windings*) of turbine-driven generators are designed to be insulated from the forging, which is at ground potential. This malfunction addresses the case when the insulation fails in one or more points, effectively *grounding* the field winding.

Of all the bad things that can happen as result of a single ground of the field winding, most probably the worst is the development of a second rotor-winding ground inside or outside the generator, causing large DC currents to flow through sections of the forging that can result in melting of sections of the forging and the winding and result in a catastrophic failure of the machine.

PROBABLE CAUSES:

- BURNING/PYROLYSIS OF ROTOR INSULATION **(M-R-04)**
- DEGRADED/DAMAGED FIELD WINDING INSULATION **(M-R-07)**
- SHORTED TURNS IN FIELD WINDING **(M-R-24)**
- Foreign object

POSSIBLE CONSEQUENCES:

- ARCING ACROSS BEARINGS AND HYDROGEN SEALS **(M-R-01)**

- BURNING/PYROLYSIS OF ROTOR INSULATION **(M-R-04)**
- DYNAMIC ROTOR UNBALANCE (MECHANICAL) **(M-R-08)**
- LOSS OF STABILITY **(M-R-17)**
- THERMAL ASYMMETRY OF ROTOR **(M-R-28)**
- VIBRATION OF ROTOR—HIGH **(M-R-30)**
- Large swings of VAR output
- Double field ground

POSSIBLE SYMPTOMS:

- GROUNDED FIELD WINDING—ALARM AND/OR TRIP **(S-R-01)**
- Temperature change of field winding (may be up or down and/or erratic in nature)
- Large swings of VARs (mainly for double grounds)
- Erratic vibration changes (mainly for double grounds)
- *Loss of excitation* alarm and/or trip (mainly for double grounds)

OPERATORS MAY CONSIDER:

1. It has been the practice in some stations to run the unit with a single field ground for long periods of time. However, there are industry surveys indicating that following that practice may carry a significant risk to developing a major failure due to a second ground.
2. A risk-minimizing policy will call for a unit trip or quickly shutdown once a ground is identified, if not removed during prompt troubleshooting.
3. For the period when the unit is running with a single ground, vibration, rotor temperatures, and MVAR output should be monitored closely. Any unexplained or sudden change in any of these parameters may indicate that the situation is unstable and consideration should be given to shutting down the unit.
4. The circuits in Figure 5.91 can be used (for units with external excitation systems), for the purpose of identifying the location of the ground as being inside or outside the rotor. If the ground is identified as being on the main generator's rotor, the References at the end of this page lead to methods for finding their approximate location along the rotor.

By inspecting both cases described below, inference can be made as to the location of the ground, that is, on the main rotor winding, or outside (e.g., brushgear, exciter, busses, cables).

- The amplitude and polarity of the measured DC voltages are required.
- All instrumentation circuits should be separated from the main field winding before the voltage measurements are performed, and reconnected immediately after the measurements are taken.

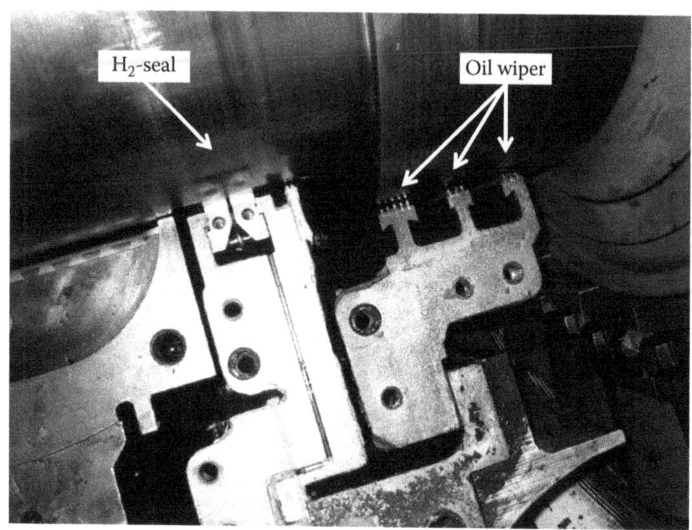

FIGURE 5.91
Example of an oil wiper arrangement for the H_2 seals and bearing.

TEST AND VISUAL INDICATIONS:

There are a number of electrical tests available for finding ground faults on field windings. Unfortunately, some grounds tend to *disappear* once the unit is standing still (no centrifugal forces exist on the winding). Therefore, some test can be done for uncovering grounds with the rotor spinning without excitation and all breakers open.

REFERENCES

1. G. Klempner and I. Kerszenbaum. *Handbook of Large Turbo-Generator Operation and Maintenance*. Piscataway, NJ: IEEE Press, 2008, chapter 6.
2. *EPRI Power Plant Reference Series*—Generators.

M-R-14: GROUNDED SHAFT (MOMENTARILY)

APPLICABILITY:

All generators

MALFUNCTION DESCRIPTION:

This malfunction addresses the situation when the shaft of the generator rotor is momentarily grounded (during operation), through a point(s) other than the grounding brushes.

Grounding of the shaft through any other point(s) than the grounding brushes defeats the purpose of the bearing insulation and grounding brushes for controlling the flow of shaft currents so that no harm comes to the rotor, in particular rotor bearings and hydrogen seals.

PROBABLE CAUSES:

- INEFFECTIVE SHAFT GROUNDING **(M-R-15)**
- RUB OF OIL WIPER **(M-R-22)**
- RUB OF HYDROGEN SEAL **(M-R-21)**
- RUB BETWEEN BEARING AND JOURNAL **(M-R-20)**

POSSIBLE CONSEQUENCES:

- ARCING ACROSS BEARINGS AND HYDROGEN SEALS **(M-R-01)**
- FAILURE OF HYDROGEN SEALS **(M-R-11)**
- FAILURE OF OIL WIPER **(M-R-12)**
- Bearing failure due to electric pitting

POSSIBLE SYMPTOMS:

- Indirect symptoms from the rubs, such as vibration
  - SHAFT CURRENT ABOVE NORMAL **(S-R-04)**
    - (if the shaft-grounding system manages to pick up this short-term event)
  - SHAFT VOLTAGE BELOW NORMAL **(S-R-06)**
    - (if the shaft-grounding system manages to pick up this short-term event)

OPERATORS MAY CONSIDER:

1. A momentarily grounded rotor is not a usually recognizable event. However, a rotor grounded in any of the friction surfaces may give rise to arcing across seals, wipers, and bearings. A well-grounded shaft (through grounding brushes) will minimize the possibility of large arcing currents flowing via the momentarily grounded bearing, seal or wiper. Therefore, if a grounding circuit is deficient or not operable:
   a. Consider troubleshooting the circuit as soon as possible, to avoid damage to seals/bearings/wipers due to electropitting.

TEST AND VISUAL INDICATIONS:

Visual inspection for arcing marks

REFERENCE

1. G. Klempner and I. Kerszenbaum. *Handbook of Large Turbo-Generator Operation and Maintenance*. Piscataway, NJ: IEEE Press, 2008, chapter 5.

M-R-15: INEFFECTIVE SHAFT GROUNDING

APPLICABILITY:

All generators

MALFUNCTION DESCRIPTION:

This malfunction addresses the situation where poor contact or none at all exists between ground and the generator shaft. This condition will not allow effective draining of shaft currents and also can cause the shaft voltage to rise.

There are a number of sources for the generation of voltage potentials in the shaft of large turbogenerators. For instance, these can be due to unbalanced capacitive coupling from the excitation system, capacitive coupling from the stator, electrostatic voltage from the turbine due to charged water droplets impacting the blades, voltage induced from asymmetrical stator core stacking, homopolar voltage from shaft magnetization, and/or large stator core faults.

It is not uncommon for shaft voltages in large generators to reach values up to 150 volts. These voltages can break down the insulation of the thin layer of oil between the bearing and rotor journal, hydrogen seal running surfaces, and the bearing wipers and shaft, resulting in the flow of *shaft currents*. Shaft currents can damage the journals, seals and bearing babbitt by an electropitting process. Bearing and seal insulation, as well as shaft-grounding devices are utilized to avoid or minimize the flow of shaft currents.

Figures 5.92 through 5.95 show a few of the various shaft-grounding methods used.

PROBABLE CAUSES:

- Worn or broken ground brushes
- Loose ground connection
- Loss of ground connection

POSSIBLE CONSEQUENCES:

- Electric arcing across the oil film in bearings, wipers, and seals cause surface damage, all the way to full failure of the seals, wipers, and bearings.
- ARCING ACROSS BEARING AND HYDROGEN SEALS (M-R-01).

POSSIBLE SYMPTOMS:

- RADIO FREQUENCY (RF) MONITOR HIGH **(S-S-13)**
- SHAFT CURRENT ABOVE NORMAL **(S-R-04)**
- SHAFT VOLTAGE ABOVE NORMAL **(S-R-05)**
- TEMPERATURE LUBE OIL AT BEARING OUTLET—HIGH **(S-R-06)**

- TEMPERATURE OF HYDROGEN SEAL ABOVE NORMAL **(S-SOS-08)**
- TEMPERATURE OF SEAL OIL ABOVE NORMAL **(S-SOS-10)**
- TEMPERTURE OF BEARING ABOVE NORMAL **(S-R-08)**
- VIBRATION OF ROTOR—HIGH **(S-R-10)**

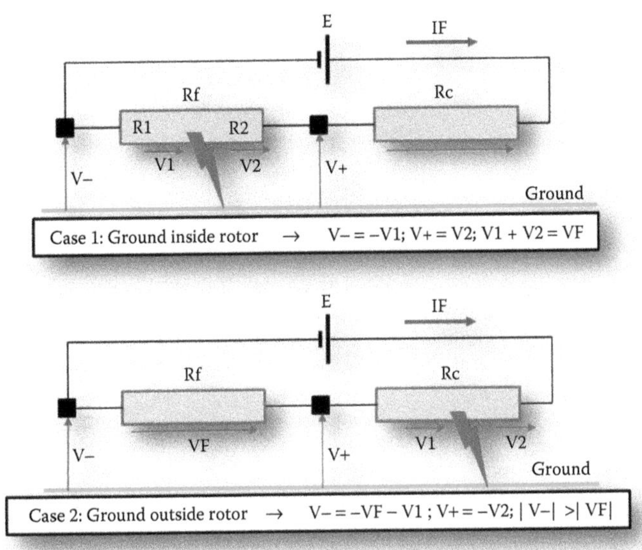

FIGURE 5.92
Circuits depicting a short-circuit to ground, either inside or outside the rotor.

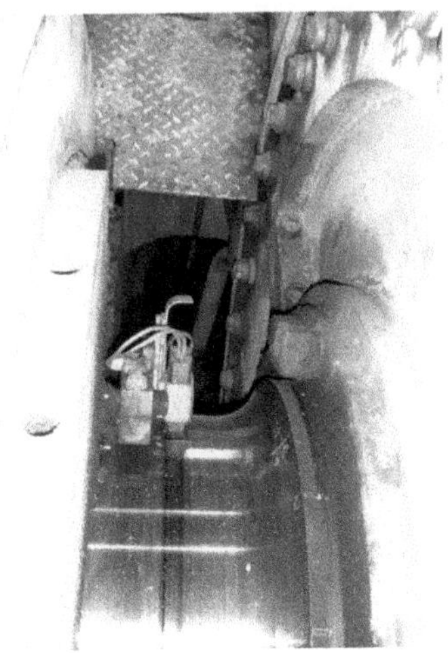

FIGURE 5.93
Shaft grounding with carbon brushes.

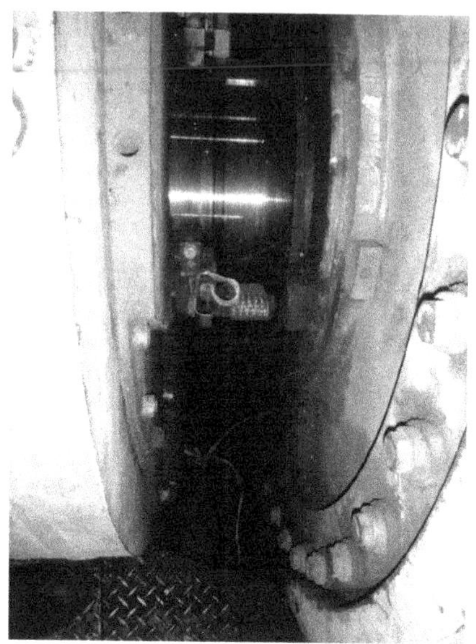

FIGURE 5.94
A shaft-grounding braided strip copper alloy.

FIGURE 5.95
A twin braided ribbon shaft-grounding method. The contact is due to gravity pulling down on the strips against the shaft.

OPERATORS MAY CONSIDER:

1. If a deficient or nonexisting grounding of the main shaft is detected, consider promptly correcting the situation by troubleshooting the circuit, and/or replacing worn brushes with new ones.
2. Extreme caution should be used when working in this area due to voltage rise on the rotor shaft.

Test and Visual Indicators:

- Continuity tests for checking ground circuit integrity
- Visual inspection of brushes and brush holders
- Inspection of running surfaces on shaft

REFERENCES

1. G. Klempner and I. Kerszenbaum. *Handbook of Large Turbo-Generator Operation and Maintenance*. Piscataway, NJ: IEEE Press, 2008, chapter 5.
2. P. Nippes, D. David, and A. Peniazev. Monitoring of Shaft Voltages and Grounding Current Brushes. In: *EPRI Motor and Generator Predictive Maintenance & Refurbishment Conference*, Orlando, FL, November 1995.
3. P. Nippes and J. Sohre. Electromagnetic Shaft Currents and Demagnetization of Rotors of Turbines and Compressors. In: *The 7th Turbomachinery Symposium*, Texas A&M University, December 1978.
4. P.I. Nippes. Magnetism and Stray Currents in Rotating Machinery. *Journal of Engineering for Gas Turbines and Power, Transactions of ASME*, 118, 225–228, January 1996.

M-R-16: INGRESSION OF OIL INTO GENERATOR

APPLICABILITY:

All generators

MALFUNCTION DESCRIPTION:

This malfunction addresses the ingression of oil into the generator. Oil can come from bearings in air- and hydrogen-cooled generators, and from hydrogen seals in hydrogen-cooled machines.

In general, it is not desirable to have oil leaking into the generator. Obviously, the magnitude of the problem depends on how much oil found its way into the machine.

PROBABLE CAUSES:

- Excessive lube-oil pressure
- Excessive H_2-seal-oil pressure
- Faulty oil wipers
- Faulty hydrogen seals
- Excessive vibrations

POSSIBLE CONSEQUENCES:

- Consequences are directly tied to the amount of oil ingression.
- Consequences tend to be long-term, unless very large quantities enter the generator, to the extent that the free rotation of the rotor is impacted (a very low probability occurrence). In this case it could cause excessive vibrations and other problems.
- A long-term issue is the weakening of the stator winding support systems.
- Contamination of the stator and rotor windings.
- In air-cooled machines, mixtures of oil and carbon dust from the brushgear can cause partial blocking of the stator cooling paths.

POSSIBLE SYMPTOMS:

- Loss of oil from lube-oil system
- Loss of oil from seal-oil system (in many cases, the lube oil and seal oil are part of the same auxiliary system)
- LIQUID-IN-GENERATOR ALARM **(S-C&F-04)**
- PRESSURE OF LUBE OIL BELOW NORMAL **(S-LOS-05)**
- TEMPERATURE OF SEAL OIL ABOVE NORMAL **(S-SOS-10)**

OPERATORS MAY CONSIDER:

1. Small quantities of oil inside the generator do not pose a significant risk. Many machines in the industry experience this type of problem, and it does not affect their reliability. In case of oil ingression in small quantities, it is prudent to keep track of the situation, by draining the oil periodically online.
2. If the quantities of oil are large or increasing, the most prudent action would be to take the unit offline for troubleshooting as soon as practically possible.
3. In all cases, bear in mind that operating a unit continuously with oil leaking into the casing may result in long-term issues that can reduce the expected life of the winding support systems, and perhaps, the winding itself.

TEST AND VISUAL INDICATORS:

- Oil in casing
- Damaged H_2 seal and/or wipers
- Oil contamination of windings and other internal components of the generator

REFERENCES

1. G. Klempner and I. Kerszenbaum. *Handbook of Large Turbo-Generator Operation and Maintenance.* Piscataway, NJ: IEEE Press, 2008, chapter 4.
2. A.J. Gonzalez et al. *Monitoring and Diagnosis of Turbine-Driven Generators.* Englewood Cliffs, NJ: Prentice Hall.

M-R-17: LOSS OF STABILITY

APPLICABILITY:

All generators

MALFUNCTION DESCRIPTION:

This malfunction addresses the situation where the magnetic coupling between the rotor and stator is lost. The reason is that the torque on the shaft of the machine becomes larger than the torque that the machine can produce in the airgap, under the present rotor excitation.

Loss of stability—also called *slipping poles*, is commonly due to operating the machine at leading power factors with a weakened rotor flux. It can also happen if the excitation is partially or totally lost due to large *field winding* problems, excitation problems, or if a major external event drastically changes grid conditions (such as a loss of one or more transmission lines connected to the power station).

The effect on the machine, in particular the stator endwindings, is similar to that of synchronizing the unit out of phase.

PROBABLE CAUSES:

- Major short-circuit in the field winding (for instance, a full-shorted coil)
- Machine fully loaded subject to a sudden major grid event (such as losing one or more transmission lines connected to the power station)
- Malfunction of the excitation system
- Wrong setting of the UEL (underexcitation limiter), allowing the field current to go too low when operating in the leading PF region

POSSIBLE CONSEQUENCES:

- Very large momentary vibrations
- Major stress of critical components:
  - Stator endwindings
  - Stator endwinding support system
  - Frame attachment to foundations
  - Unit alignment
  - Possible damage to the long blades of low pressure turbines, in particular those driving large four-pole generators
  - DYNAMIC ROTOR UNBALANCE (MECHANICAL) **(M-R-08)**
  - RUB BETWEEN BEARING AND JOURNAL **(M-R-20)**
  - RUB OF HYDROGEN SEAL **(M-R-21)**
  - RUB OF OIL WIPER **(M-R-22)**
  - VIBRATION OF ROTOR—HIGH **(M-R-30)**

POSSIBLE SYMPTOMS:

- Generators may trip by *loss-of-field* relay
- Generators may trip by *out-of-step* relay
- High and sudden rotor vibrations
- MW and MVAR swings

OPERATORS MAY CONSIDER:

1. Slipping poles is a serious operational event, not too different from synchronizing out of phase. And, the same way that the severity of the out-of-phase synchronization depends on the closing angle, among other things, the severity of a loss-of-stability event depends on the load conditions at the time of the event, the length of the event (how many poles slipped), did the unit recover of kept slipping poles till it tripped, and so on.
2. Operators can use ***DFR***[5] traces to analyze the severity of the event. In any case, it would be prudent to be conservative about possible damage to the endwindings, keybar bolts, and so on. Thus, a partial visual inspection and a number of tests may be desirable before reconnecting to the grid.
3. After the event and with the unit back on line, monitor vibration, temperatures, and hydrogen leaks in hydrogen-cooled generators, in particular hydrogen-in-SCW in machines with directly water-cooled stators. If a *rotor (airgap) flux monitor* (RFM[6]) is installed, check for shorted turns and compare to pre-event values.

TEST AND VISUAL INDICATIONS:

Visual inspections should focus mainly on endwinding support systems and coils, as well as frame alignment, bearings, seals, and the integrity of the couplings.

REFERENCE

1. G. Klempner and I. Kerszenbaum. *Handbook of Large Turbo-Generator Operation and Maintenance*. Piscataway, NJ: IEEE Press, 2008, chapter 4.

[5] DFR—Digital fault recorder.
[6] RFM—Rotor flux monitor, also known as airgap flux monitor.

M-R-18: MECHANICAL FAILURE OF ROTOR COMPONENT

APPLICABILITY:

All generators

MALFUNCTION DESCRIPTION:

This malfunction addresses the situation where a rotor component is experiencing (has experienced) a mechanical failure. Mechanical failures for the purpose of this definition are significant cracks or broken structural components. More minor indications, (small cracks, surface pitting, imperfections) do not qualify as a *mechanical failure*.

Typical examples of mechanical failures are cracks on shafts, wedges, teeth, retaining rings, zone rings, centering rings, fan hubs, fan blades, sliprings; broken bolts, wedges, and so on. An extreme example would be a fan blade or a radial stud being thrown (see Figures 5.96 through 5.98).

Typically, a mechanical failure can cause vibrations and/or major failure (sometimes catastrophic) of the generator.

PROBABLE CAUSES:

Any cracked or broken structural component

POSSIBLE CONSEQUENCES:

- VIBRATION OF ROTOR—HIGH (M-R-30)
- Major failure up to catastrophic one (e.g., failure of a retaining ring or rotor wedge)

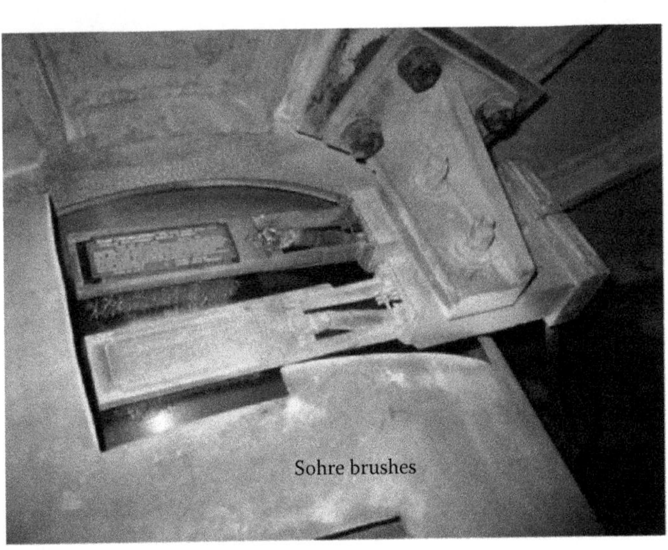

FIGURE 5.96
Twin shaft-grounding brushes arrangement.

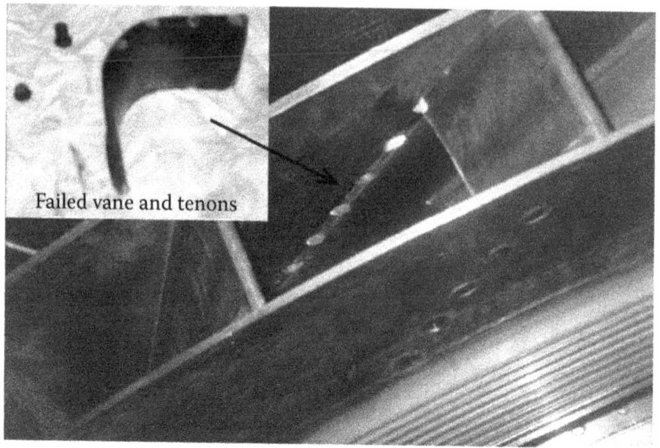

FIGURE 5.97
Failed centrifugal fan vane and the resulting missile.

FIGURE 5.98
A typical radial stud and main lead connection in good condition.

POSSIBLE SYMPTOMS:

- VIBRATION OF ROTOR—HIGH **(S-R-10)**
- Trip due to a major failure

OPERATORS MAY CONSIDER:

1. Failure of a structural rotor component can happen suddenly, or can happen slowly by fatigue. Only analysis of the vibrations and their trending may identify the nature of the fault, plus fracture mechanics investigation, and postfailure.

2. Identification of what specific component is failing may be impossible without disassembly and visual inspection. A cracked shaft will lend itself to identification as the source of the vibrations, if the right test is done (correlation of vibration vs. watts loading of the generator).
3. Mechanical failures have all the potential to leading to major loss. Thus, they ought to be dealt diligently and promptly.
4. A THERMAL SENSITIVITY TEST **(S-R-29)** is a good tool for discriminating mechanical from electrical sources of vibration. However, to perform this test, the unit must be able to operate at load. Therefore, if a developing situation arises where the vibrations are showing an upward trend, operators may consider doing the test sooner than later.

TEST AND VISUAL INDICATIONS:

Visual inspection and NDE of rotor components

REFERENCE

1. G. Klempner and I. Kerszenbaum. *Handbook of Large Turbo-Generator Operation and Maintenance*. Piscataway, NJ: IEEE Press, 2008, chapters 4 and 9.

M-R-19: MISALIGNMENT OF HYDROGEN SEAL

APPLICABILITY:

Hydrogen-cooled generators

MALFUNCTION DESCRIPTION:

This malfunction addresses the situation where a hydrogen seal is misaligned, that is, it is not sitting perpendicular to the shaft. Misalignment is usually due to installation and re-assembly errors or an operational event.

PROBABLE CAUSES:

- Installation and/or reassembly issues
- Operation mishap (e.g., large grid disturbance)

POSSIBLE CONSEQUENCES:

- FAILURE OF HYDROGEN SEALS **(M-R-11)**
- HYDROGEN ESCAPING THE GENERATOR **(M-H-08)**
- INGRESSION OF OIL INTO GENERATOR **(M-R-16)**
- RUB OF HYDROGEN SEAL **(M-R-21)**
- TEMPERATURE OF HYDROGEN SEAL ABOVE NORMAL **(M-R-27)**
- VIBRATION OF ROTOR—HIGH **(M-R-30)**

POSSIBLE SYMPTOMS:

- HYDROGEN MAKE-UP RATE ABOVE NORMAL **(S-H-05)**
- HYDROGEN PRESSURE INSIDE THE GENERATOR BELOW NORMAL **(S-H-06)**
- LIQUID-IN-GENERATOR ALARM **(S-C&F-04)**
- TEMPERATURE OF HYDROGEN SEAL ABOVE NORMAL **(S-SOL-08)**
- TEMPERATURE OF SEAL OIL ABOVE NORMAL **(S-SOL-10)**
- VIBRATION OF ROTOR—HIGH **(S-R-10)**

OPERATORS MAY CONSIDER:

1. Depending on the seriousness of the symptoms, operator action may require immediate shutdown and troubleshooting, or monitoring the *seal-oil temperatures, hydrogen make-up rate*, and *rotor vibration* until a scheduled repair.
2. Seal misalignment has the potential to cause significant material losses and also serious impact on safety of those working in close vicinity to the generator, thus it must be given all the seriousness it deserves.

TEST AND VISUAL INDICATIONS:

- Measurement of hydrogen leakage from the generator (generally by the H_2 make-up rate)
- Damage to seal and shaft running surfaces
- Seal oil in generator

REFERENCES

1. G. Klempner and I. Kerszenbaum. *Handbook of Large Turbo-Generator Operation and Maintenance*. Piscataway, NJ: IEEE Press, 2008, chapters 2, 5, and 9.
2. IEEE Std 1129–2014. IEEE Guide for Online Monitoring of Large Synchronous Generators (10 MVA and Above).

M-R-20: RUB BETWEEN BEARING AND JOURNAL

APPLICABILITY:

All generators

MALFUNCTION DESCRIPTION:

This malfunction addresses the situation where a *rub* occurred between a bearing (babbitt) and its journal (Figure 5.99). The rub happens when the movement of the journal against the bearing squeezes all the oil from in between them, allowing the metals to contact each other directly while the rotor is rotating, resulting in some damage (most often to the

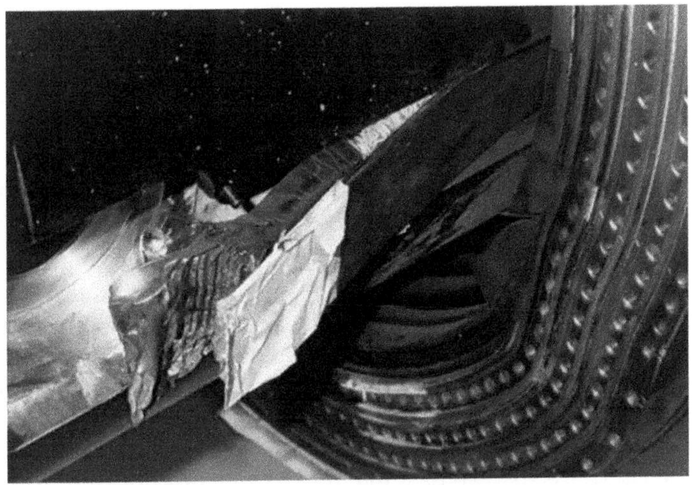

FIGURE 5.99
The result of a mechanical failure of a radial stud and the connection being ripped open.

bearing's babbitt, which is made of softer material than the journal). A major rub, where significant babbitt material is displaced, is called a *wipe*, and the result is called a *wiped bearing*.

PROBABLE CAUSES:

- ARCING BETWEEN BEARINGS AND HYDROGEN SEALS (M-R-01)
- BEARING MISALIGNMENT **(M-R-02)**
- High inlet lube-oil temperature reducing oil viscosity

Insufficient flow of lube oil:

- BLOCKAGE OF LUBE-OIL FLOW (M-LOS-01)
- Large grid disturbance
- VIBRATION OF ROTOR—HIGH (M-R-30)

Out-of-step synchronization or:

- LOSS OF STABILITY (M-R-17)

POSSIBLE CONSEQUENCES:

- VIBRATION OF ROTOR—HIGH (M-R-30)
- Damage to journal
- Damage to bearing housing
- Shaft current-induced pitting of journal and/or babbitt
- Hydrogen escaping the generator
- Extensive bearing failure leading to *seizing* and unit trip with damage to generator rotor, turbines, couplings, and so on

POSSIBLE SYMPTOMS:

- TEMPERATURE OF LUBE OIL AT BEARING OUTLET—HIGH **(S-LOS-06)**
- TEMPERATURE OF BEARING ABOVE NORMAL **(S-R-08)**
- VIBRATION OF ROTOR—HIGH **(S-R-10)**
- Sudden hydrogen leaks. There are several *malfunctions* reflecting this case.

OPERATORS MAY CONSIDER:

1. A rub/wipe is a significant occurrence. Depending on the magnitude of the event, it could be managed by bringing the unit offline for repairs in a scheduled manner, or it may require immediate attention.
2. In dealing with the situation, operators must keep in mind that the cause for the rub or wipe may still be present, and lead to a quick bearing failure with large financial consequences. Thus, this situation should be treated with urgency.
3. If the unit is not brought offline immediately, temperatures and lube-oil flow should be monitored continuously, as well as vibrations and hydrogen pressure as well as the rate of hydrogen leak from the casing.

TEST AND VISUAL INDICATIONS:

- Visual inspection of bearing and journals
- Oil chemical analysis (for metal contents)

REFERENCES

1. G. Klempner and I. Kerszenbaum. *Handbook of Large Turbo-Generator Operation and Maintenance*. Piscataway, NJ: IEEE Press, 2008, chapters 2 and 11.
2. EPRI Technical Report 1026557. Bearing Failures in Power Plant Rotating Equipment, October 2012.
3. EPRI Technical Report 1026566. Field Guide: Bearing Damage Mechanisms, November 2012.
4. EPRI Technical Report CS-4555. Guidelines for Maintaining Steam Turbine Lubrication, July 1986.
5. EPRI Technical Report 1026605. Generator Maintenance Applications Center: Combined-Cycle Combustion Turbine Lube Oil System Maintenance Guide, October 2012.

M-R-21: RUB OF HYDROGEN SEAL

APPLICABILITY:

Hydrogen-cooled generators

MALFUNCTION DESCRIPTION:

This malfunction is about the situation where the hydrogen seals contact the shaft running surface, in a way that damage occurs to either surfaces, or both.

A rub may be a momentary one-time event, or may be repetitive in nature. A small one-time rub may not cause a long-term problem. A continuous or repetitive rub will result in damage that may require prompt repairs.

Figure 5.100 shows an example of radial hydrogen-seal arrangement. Figure 5.101 shows the result of a seal-to-shaft rub.

PROBABLE CAUSES:

- MISALIGNMENT OF HYDROGEN SEAL **(M-R-19)**
- VIBRATION OF ROTOR—HIGH **(M-R-30)**
- Wrong seal clearances
- Temperature of seal oil too high
- Seal oil pressure too high
- Wrong seal installation

POSSIBLE CONSEQUENCES:

- FAILURE OF HYDROGEN SEALS **(M-R-11)**
- VIBRATION OF ROTOR—HIGH **(M-R-30)**
- Increase in shaft currents
- Contamination with metal particles of seal oil

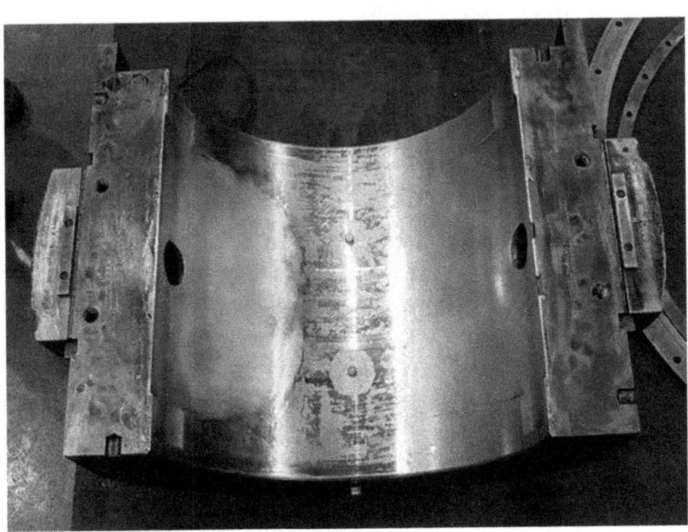

FIGURE 5.100
Bearing rub.

FIGURE 5.101
Typical hydrogen seal arrangement.

POSSIBLE SYMPTOMS:

- TEMPERATURE OF HYDROGEN SEAL ABOVE NORMAL **(S-SOS-08)**
- VIBRATION OF ROTOR—HIGH **(S-R-10)**

OPERATORS MAY CONSIDER:

1. Sometimes a single rub can go unnoticed. However, when the seal temperatures increase significantly, and the vibration levels also go up, a seal rub could demand action from the operator to solve the situation, given the possibility of a failure of the hydrogen seal.
2. Analysis of the vibration signature may identify the cause as a seal rub. Thus, measuring vibrations over a wide range of frequencies is advantageous in troubleshooting a rub.
3. A seal rub can be dependent on load, that is, reducing load may eliminate the rub and provide breathing space to the operator to plan for the best approach in solving the problem.
4. A THERMAL SENSITIVITY TEST (M-R-29) (TST) is a good way in narrowing the search for the causes of elevated machine vibrations.

TEST AND VISUAL INDICATIONS:

- Measuring clearances
- Measuring misalignment
- Damaged seal and/or shaft running surfaces

REFERENCE

1. G. Klempner and I. Kerszenbaum. *Handbook of Large Turbo-Generator Operation and Maintenance*. Piscataway, NJ: IEEE Press, 2008, chapters 2, 3, 5, and 9.

M-R-22: RUB OF OIL WIPER

APPLICABILITY:

Hydrogen-cooled generators

MALFUNCTION DESCRIPTION:

This malfunction addresses the situation where a seal-oil wiper touches the shaft momentarily during operation, causing what it is called: *a wiper rub*.

A rub may be a momentary one-time event, or may be repetitive in nature. A small one-time rub may not cause a long-term problem. A continuous or repetitive rub will—most probably—result in damage that requires repairing the unit. Figure 5.100 (in the previous malfunction description) shows a typical radial oil-wiper arrangement, while Figure 5.102 shows a damaged wiper.

PROBABLE CAUSES:

- VIBRATION OF ROTOR—HIGH (**M-R-30**)
- Wrong wiper clearances
- Wrong wiper installation
- Misalignment of the wiper
- Seal-oil temperature and/or pressure too high
- Major grid disturbance

POSSIBLE CONSEQUENCES:

- FAILURE OF OIL-WIPER (**M-R-12**)
- INGRESSION OF OIL INTO GENERATOR (**M-R-16**)
- VIBRATION OF ROTOR—HIGH (**M-R-30**)

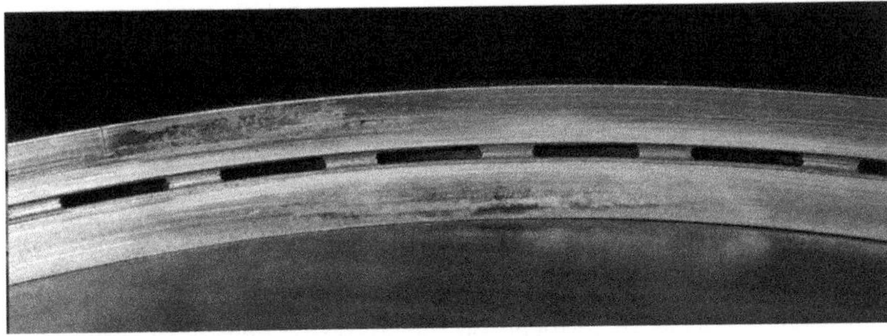

FIGURE 5.102
A minor hydrogen-seal rub.

- Seal-oil contamination with metal particles
- Electric arcing between wipers and shaft:
- ARCING ACROSS BEARINGS AND HYDROGEN SEALS (M-R-01)

POSSIBLE SYMPTOMS:

- TEMPERATURE OF BEARING ABOVE NORMAL **(S-R-08)**
- TEMPERATURE OF LUBE OIL AT BEARING OUTLET—HIGH **(S-LOS-06)**
- VIBRATION OF ROTOR—HIGH **(S-R-10)**

Increased *RF activity* from arcing between wipers and shaft:

- RADIO FREQUENCY (RF) MONITOR HIGH **(S-S-13)**

OPERATORS MAY CONSIDER:

1. Rubs of seals and/or wipers tend to manifest themselves as high vibrations of the rotor.
2. Vibrations may be erratic in nature.
3. Vibrations may be load dependent.
4. Vibration caused by rubs are in general of frequencies higher than *running* frequency.
5. Depending on the severity of the rub (and resulting vibrations), the unit may have to be shut down in short notice, or may be operated until such a time a scheduled repair can be made.
6. Monitor carefully any change in vibration and seal-oil temperature, as well as *liquid-in-generator* alarms.

TEST AND VISUAL INDICATIONS:

- Measuring clearances
- Measuring misalignment
- Damaged wiper and/or shaft running surfaces

REFERENCE

1. G. Klempner and I. Kerszenbaum. *Handbook of Large Turbo-Generator Operation and Maintenance*. Piscataway, NJ: IEEE Press, 2008, chapter 11.

M-R-23: SHAFT VOLTAGE—HIGH

APPLICABILITY:

All generators

MALFUNCTION DESCRIPTION:

This malfunction relates to the situation when unusually high voltages are impressed upon, or induced on the shaft of a generator rotor.

There are a number of possible sources for the generation of voltage potentials in the shaft of large turbogenerators. For instance, due to unbalanced capacitive coupling from the excitation system, capacitive coupling from the stator, electrostatic voltage from the turbine due to charged water droplets impacting the blades, voltage induced from asymmetrical stator core stacking, homopolar voltage from shaft magnetization, and large core faults.

It is not uncommon for shaft voltages in large generators to reach values of up to about 150 volts. These voltages can break the insulation of the thin layer of oil between bearing and journal, or between hydrogen seals and wipers and shaft, resulting in the flow of *shaft currents*. Shaft currents can damage the journals, seals, and babbitt by an electropitting process. Bearing and H_2-seal insulation, as well as shaft-grounding devices, are utilized to avoid or minimize the flow of shaft currents.

PROBABLE CAUSES:

- GROUNDED FIELD WINDING **(M-R-13)**
- INEFFECTIVE SHAFT GROUNDING **(M-R-15)**
- INTERLAMINAR INSULATION FAILURE **(M-C&F-13)**
- LOCAL CORE FAULT **(M-C&F-15)**
- SHORTED TURNS IN FIELD WINDING **(M-R-24)**
- Shaft magnetization.
- Other mechanisms described in Section *Malfunction Description*. These mechanisms are not related to any failure mode, and are not given to change by operator intervention. These are inherent to the specific design of the unit, and are handled by the bearing/seal insulation and shaft-grounding devices. Example of those issues: excitation system unbalance; capacitive coupling; unsymmetrical core stacking; steam turbine static charges.

POSSIBLE CONSEQUENCES:

Electric arcing across the oil film in bearings, wipers, and seals cause surface damage, all the way to full failure of the seals, wipers, and bearings:

- ARCING ACROSS BEARINGS AND HYDROGEN SEALS **(M-R-01)**

POSSIBLE SYMPTOMS:

- RADIO FREQUENCY (RF) MONITOR HIGH **(S-S-13)**
- SHAFT CURRENT ABOVE NORMAL **(S-R-04)**
- SHAFT VOLTAGE ABOVE NORMAL **(S-R-06)**

Significant stator core problems (indicated by core local or global temperature rise):

- LOCAL CORE TEMPERATURE ABOVE NORMAL **(S-C&F-05)**
- TEMPERATURE OF CORE ABOVE NORMAL **(S-C&F-06)**

OPERATORS MAY CONSIDER:

1. Shaft voltages are always present in electric machines. As stated above, voltages up to 150 volts are not uncommon. What is important is to monitor the resulting shaft currents. An increase in shaft currents means that the bearing/seal insulation is deficient, and/or the grounding brushes are not clamping the shaft voltages effectively. If any of these occur, the problem should be addressed ASAP. If left as is, bearing and/or seal failure can eventually happen, and depending on the severity of the situation, can happen quickly.
2. A machine that is experiencing a significant increase in shaft voltage amplitude may be undergoing a serious stator core failure. Therefore, this type of voltage increases should be investigated diligently.

TEST AND VISUAL INDICATIONS:

- Bearing/H_2-seal insulation testing
- Visual inspection of babbitt and journal
- EL-CID and/or flux tests of stator core

REFERENCE

1. G. Klempner and I. Kerszenbaum. *Handbook of Large Turbo-Generator Operation and Maintenance.* Piscataway, NJ: IEEE Press, 2008, chapter 5.

M-R-24: SHORTED TURNS IN FIELD WINDING

APPLICABILITY:

All generators

MALFUNCTION DESCRIPTION:

This malfunction addresses the situation where the insulation fails between two or more contiguous turns of the rotor winding. These are called *shorted turns.*

A field (rotor) winding may have one to many shorted turns in one location, or in a number of locations. This includes both the winding slots and the endwindings.

The behavior of a rotor due to having one or more shorted turns is not the same in every case. It depends on the number of rotor poles, number of shorts, location of the shorts, and the nature of the short (i.e., solid short, high-resistance short, intermittent short). Also, shorted turns that are closer to the pole face of a two-pole rotor are more likely to produce higher vibration due to the lack of symmetry in the thermal distribution of temperature in the rotor.

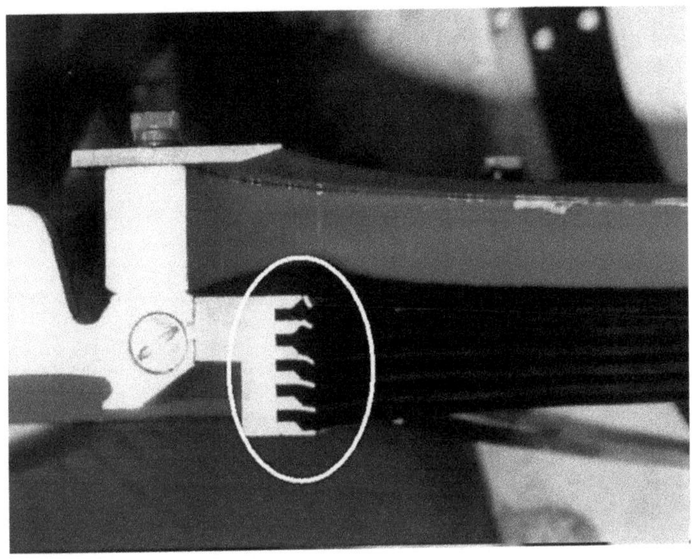

FIGURE 5.103
Oil-wiper rub damage.

The prognosis about how shorted turns in a particular generator will lead to future degradation also depends heavily on all the factors mentioned above, plus the nature of the insulation degradation. For a detailed description of this phenomena, look at the *References* section, below.

Figures 5.103 and 5.104 show two different mechanisms (out of many more) for the formation of shorted turns.

PROBABLE CAUSES:

- *Copper dusting* due to long periods on turning gear, coupled with a coil design susceptible to this type of problem
- Degraded insulation due to
  - Mechanical wear
  - Electrical stress (such as from solid-state excitation and/or high forced excitation events)
  - Overheating of the rotor
  - Excessive vibrations
  - Abnormal coil movement in the slot
  - Abnormal coil clearances
  - Aging
  - Manufacturing defects
- Abnormal grid events

FIGURE 5.104
Shorted turns in the endwinding region due to copper conductor deformation.

POSSIBLE CONSEQUENCES:

Some industry studies give the number of generators operating with shorted turns in their rotor windings to be about 50% of all running units. Most of them operate without any problem. In many cases the operators don't know that their units have one or more shorted turns, unless the machines have ***RFM***s[7] installed and are being monitored. However, there are cases where shorted turns affect the normal operation of the machine to different degrees, and in some cases, shorted turns develop into more serious faults, such as ground faults. Some of the consequences of operating with shorted turns may be:

- Elevated rotor vibration
- Elevated stator and frame vibration in four pole machines
- Rotor vibration that is sensitive to VAR loading
- In extreme cases, inability of the generator to supply significant amount of reactive power at lagging power factors
- Local overheating (and possible burning) of the rotor insulation
- Ground faults of the field winding

POSSIBLE SYMPTOMS:

RFM showing the existence of shorted turns:

- SHORTED TURNS IN FIELD WINDING **(S-R-07)**

[7] RFM—Rotor (airgap) flux monitor.

High rotor (and frame/core) vibration:

- HIGH VIBRATION OF CORE AND/OR FRAME **(S-C&F-03)**
- VIBRATION OF STATOR ENDWINDING HIGH **(S-S-21)**
- VIBRATION OF ROTOR—HIGH **(S-R-10)**
- Loss of capability to deliver lagging reactive power (when many shorted turns are present)
- Field ground can be a direct result from a shorted turn)

OPERATORS MAY CONSIDER:

1. As explained above, many units are known to operate without any problems with one or more shorted turns. In any case, given the potential to developing into larger problems, the situation should be investigated to evaluate the risks, and monitored for trending purposes and long-term risk evaluation.
2. If rotor vibrations are elevated, the unit should be scheduled for shutdown and troubleshooting, based on the severity of the vibration, and how the vibration is trending.
3. Sudden step on rotor/frame vibration magnitudes should be investigated.
4. If troubleshooting is required, and a RFM probe is not installed, there are a number of online and offline tests that can be made to ascertain the existence of shorted turns.
5. If possible, before going offline, performing a THERMAL SENSITIVITY TEST **(M-R-29)** will go a long way in diagnosing the source of elevated vibrations.

TEST AND VISUAL INDICATIONS:

- Offline tests such as the RSO, pole-drop measurements, and others
- Visible damage to insulation

REFERENCE

1. G. Klempner and I. Kerszenbaum. *Handbook of Large Turbo-Generator Operation and Maintenance.* Piscataway, NJ: IEEE Press, 2008, chapters 5, 9, and 11.

M-R-25: TEMPERATURE OF BEARING ABOVE NORMAL

APPLICABILITY:

All generators

MALFUNCTION DESCRIPTION:

This malfunction addresses bearing temperatures above *normal.* By *normal,* it means temperatures that are higher than historically reported for the bearing in question, under similar load conditions.

PROBABLE CAUSES:

There are many factors affecting the operation of sleeve and thrust bearings in large turbine-driven generators, for instance, angular misalignment, linear misalignment, concentricity, journal and/or babbitt deformations, oil viscosity, oil temperature, oil pressure, oil impurities, thermal expansion, and so on. Temperature above normal may be the result of a bearing issue, or due to abnormal supply of oil.

- ARCING ACROSS BEARINGS AND HYDROGEN SEALS **(M-R-01)**
- BEARING MISALIGNMENT **(M-R-02)**
- BLOCKAGE OF LUBE-OIL FLOW **(M-LOS-01)**
- EFFICIENCY OF LUBE-OIL COOLERS DEGRADED **(M-LOS-02)**
- LUBE-OIL IMPURITIES **(M-LOS-10)**
- RUB BETWEEN BEARING AND JOURNAL **(M-R-20)**

POSSIBLE CONSEQUENCES:

- Bearing failure
- Damage to journal
- Damage to other rotor components if the bearing fails in operation

POSSIBLE SYMPTOMS:

- RADIO FREQUENCY (RF) MONITOR HIGH **(S-S-13)**
- TEMPERATURE OF LUBE OIL AT BEARING OUTLET ABOVE NORMAL **(S-LOS-06)**
- VIBRATION OF ROTOR—HIGH **(S-R-10)**
- Impurities in oil
- Arcing signs on babbitt and/or journal
- Wrong operation of lube-oil system
- High lube-oil pressure
- Excessive moisture in the lube oil

OPERATORS MAY CONSIDER:

1. Bearing temperatures above *normal* requires immediate operator attention. If the temperature is below maximum allowable:
   a. Monitor bearing temperatures
   b. Monitor and compare lube-oil inlet and outlet temperatures
   c. Look for lube-oil system problems
   d. Look for shaft current issues
2. If the bearing temperature approaches or breaches the maximum allowable temperature, consider bringing the unit offline for troubleshooting.

TEST AND VISUAL INDICATIONS:

- Alignment measurements
- Damage to the babbitt
- Blockage of oil passages
- Impurities in oil (metallic particles are of great concern)
- Problems with the lube-oil system

REFERENCES

1. G. Klempner and I. Kerszenbaum. *Handbook of Large Turbo-Generator Operation and Maintenance.* Piscataway, NJ: IEEE Press, 2008, chapters 4 and 5.
2. EPRI Technical Report 1026557. Bearing Failures in Power Plant Rotating Equipment, October 2012.
3. EPRI Technical Report 1026566. Field Guide: Bearing Damage Mechanisms, November 2012.
4. EPRI Technical Report CS-4555. Guidelines for Maintaining Steam Turbine Lubrication, July 1986.
5. EPRI Technical Report 1026605. Generator Maintenance Applications Center: Combined-Cycle Combustion Turbine Lube Oil System Maintenance Guide, October 2012.

M-R-26: TEMPERATURE OF FIELD WINDING ABOVE NORMAL

APPLICABILITY:

All generators

MALFUNCTION DESCRIPTION:

This malfunction addresses the situation when the rotor (*field*) winding reaches temperatures above *normal*. In the context of this description, *normal* means temperatures that were measured on the same machine previously, under same operating conditions. Also, *normal* means temperatures within the insulation *class rise* limits.

Field-winding temperatures are not measured directly, but indirectly from dividing V_{field} by I_{field}, on the brushgear for external excitation, or via auxiliary sliprings in shaft-mounted brushless excitation. The problem with this type of measurement is that it is global; that is, a localized temperature increase due to a fault will most probably be masked by the rest of the winding's temperature.

PROBABLE CAUSES:

- OPERATION OUTSIDE CAPABILITY CURVE ONLINE **(M-E-11)**
- OVEREXCITATION ONLINE **(M-E-13)**

Problem with global cooling:

- AIR OR HYDROGEN HEAT-EXCHANGER COOLING-WATER SUPPLY OFF **(M-H-01)**

- BLOCKAGE OF AIR COOLERS **(M-H-02)**
- BLOCKAGE OF HYDROGEN COOLERS **(M-H-03)**
- HYDROGEN PRESSURE INSIDE GENERATOR BELOW NORMAL **(M-H-09)**
- INADEQUATE FLOW OF COOLING WATER TO AIR OR H_2 COOLERS **(M-H-10)**
- INLET TEMPERATURE OF AIR OR HYDROGEN COOLING WATER—HIGH **(M-H-11)**
- INSUFFICIENT PERFORMANCE OF AIR OR HYDROGEN HEAT EXCHANGERS **(M-H-12)**
- TEMPERATURE OF COLD GAS ABOVE NORMAL **(M-H-17)**

Shift of blocking or insulation large enough to affect the global temperature of the rotor, or other debris blocking cooling paths:

- BLOCKAGE OF ROTOR VENTILATION PATH **(M-R-03)**
- In those very few directly water-cooled rotors in existence, malfunction of the rotor cooling water (RCW) system must be considered in addition to the aforementioned
- SHORTED TURNS IN FIELD WINDING **(M-R-24)**

POSSIBLE CONSEQUENCES:

Loss of expected life of the insulation:

- BURNING/PYROLYSIS OF ROTOR INSULATION **(M-R-04)**
- DEGRADED/DAMAGE FIELD WINDING ISULATION **(M-R-07)**
- GROUNDED FIELD WINDING **(M-R-13)**
- SHORTED TURNS IN FIELD WINDING **(M-R-24)**
- THERMAL ASYMMETRY OF ROTOR **(M-R-28)**
- During certain type of faults, after the unit received a shutdown command and is offline, it is possible that high current will continue to flow in the field winding at low speeds (even standing still), that is, at poor rotor cooling conditions. The rotor thermal duty can be exceeded and heavy damage can occur (rotors are manufactured to withstand the overcurrent capabilities stipulated by IEEE C50.13, but they are true only at rated speed and obviously much lower at decreased speeds).

POSSIBLE SYMPTOMS:

- Overexcitation alarms
- Overexcitation limits
- HYDROGEN PRESSURE INSIDE GENERATOR BELOW NORMAL **(S-H-06)**
- TEMPERATURE OF FIELD WINDING ABOVE NORMAL **(S-R-09)**

Alarms indicating malfunction of the hydrogen or air cooling system:

- COOLING-GAS HEAT-EXCHANGER DIFFERENTIAL TEMPERATURE—HIGH **(S-H-01)**
- DEGRADED HYDROGEN PURITY **(S-H-02)**
- ELEVATED HYDROGEN DEW POINT **(S-H-03)**
- PRESSURE OF HYDROGEN SUPPLY—LOW **(S-H-09)**
- TEMPERATURE OF COLD GAS ABOVE NORMAL **(S-H-10)**

Liquid-in-generator alarms in machines with rotor directly cooled by water (the liquid being RCW):

- LIQUID-IN-GENERATOR ALARM **(S-C&F-04)**
- VIBRATION OF ROTOR—HIGH **(S-R-10)**

Shorted-turn development from online (or offline) monitoring of the gas-gap by the RFM:

- SHORTED TURNS IN FIELD WINDING **(S-R-07)**

OPERATORS MAY CONSIDER:

With respect to rotor winding temperatures, the key goal is maintaining them within the allowable range. Keep in mind that insulation follows the Arrhenius law (i.e., the expected life of the insulation is halved for a temperature increase of 8°C–10°C). Thus, allowing the temperature to exceed OEM and limits on the standards will rapidly shorten the life of the insulation, not to mention leading to catastrophic failure due to burning/pyrolysis, if the temperatures are allowed to go too high.

1. The degree of urgency for operator response depends on the magnitude of the problem; that is, for temperatures above normal but within limits, there is more time to investigate the source of the problem without having to reduce load, than the case when the temperature limits given by the OEM and/or standards are exceeded.
2. As stated above, the Arrhenius law means the expected life of the insulation is halved by every 8°C–10°C increase of temperature.
3. Keep in mind the problem may be shown as being minor as measured by the global temperature increase of the rotor winding, but actually, it may be a serious localized fault.
4. Temperature rises combined with increase in vibration, a ground-fault or new shorted turns are a sign of a serious problem.

TEST AND VISUAL INDICATIONS:

Visual cues on insulation discoloration, problems with heat exchangers, dislocated insulation/blocking, clogged vents, and so forth.

REFERENCE

1. G. Klempner and I. Kerszenbaum. *Handbook of Large Turbo-Generator Operation and Maintenance.* Piscataway, NJ: IEEE Press, 2008, chapters 1, 4, 5, 9, and 10.

M-R-27: TEMPERATURE OF HYDROGEN SEAL ABOVE NORMAL

APPLICABILITY:

Hydrogen-cooled generators

MALFUNCTION DESCRIPTION:

This malfunction addresses a situation where the temperature of a hydrogen seal is above *normal*. By *normal*, it means a temperature that is higher than on previous occasions under similar load conditions, or that it is at or above the maximum limit temperature provided by the OEM.

PROBABLE CAUSES:

- Inadequate cooling of the seal oil (high seal-oil inlet temperature)
- Insufficient flow of seal oil
- Seal-oil pressure too high
- Failure of the seal insulation
- Failure of the shaft-grounding circuit
- Shaft-seal rub:
  - Wrong seal dimensions
  - Excessive rotor/frame vibrations
  - Seal misalignment
- ARCING ACROSS BEARINGS AND HYDROGEN SEALS **(M-R-01)**
- MISALIGNMENT OF HYDROGEN SEAL **(M-R-19)**
- RUB OF HYDROGEN SEAL **(M-R-21)**
- VIBRATION OF ROTOR—HIGH **(M-R-30)**

POSSIBLE CONSEQUENCES:

- FAILURE OF HYDROGEN SEAL **(M-R-11)**

POSSIBLE SYMPTOMS:

- TEMPERATURE OF HYDROGEN SEAL ABOVE NORMAL **(S-SOS-08)**
- TEMPERATURE OF SEAL OIL ABOVE NORMAL **(S-SOS-10)**

OPERATORS MAY CONSIDER:

1. Hydrogen-seal temperatures above *normal* should be investigated immediately.
2. Temperatures should be measured at close intervals and trended.
3. If the temperatures get close to, or are above limit values provided by the OEM, consider a unit shutdown for troubleshooting. The machine load point has little effect on seal temperature, unless it affects vibrations, or the shaft is misaligned with respect to the turbine(s) and/or rotating exciter.

TEST AND VISUAL INDICATIONS:

- Damage to the seals and/or running surfaces.
- Problems with seal-oil system.
- Blockage of the seal-oil flow path

REFERENCE

1. G. Klempner and I. Kerszenbaum. *Handbook of Large Turbo-Generator Operation and Maintenance*. Piscataway, NJ: IEEE Press, 2008, chapter 5.

M-R-28: THERMAL ASYMMETRY OF ROTOR

APPLICABILITY:

All generators

MALFUNCTION DESCRIPTION:

This malfunction addresses the situation where a rotor has developed a thermal asymmetry.

A thermal asymmetry can manifest itself in various forms, but overwhelmingly, thermal asymmetry of a synchronous generator rotor will take the form of uneven temperature distribution around the circumference of the forging. This type of asymmetry tends to bow the rotor, resulting in an unbalanced mass around the axis of rotation, leading to elevated vibrations.

Invariably, rotor thermal asymmetries are load-dependent. Actually, they depend on the magnitude of the rotor field current. Nevertheless, some types of asymmetries are more or less dependent than others on rotor field current magnitudes, all else being equal. For instance, a blocked rotor cooling passage will, in general, be somewhat less dependent on the rotor field current than shorted turns.

There are few instances where a thermal asymmetry of the rotor is created by a cause external to the rotor. For instance, negative-sequence currents resulting in the rotor may not be evenly distributed around the machine, producing uneven heating and bowing of the rotor. However, this is mostly a temporary condition.

Figure 5.105 shows how the thermal asymmetry will augment the unbalanced magnetic pull, in the case of shorted turns being the cause for the thermal asymmetry, with the result being increased rotor vibrations.

PROBABLE CAUSES:

- BLOCKAGE OF ROTOR VENTILATION PATH **(M-R-03)**
- GROUNDED FIELD WINDING **(M-R-13)**
- NEGATIVE-SEQUENCE CURRENTS **(M-S-42)**
- RUB BETWEEN BEARING AND JOURNAL **(M-R-20)**
- RUB OF HYDROGEN SEAL **(M-R-21)**

FIGURE 5.105
Shorter turn due to erosion of the interturn insulation.

- RUB OF OIL WIPER (M-R-22)
- SHORTED TURNS IN FIELD WINDING (M-R-24)
- Shifted rotor winding coils
- Shifted endwinding blocking
- Blocked cooling passages (oil, dust, shifted insulation, etc.)
- And so on

POSSIBLE CONSEQUENCES:

- HIGH VIBRATION OF CORE ANS/OR FRAME (M-C&F-10)
- VIBRATION OF ROTOR—HIGH (M-R-30)
- VIBRATION OF STATOR ENDWINDING—HIGH (M-S-56)

POSSIBLE SYMPTOMS:

- SHORTED TURNS IN FIELD WINDING **(S-R-07)**
- TEMPERATURE OF BEARING ABOVE NORMAL **(S-R-08)**
- TEMPERATURE OF HYDROGEN SEAL ABOVE NORMAL **(S-SOL-08)**
- VIBRATION OF ROTOR—HIGH **(S-R-10)**

OPERATORS MAY CONSIDER:

1. The most common result of thermal unbalance of the rotor is elevated *rotor vibrations*. However, there are many causes for thermal asymmetries. Therefore, once vibrations are recognized

as increasing or beyond *normal* amplitudes, the first activity should be searching for the source (if the vibrations and/or other machine parameters do not require immediate shutdown).

2. Thermal asymmetries may result from hydrogen/wiper/bearing rubs, but other than those, they will be load and rotor field current dependent.
3. The THERMAL SENSITIVITY TEST (M-R-29) is an excellent method for identifying the origin of vibrations as being mechanical or electrical in nature. Once this identification is made, elucidation of the source of vibrations becomes easier.
4. It is wise to closely monitor rotor temperature, other machine temperatures, vibration levels, MW, MVAR, and field current. In addition to conveying information about the status of the machine, they can provide valuable information on the source of the problem.

TEST AND VISUAL INDICATORS:

- RSO (test done offline or on turning gear)
- Other offline tests searching for shorted turns (such as *pole balance* or *volt drop*) or rotor field grounds
- Visible cues of anomalies in the rotor winding, blocking, and so on
- TST[8]

REFERENCE

1. G. Klempner and I. Kerszenbaum. *Handbook of Large Turbo-Generator Operation and Maintenance*. Piscataway, NJ: IEEE Press, 2008, chapter 11.

M-R-29: THERMAL SENSITIVITY TEST

APPLICABILITY:

All generators

BACKGROUND INFORMATION:

Remark: The *THERMAL SENSITIVITY TEST (M-R-29)* is actually *not* a *malfunction*. It is for convenience-sake listed under malfunctions because it is the only test described in this book. It is designed to identify sources of vibration in the machine by devising a number of tests that check how the vibrations respond to certain parameter changes in the machine.

The *TST* is performed when vibrations of unknown origin afflict the generator. The purpose of the test is to narrow the search for the cause of the vibrations by ascertaining if they are due to changes in the

[8] TST—Thermal sensitivity test.

magnitude of the rotor field current (I_f) in the field winding, temperature of the hydrogen gas, real power output, or from some other cause which may even be strictly mechanical.

Generator rotor thermal sensitivity is a phenomenon that may occur in the generator rotor causing the rotor vibrations to change as the rotor field current is changed. This can occur in the generator fields of all manufacturers.

Thermal sensitivity can be caused by uneven temperature distribution circumferentially around the rotor, or by axial forces that are not distributed uniformly in the circumferential direction. The primary driver of this second cause is the large difference in coefficients of thermal expansion between the copper coils and the steel alloy rotor forging and components. If the rotor winding is not balanced both electrically and mechanically around its circumference, it will become unevenly stressed, which can cause it to bow and vibrations to change.

Vibrations due to a thermally sensitive rotor are mostly at running speed. The vibration origin can be further discriminated as being *reversible or repetitive* and *irreversible*. Reversible vibrations are those in which the vibration vector, when plotted on a polar graph, will not shift. Irreversible vibrations will show a shifting vector. These last ones are the most onerous, as they cannot be balanced for the long run. Almost invariably they result in total winding removal and a full rewind. To capture the presence of these vibrations, the *Constant MW* part of the test is done in both directions, that is, with the rotor field current changed both ways. Figures 5.106 through 5.108 show what to expect in both cases. In the figures, the test is started in point A and ends in point G.

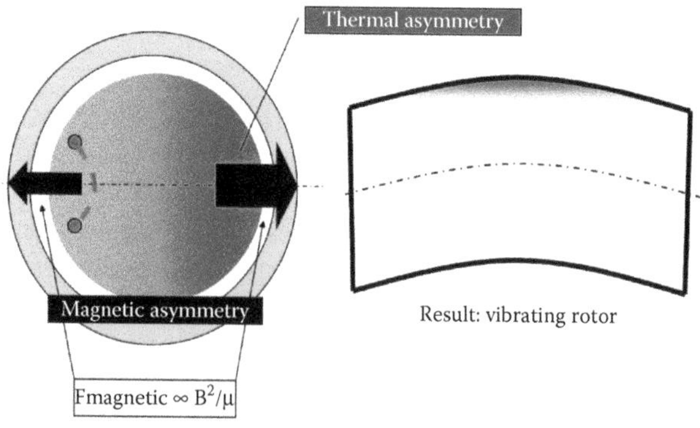

FIGURE 5.106
Thermal asymmetry of the rotor and unbalanced magnetic pull shown as the result of a shorted-turn in the winding.

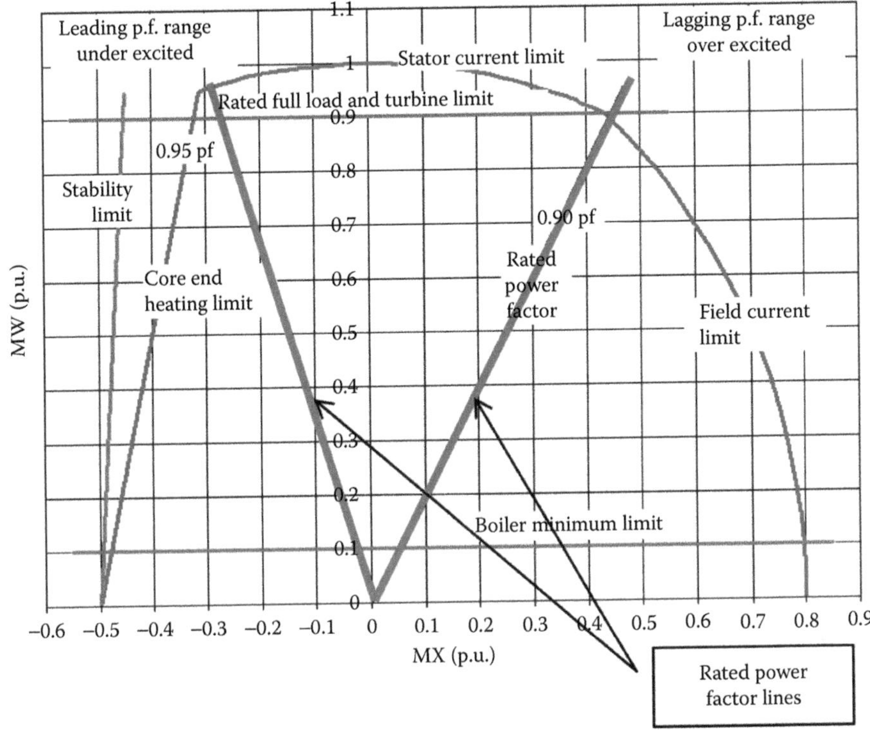

FIGURE 5.107
Capability curve (*English view*) of a typical machine.

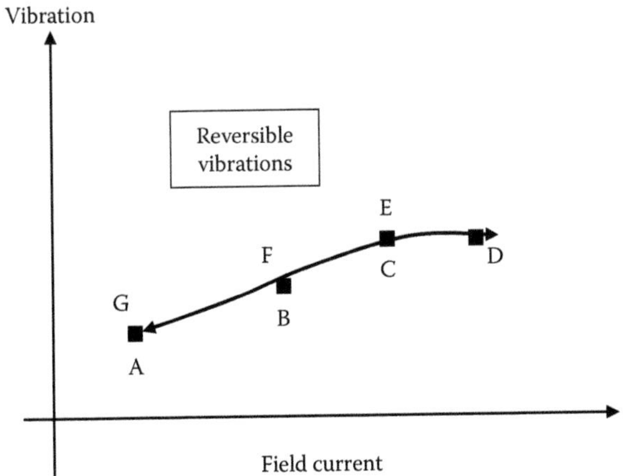

FIGURE 5.108
Reversible type of vibrations.

Field-related vibrations might indicate the presence of one or more of the following:

- Shorted turns in the rotor field winding (vibration vector is reversible vs. *If* for a given number of shorted turns)
- A ratcheting effect, that is, coils shift to one side of the rotor (vibration vector tends to be irreversible)
- *Sticky coil* (vibration vector tend to be reversible, but vibration changes might be erratic with rotor field current changes)
- Crease in the retaining ring insulation liner under the retaining ring (vibration vector may be reversible in the short run but might be of a ratcheting nature in the long run)
- Partial blockage of gas paths inside the rotor (vibration vector is reversible in the short and medium turn, and irreversible if insulation keeps moving—reversible during the test)
- Rotor stiffness dissymmetry (reversible vibrations)
- Wedges that were partially replaced during overhaul, and are not of identical dimensions—resulting in uneven fit along the circumference (vibrations tend to be reversible)
- Insulation that has broken and shifted in the slots (reversible or irreversible during test, depending on severity of insulation damage)
- Nonhomogenous forging (vibration vector is reversible—vibration pattern will show up when unit is new or after repair in the forging, e.g., welding, machining, etc.)

If the unit does not indicate a rotor field current dependency, then most probably vibrations are of mechanical nature.

ROTOR POSSIBLE CAUSES:

- Rotor mechanical unbalance due to a mass shift
  - Retaining ring movement
  - Balancing weights shift
  - Wedges shift
  - Fan shift
  - Shift of blocking under the retaining rings
  - Any other mass shift
- Crack in the rotor forging or on one of its components
- Loose components

STATOR/FRAME POSSIBLE CAUSES:

- Uneven heating of the stator coils, core, and frame
- Bearing misalignment
- H_2-seals rub
- Coupling misalignment

- Loose footing
- Loose components
- Frame twisted

TYPICAL THERMAL SENSITIVITY TEST:

In order to segregate between the various sources, the following TST is performed. The test has three main parts.

The following variables are recorded during the test (others may be added to the list).

- MW
- MVAR
- Terminal volts
- Terminal current
- Field voltage
- Field current
- Vibration of all bearings of interest
- Temperatures of interest
- Time (hours; minutes) to stamp and correlate all readings

It is important to verify that the unit remains within its capability curve under the entire duration of the test. For that purpose, it is convenient to plan the load points ahead of the test, though it is not critical whether the actual load points selected during the test closely match those in the plan.

1. *Constant MW and MVAR*: The aim of this test is to identify a rotor that is thermally sensitive to forging asymmetries (due to the forging process or to machining, welding, etc.), or to stator-frame issues. During this part of the test, the unit's *active and reactive power is held constant* (AVR manual). The temperature of the generator's hydrogen is changed in steps by controlling flow of cooling water to the hydrogen heat exchangers. Stator water temperature is not changed for directly water-cooled stators. It is important to ascertain the unit remains within its temperature operating limits. Readings are taken when the temperatures are stable, and the time is noted (hour; minute). This test is done at a reduced load point, to *make room* for temperature increases during the test.
2. *Constant MW*: During this part of the test, the generator's gross MW output *must remain constant*. With MW constant, the field current (I_f) is changed (thus, the VARs also change), and readings are taken of the variable parameters. The test is repeated for a number of rotor field current values. After a new I_f is set, the machine is run for a given time to allow temperatures to stabilize, and then the

readings are noted and the time is noted. The rotor field current is changed in both directions, to capture the *ratcheting* problems.

3. *Constant field current* (I_f): During this part of the test, *the* I_f *must remain constant*—the AVR must be set to manual. Under this condition, the MW of the generator is changed in steps. After every change the unit is allowed to stabilize, and readings of the variables of interest are taken and the time is noted.

WORKING WITH THE MW–MVAR CAPABILITY CURVE AND WITH THE V-CURVES:

Operators can make the planning of the test easy by working with both the MW–MVAR capability curve, and the V-curve.

On the MW–MVAR capability graph, any line perpendicular to the MW axis represents constant output.

On the V-curve, any line perpendicular to the rotor field current axis represents loads with the same rotor field current.

What to do:

1. On the MW–MVAR *capability graph*, select a number of load points on a line perpendicular to the MW axis, such that the MW output is about 50%–75% of maximum. The most extreme points should cover the area from leading PF to lagging PF. Make sure that in the leading PF zone, the load point chosen is above the stability limit of the machine. These points represent the *constant MW* part of the test. Decide which one of the above points will start the test, and which one will finish.
2. Move to the V-curve, and pencil on the graph the last load point of 1), above. Then draw a line perpendicular to the rotor field current axis, and mark on it a number of load points. These will mark the *constant rotor field current* part of the test.
3. When carrying out the test, allow temperatures to stabilize in each point before recording all pertinent parameters.
4. For the constant MW–MVAR test, chose a load point of about 50% of load, and a power factor close to unity. Then very slowly change cooling and wait for temperature stabilization before taking readings

REFERENCES

1. G. Klempner and I. Kerszenbaum. *Handbook of Large Turbo-Generator Operation and Maintenance*. Piscataway, NJ: IEEE Press, 2008, chapter 11.
2. R.J. Zawoysky and W.M. Genovese. Generator Rotor Thermal Sensitivity—Theory and Experience. *GE Power Systems*, GER-3809.

3. A. Febriyanto and H.T. Sandewan. Rotor Thermal Sensitivity Experience on 143.4 MVA Hydrogen-Cooled Turbo Generator with Direct-Cooled Rotor Design. In: *2011 International Conference on Electrical Engineering and Informatics*, Bandung, Indonesia, July 2011.

M-R-30: VIBRATION OF ROTOR—HIGH

APPLICABILITY:

All generators

MALFUNCTION DESCRIPTION:

This malfunction addresses the situation where the rotor vibration is high compared with the normal vibrations as given by the OEM and/or the standards applicable to that machine.

There are many possible sources of rotor vibration, and their elucidation often requires significant effort and time.

In general, rotor vibrations can be mechanical or electromagnetic in origin. Electromagnetic sources can be straightforward (such as the unbalanced magnetic pull due to shorted turns in the rotor), or can be masked by thermal bows (such as temperature uneven distribution due to shorted turns, or cooling vents being blocked). The most commonly used terminology distinguishes between those that depend mainly on torque (*mechanical causes*), and those that depend on reactive power output (*electrical causes*). Some other vibrations appear to be mainly dependent on the temperature of the machine—thus to its cooling or lack thereof.

PROBABLE CAUSES:

In the following section, one can find a number of possible causes for rotor vibration:

- THERMAL SENSITIVITY TEST **(M-R-29)**

POSSIBLE CONSEQUENCES:

- DYNAMIC ROTOR UNBALANCE (MECHANICAL) **(M-R-08)**
- HIGH VIBRATION OF CORE AND FRAME **(M-C&F-10)**
- RUB BETWEEN BEARING AND JOURNAL **(M-R-20)**
- RUB OF HYDROGEN SEAL **(M-R-21)**
- RUB OF OIL WIPER **(M-R-22)**
- VIBRATION OF STATOR ENDWINDING HIGH **(M-S-56)**
- Poor commutation (the brushes *chatter* on the collector rings, making and breaking contact, resulting in arcing and possible damage to the rings)

POSSIBLE SYMPTOMS:

- VIBRATION OF ROTOR—HIGH **(S-R-10)**

Hydrogen leakage rate increase (hydrogen escapes through seals):

- HYDROGEN MAKE-UP RATE ABOVE NORMAL **(S-H-05)**
- HYDROGEN PRESSURE INSIDE GENERATOR BELOW NORMAL **(S-H-06)**

OPERATORS MAY CONSIDER:

1. Any step increase or trending up of rotor vibrations should immediately capture the attention of the machine operators. Vibration increase can be due to a *manageable* problem, or they may be part of a developing runaway situation. The only way of ascertaining which one is by quickly starting an effort of identification of the root causes for the vibration.
2. It is universally accepted that when the unit is in operation, the effort of finding the causes for high vibration—difficult as they may be, is significantly lower to that of finding the causes with the unit offline. Therefore, operators should be diligent in pursuing this effort while the level of vibration hasn't reached a magnitude that requires to shutdown the unit promptly.
3. Finding out the causes of vibrations most often than not requires taking specialized vibration measurements with specialized instrumentation, beyond that taken every day. For instance, vibration phase in addition to vibration magnitude might be helpful, as well as cross-reference to MW, MVAR, and field-current magnitudes.
4. Keep in mind that although the unit may be running with vibrations below the maximum given limits, if they are significantly above *normal*, they have the potential to creating long-term deterioration of main components.
5. The THERMAL SENSITIVITY TEST (M-R-29) may be very helpful in narrowing the search for the cause of vibrations. This is an online test, so if there are expectations the situation will deteriorate to the extent the machine will have to be taken offline, the operators may want to do this test early on.

TEST AND VISUAL INDICATIONS:

There are too many causes providing visual clues, to enumerate. Refer to the reference provided to carry out an in-depth search.

REFERENCES

1. G. Klempner and I. Kerszenbaum. *Handbook of Large Turbo-Generator Operation and Maintenance.* Piscataway, NJ: IEEE Press, 2008, chapters 4, 5, 11, and 12.

2. R.J. Zawoysky and W.M. Genovese. Generator Rotor Thermal Sensitivity—Theory and Experience. *GE Power Systems*, GER-3809.
3. A. Febriyanto and H.T. Sandewan. Rotor Thermal Sensitivity Experience on 143.4 MVA Hydrogen-Cooled Turbo Generator with Direct-Cooled Rotor Design. In: *2011 International Conference on Electrical Engineering and Informatics*, Bandung, Indonesia, July 2011.
4. J. McDowell and P. Swan. Identification of a Unique Vibration Problem. *Orbit Magazine*, May 1992.

M-R-31: WATER LEAK IN A WATER-COOLED ROTOR

APPLICABILITY:

Generators with water inner-cooled rotors

MALFUNCTION DESCRIPTION:

Over the years, a relatively small (compared to the conventional type of rotor cooling) generators were built with rotor inner-cooled by water. This malfunction addresses those machines. Figure 4.16 shows that in a typical water-cooled rotor the pressure of the RCW can be above 1000 Psi ($\approx$ 6900 kPa). This pressure is an order of magnitude above the pressure of the hydrogen in the casing, resulting in water ingressing the generator whenever a leak develops in the rotor. This situation is different than that of the stator, where the SCW is always at a lower pressure than the hydrogen, so that leaks result in hydrogen leaking into the SCW, keeping the internals of the generator water-free, for the most. The greatest advantages of the water-cooled rotor are

- More effective cooling
- Smaller rotor dimensions
- Lower windage losses
- Major disadvantages are:
    - Increased likelihood of water ingression to the generator and consequential damage to components
    - Increased complexity (additional water skid, instrumentation and control systems)
    - Increased maintenance

PROBABLE CAUSES:

The causes for leaks in the rotor are almost always due to degradation of the piping connections (the rotor field copper conductors are part of the piping system).

POSSIBLE CONSEQUENCES:

- Longer outages
- Water damage to other components inside the generator

POSSIBLE SYMPTOMS:

- INSULATION RESISTANCE LOW IN WATER-COOLED ROTORS **(S-R-02)**
- LIQUID-IN-GENERATOR ALARM **(S-C&F-04)**

OPERATORS MAY CONSIDER

While managing rotor grounds as with any other rotor, in the case of water-cooled rotors operators ought to ascertain that no significant RCW is ingressing the generator. Liquid detectors can provide indications that water is present in the machine.

TEST AND VISUAL INDICATIONS:

Megger test of rotor winding. However, electrical testing of rotor windings has its own challenges, not too different than testing stator water cooled windings. Each case must be evaluated on its own.

REFERENCE

1. S. Eriksson. *The Swedish Development of Turbogenerators with Directly Water-Cooled Rotors*. IEEE.

5.5 Excitation System Malfunctions

M-E-01: ACCELERATED BRUSH WEAR

APPLICABILITY:

All generators with collector rings (sliprings) and brushgear fed from static or rotating exciters

MALFUNCTION DESCRIPTION:

Accelerated brushwear is a condition where the brushes are wearing out faster than normal and often in an abnormal manner.

The brushes can still be operating well and it is simply a matter of condition than cause that makes the brush surface to look normal but the wear rate is simply much faster. On the other hand, the excessive wear rate can also be due to some conditions that causes the brush surface to be damaged (Figure 5.109).

It is often difficult to diagnose the root cause for accelerated brushwear, because there are literally a couple of dozen conditions that can cause brushwear issues. Recognizing that there is excessive wear is simply a matter of determining the time frame between required

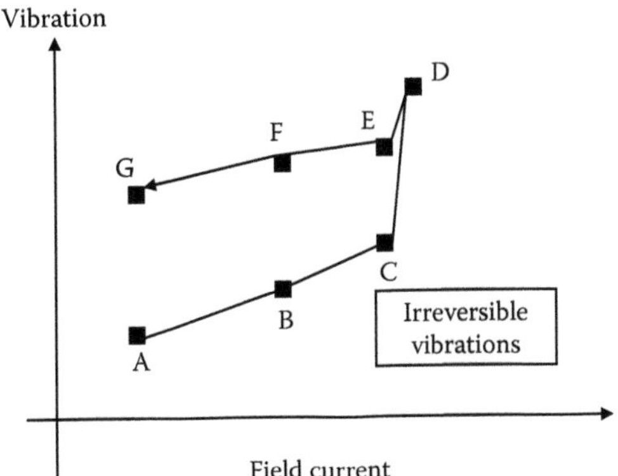

FIGURE 5.109
Irreversible type of vibrations.

brush changes. However, as stated above, determining the root cause is another matter.

PROBABLE CAUSES:

There are many possible causes for accelerated brushwear. The following are some of the most likely:

- OVEREXCITATION ONLINE **(M-E-13)**
- VIBRATION OF ROTOR—HIGH **(M-R-30)**
- Misaligned brushes in the brush holder
- Misaligned brush holders
- Poor atmospheric conditions (i.e., excessively dry air, contaminated cooling air, etc.)
- Damaged collector ring surface
- Inadequate cooling of the brushes and collector rings
- Brush spring pressure too high
- Brush spring pressure too low
- Brush sparking
- Over- or underbrushing (i.e., the current density per brush is too high or too low)
- Brushes of different composition/grade creating unbalance in current distribution between brushes

POSSIBLE CONSEQUENCES:

When brushgear operation is compromised, the consequences can range from simply requiring a change of brushes to new ones, all the way up to a catastrophic failure (Figure 5.110).

FIGURE 5.110
Accelerated and abnormal brushwear.

POSSIBLE SYMPTOMS:

- FIELD CURRENT TOO HIGH ONLINE **(S-E-04)**
- HIGH RATE OF BRUSH WEAR **(S-E-06)**
- SPARKING ON SLIPRINGS **(S-E-14)**

OPERATORS MAY CONSIDER

1. Starting troubleshooting the problem by looking for the possible symptoms and root causes.
2. Decreasing the time between brush changes.

TEST AND VISUAL INDICATIONS:

- Determine the brush change-out frequency for comparison to the normal change-out time.
- Inspect for visual signs of any of the listed symptoms or causes above.

REFERENCE

1. G. Klempner and I. Kerszenbaum. *Handbook of Large Turbo-Generator Operation and Maintenance.* Piscataway, NJ: IEEE Press, 2008, chapter 9.

M-E-02: BLOCKAGE OF BRUSHGEAR COOLING AIR

APPLICABILITY:

All externally excited generators (via sliprings and brushes)

MALFUNCTION DESCRIPTION:

This malfunction relates to the situation where the path of the cooling air in and out of the brushgear compartment is partially or totally blocked.

In general, the flow of cooling air to the brushgear compartment is driven by a shaft-mounted fan. The air flows into the compartment through an air filter. In many large units, the brushgear is enclosed in a much larger housing, normally called by a number of names, such as *exciter housing*. In some of those cases, air filters are not installed.

The temperature of the air drawn into the brushgear by the brushgear fan varies over the day, the season, and between units, and depending on their location at the same plant. Therefore, the temperature of the outlet air exiting the brushgear will also change during the day and season, even at the same MVA output of the generator. Thus, a better measure of the cooling of the brushgear is the differential air temperature across it, and may be an indicator of the condition of the filter, among other things.

In general, it is rare for temperature pickups to be installed on the brushgear. Operators rely on periodic maintenance of the filters. However, under certain circumstances the filters may get plugged before the next filter maintenance is scheduled, or a planned maintenance activity is not carried out as scheduled.

A more direct indicator ascertaining the quality of the flow of air through the brushgear is the pressure drop across the brushgear filter—in those cases one is installed. However, it is not common practice to monitor this pressure differential.

POSSIBLE CONSEQUENCES:

Overheating of the brushgear can lead, first, to excessive wear of the brushes and/or sliprings, and finally, to damage of the insulation between sliprings and shaft or other insulation in the brushgear, causing short-circuits between polarities and/or ground faults.

- DAMAGE/DEGRADATION OF THE BRUSHGEAR (M-E-03)

POSSIBLE SYMPTOMS:

In those rare cases where the brushgear is instrumented—temperature indications:

- HIGH RATE OF BRUSH WEAR **(S-E-06)**
- PRESSURE DROP ACROSS BRUSHGEAR FILTER—HIGH **(S-E-12)**
- TEMPERATURE OF SLIPRINGS/BRUSHGEAR ABOVE NORMAL **(S-E-16)**

OPERATORS MAY CONSIDER:

1. Start immediately troubleshooting the problem, including replacing clogged filters or removing airflow blockage.
2. If the troubleshooting is not expected to happen in short term, or the root cause is not readily identified, reducing load to lower field current may help keep the temperature of the brushgear minimized to an acceptable level.
3. Keeping track of temperature of the brushes and collector rings with an infrared gun, and comparing with normal temperatures. If the temperatures are too high, mitigate the situation by lowering field current or shutting down the unit.

TEST AND VISUAL INDICATIONS:

Visual inspection for plugged cooling path/filter

REFERENCES

1. G. Klempner and I. Kerszenbaum. *Handbook of Large Turbo-Generator Operation and Maintenance.* Piscataway, NJ: IEEE Press, 2008, chapter 9.
2. EPRI TR-107137. Main Generator Online Monitoring and Diagnostics, Sections 3.5.5 through 3.5.7.

M-E-03: DAMAGE/DEGRADATION OF THE BRUSHGEAR

APPLICABILITY:

All generators with external excitation

MALFUNCTION DESCRIPTION:

This malfunction addresses the situation where the brushgear in general (including the collector rings of the main generator), are damaged or otherwise degraded to the extent that commutation is poor.

Good commutation is based on excellent quality of the *patina*, which itself depends on a number of factors. Although simple in appearance, the physics of the patina production and commutation efficiency is rather complex, requiring ample experience for troubleshooting, if and when needed. Before carrying out any change to the brushes (for instance, their *grade*), or to the sliprings (e.g., the *roughness number* of their surface), one must consult with the OEM and/or experts with the proper knowledge, given that the fine balance between elements and conditions that make for a good commutation can easily unravel.

Basic Theory on Patina Formation:

During the process of the conduction of current between the brushes and the slipring, a thin film of about 2×10^{-6} mm is formed on the slipring. This film (called *patina*), which reduces the friction between the brush and slipring surfaces and offers a low resistance path to the current, is critical for the operation of the brushgear. The patina is made of oxide of the slipring material, moisture in the air, airborne contamination, and graphite.

The current flows between the brush and the slipring through a relatively small number of contact points, which share the total current flowing through the brush. The contact points are constantly being eroded and newly formed. Thus, the current flow constantly changes location across the face of the brush. The patina is formed due to the constant erosion of brush and slipring material, and the high temperatures at the contact-conducting points.

The patina, formed as explained above, is essential in maintaining a good contact between the running surfaces, with very low friction. The condition of the patina depends on many variables, key among them are

Temperature of the slipring: At higher temperatures the slipring material will oxidize quicker. The best temperature range for proper oxidation of the ring appears to be 60°C to 90°C (140°F to 194°F). Most of the temperature rise is due to friction, and only about 10% is due to the electrical losses generated by the current. It is important that the temperature is homogeneous across the width of the slipring. This homogeneity depends on the proper cooling arrangement, and the correct distribution of currents among the brushes.

Current: The ionized metal gas from the flow of current also results in oxidation of the running surfaces. This effect is somewhat different in the positive and the negative polarities.

Contaminants: Oil, dust, smoke, dirt, gasses, and so on will either reduce or enhance the formation of an oxide layer.

Humidity: Moisture is a major factor the composition of the patina. The best patina formation versus moisture occurs when the humidity is in the range 4.5–25 g/m³. The graph in Figure 5.111 (after Morgan Industrial Carbon) provides a sense of the effect of humidity and temperature on the life of the brush.

Polarity effect: The patina formation depends on the direction of the current (from brushes to or from the slipring). As a result, *the*

FIGURE 5.111
The entire brushgear equipment has failed in a catastrophic manner due to brushwear being excessive and lack of attention to it.

positive slipring tends to wear somewhat faster than the negative one. Depending on the patina, and the composition of the slipring, the wear rates for both polarities can vary for as much as 5 times. Therefore, it is no uncommon for operators to swap polarities every few years, to even out the wear of the sliprings, and doing that, defer major slipring refurbishment (replacement).

Grooving of the sliprings: Helical grooving of the slipring results in a more even distribution of the currents among the brushes. Also, helical grooving allows some of the gas formed under the brushes to escape, with benefit to the overall behavior of the contact area. Grooving can come in different shapes, and it is not uncommon to combine helical grooving with radial cooling vents.

Other issues, such as optimal brush pressure, angle of the brush against the slipring, and so on are covered in the References noted below.

PROBABLE CAUSES:

Insufficient cooling of the brushgear:

- BLOCKAGE OF BRUSHGEAR COOLING AIR **(M-E-02)**
- TEMPERATURE OF INLET COOLING AIR TO BRUSHGEAR—HIGH **(M-E-20)**
- TEMPERATURE OF SLIPRINGS/BRUSHGEAR ABOVE NORMAL **(M-E-21)**
- Debris from a broken carbon brush
- Contamination from airborne chemicals

- Insufficient brush pressure
- Mixing brush grade
- Brushes too tight or too loose inside their brush-holder
- Changing too many brushes at the same time without first *sitting* them properly
- High rotor vibration
- High humidity
- High slipring runout
- Having the brushes sitting on sliprings standing still for long periods of time

POSSIBLE CONSEQUENCES:

- TEMPERATURE OF SLIPRINGS/BRUSHGEAR ABOVE NORMAL **(M-E-21)**
- Ground faults in the brushgear:
- GROUNDED FIELD WINDING **(M-R-13)**
- Damage to sliprings

POSSIBLE SYMPTOMS:

- In those cases where the brushgear compartment is instrumented: temperatures above normal
- High temperature readings seen via infrared cameras
- Number of broken brushes above normal
- HIGH RATE OF BRUSH WEAR **(S-E-06)**
- SPARKING ON SLIPRINGS **(S-E-14)**

OPERATORS MAY CONSIDER:

1. No commutation is perfect. They all have losses and the heat thus generated must be removed by an effective cooling system. However, in well-commutated machines it is very rare to experience a forced outage or load reduction because of poor commutation, while in machines with inferior or degraded commutation, this may happen, and certainly brushes wear off in short notice.
2. Commutation problems should be addressed as soon as possible, if they result in markedly elevated temperatures of the sliprings, and/or considerable sparking.
3. In some cases the magnitude of the problem requires reducing load, so that the temperatures are maintained within normal values.
4. In some cases the root of the problem may be exogenous to the machine (like chemical contamination). In others, it may be a result of other problem, like high rotor vibration.
5. Operators ought to *control* temperatures (*load reduction* is one way), and verify that there is no risk to the physical integrity of the slipring and brushgear components, while investigating the root cause of the problem and troubleshooting it.

6. Keep in mind that bad commutation can result in faults such as DC ground faults involving sliprings and/or brushgear, potentially with great cost to equipment and loss of production.

TEST AND VISUAL INDICATIONS:

- Broken brushes
- Carbon debris
- Quickly worn brushes
- Quickly worn sliprings
- Damage sliprings/brushes/brushgear insulation
- Discoloration of the sliprings
- Discoloration of the pigtails of the brushes

REFERENCES

1. G. Klempner and I. Kerszenbaum. *Handbook of Large Turbo-Generator Operation and Maintenance*. Piscataway, NJ: IEEE Press, 2008, chapter 9.
2. C. Maughan. Maintaining Carbon-Brush Collectors, CCJ-Online, 2014.
3. Sliprings and Carbon Brushes on Turbo-Alternators. Morgan Industrial Carbon, pdf online.

M-E-04: FAILURE OF THE EXTERNAL STATIC EXCITATION

APPLICABILITY:

All generators with external static excitation

MALFUNCTION DESCRIPTION:

This malfunction describes the situation where the static excitation on machines excited via sliprings and brushes fails, partially or totally.

PROBABLE CAUSES:

- Loss of power to the excitation transformer
- Faulty solid-state power rectifier
- Faulty firing circuit board
- Faulty capacitor in the power circuit of the excitation
- Faulty excitation breaker
- Short-circuit on the AC or DC side of the excitation

POSSIBLE CONSEQUENCES:

- FIELD CURRENT TOO LOW OFFLINE **(M-E-05)**
- FIELD CURRENT TOO LOW ONLINE **(M-E-06)**

POSSIBLE SYMPTOMS:

- Alarms from the excitation panel
- FIELD CURRENT TOO LOW ONLINE **(S-E-05)**

OPERATORS MAY CONSIDER:

Refer to actions in the following:

- FIELD CURRENT TOO LOW OFFLINE (M-E-05)
- FIELD CURRENT TOO LOW ONLINE (M-E-06)

TEST AND VISUAL INDICATIONS:

Refer to actions in the following:

- FIELD CURRENT TOO LOW OFFLINE (M-E-05)
- FIELD CURRENT TOO LOW ONLINE (M-E-06)

REFERENCE

1. G. Klempner and I. Kerszenbaum. *Handbook of Large Turbo-Generator Operation and Maintenance*. Piscataway, NJ: IEEE Press, 2008, chapter 3.

M-E-05: FIELD CURRENT TOO LOW OFFLINE

APPLICABILITY:

- All generators *offline*
- AVR in *manual* or *automatic*

MALFUNCTION DESCRIPTION:

This malfunction describes the case where the unit is running at full speed, but the excitation current is too low to achieve rated voltage.

PROBABLE CAUSES:

- Malfunction of the AVR
- Malfunction of the permanent magnet generator (PMG) or high-frequency generator (HFG) or field flashing circuit (FFC)
- Malfunction of the exciter rectifier
- Discharge resistor still connected to the field winding
- External short-circuit on the *field* circuit
- External open excitation circuit
- Commutation brushes left in the *lifted* position
- Loss of external excitation transformer on static excitation systems
- Significant number of SHORTED TURNS IN FIELD WINDING (M-R-24)
- GROUNDED FIELD WINDING (M-R-13)

POSSIBLE CONSEQUENCES:

Synchronization with the grid not possible due to low terminal voltage.

POSSIBLE SYMPTOMS:

- Terminal voltage too low
- Sync-check relay not allowing breaker closure
- Excitation malfunction alarms

OPERATORS MAY CONSIDER:

1. Operator ought to quickly ascertain what the root cause of the trouble is, and if this can't be done in short notice, the most prudent course of action may be to shut down the excitation (and perhaps bring the generator to rest).
2. AVR control problems may just require switching to another AVR (if a backup is part of the excitation scheme). This normally happens automatically, but it can be done manually.
3. Rectifier, FFC, and PMG/HFG problems may require the unit to be shutdown, but little or no collateral damage will occur.
4. Short-circuits on the DC current circuit, inside or outside the machine, may end up being very costly.
5. If in addition to low excitation, the machine shows abnormally high vibrations, field grounds or large number of shorted turns or shorted coils may be assumed

TEST AND VISUAL INDICATIONS:

- Testing if the AVR is properly aligned
- Testing AVR functionality
- Visual inspection of brushes on externally excited machines
- Testing PMG/HFG/FCC systems
- Series resistance tests on field winding and excitation circuits
- Visual inspection of brushgear, cables, and busses
- Megger tests of rotor field and excitation conductors
- Testing for large number of shorted turns or shorted coils
- Testing rectifier (external or shaft-mounted)

REFERENCE

1. G. Klempner and I. Kerszenbaum. *Handbook of Large Turbo-Generator Operation and Maintenance.* Piscataway, NJ: IEEE Press, 2008, chapters 3, 9, and 11.

M-E-06: FIELD CURRENT TOO LOW ONLINE

APPLICABILITY:

All generators when on line

MALFUNCTION DESCRIPTION:

This malfunction describes the situation where the excitation current is too low while the generator in online.

The *minimum value of rotor field current* is limited by the maximum *leading* VAR capability of the generator, and/or the stability limit of the machine. These limits are different for different load points in the leading power factor hemisphere of the capability curve. Thus, it is convenient to focus on the *VAR limits*, and the excitation current limits are derived therefrom. See Figure 5.112 for an example of a *capability characteristic.*

In the figure above, the stability limit line is shown just outside the leading VAR region of the capability curve, but it can also be located inside, between the MW axis and the leading power factor edge of the capability curve.

One should keep in mind that it is not too rare to find a machine sustaining core-end damage due to being operated well in its leading power factor region, though still within the capability curve. This is more so in older machines, where the interlaminar insulation has deteriorated over many years of operation.

PROBABLE CAUSES:

- Operator error
- AVR malfunction
- Excitation malfunction
- Loss of excitation's power supply
- Sustained short-circuit on the DC field current leads/busses

POSSIBLE CONSEQUENCES:

- Automatic trip of the generator by the *loss-of-field* relay
- LOSS OF STABILITY (M-R-17) (i.e., *slipping poles*)
- Lowering of terminal voltage below 95% of nominal volts

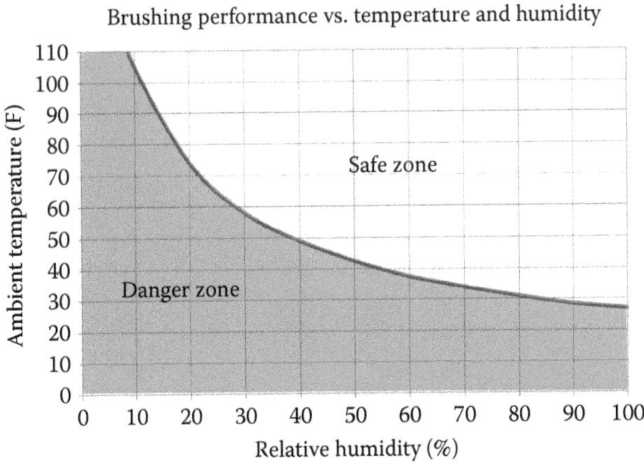

FIGURE 5.112
Humidity and its effect on brush life.

Overfluxing of the core ends:

- TEMPERATURE OF CORE END ABOVE NORMAL (M-C&F-21)

POSSIBLE SYMPTOMS:

- TEMPERATURE OF CORE END ABOVE NORMAL **(S-C&F-06)**
- Trip/alarm of the *loss-of-field* relay
- Terminal voltage below 95% of rated
- Machine becoming unstable
- Alarm indicating activation of the *minimum field current limiter* in the AVR

OPERATORS MAY CONSIDER:

1. *Field current too low online* means the unit is beyond its limits in the *leading* power factor region. Thus, if it hasn't tripped yet by its protection system, it should be taken immediately back into its allowable operating area by operator action.
2. Keep in mind that *slipping poles* may, in addition to loss of production by a sudden trip, cause significant damage to the generator and perhaps the turbine(s). Thus, if this has happened, consideration ought to be given to a serious inspection of possible affected components, such as endwindings and endwinding support systems, couplings, and footing (bolts).
3. Long operation in leading power factors may result in overheating of the core ends, with serious consequences to the integrity of the interlaminar insulation. Thus, monitoring closely the *core-end and flux-shield/shunt temperatures* for a number of weeks, post event, is extremely important to ascertain that a runaway situation does not exist (whereas the interlaminar insulation failed creating more loses and more insulation degradation in a vicious circle).

TEST AND VISUAL INDICATIONS:

- Postevent tests should focus on visual and NDE inspections of those areas prone to damage
- Flux tests may find problems in the core-end area, if these are suspected of having occurred

REFERENCE

1. G. Klempner and I. Kerszenbaum. *Handbook of Large Turbo-Generator Operation and Maintenance*. Piscataway, NJ: IEEE Press, 2008, chapters 1, 4, and 8.

M-E-07: FREQUENCY BELOW NOMINAL—ONLINE

APPLICABILITY:

All generators when *synchronized* to the grid

MALFUNCTION DESCRIPTION:

This malfunction addresses the situation when the grid frequency is below nominal value (60 Hz in North America/50 Hz in Europe, Asia, and most other places in the world). Given that a synchronous generator runs at a speed directly proportional to the frequency of the grid, this problem is covered under:

- ROTOR SPEED BELOW NOMINAL—ONLINE (M-E-17)

REFERENCE

1. G. Klempner and I. Kerszenbaum. *Handbook of Large Turbo-Generator Operation and Maintenance*. Piscataway, NJ: IEEE Press, 2008, chapters 4 and 6.

M-E-08: INSUFFICIENT COOLING OF SOLID-STATE EXCITATION CABINETS

APPLICABILITY:

All generators excited by solid-state exciters through brushes and sliprings

MALFUNCTION DESCRIPTION:

This malfunction addresses those situations where the cubicles containing the solid-state power devices of the excitation, are not properly cooled.

Figure 5.113 shows a typical cabinet containing solid-state power devices of an externally excited generator.

In some arrangements the source excitation transformer may be located adjacent to the rectifying cabinets (as in the figure above, to the left). In others it may be removed.

The rectification process releases significant amounts of heat. In large solid-state excitation systems, the most efficient manner for removing the heat generated by the rectifying elements (diodes or thyristors) is by cooling them directly with water (fed through the DC and AC busses). The rest of the cabinet is cooled by forced air. In other arrangements, the entire system is cooled by forced air.

Although the excitation cabinets may be somewhat removed from the generator, and thus may be seen as also removed in importance, the fact is that the entire operation of the generator depends on this static excitation being able to deliver the required field voltage and current. Thus,

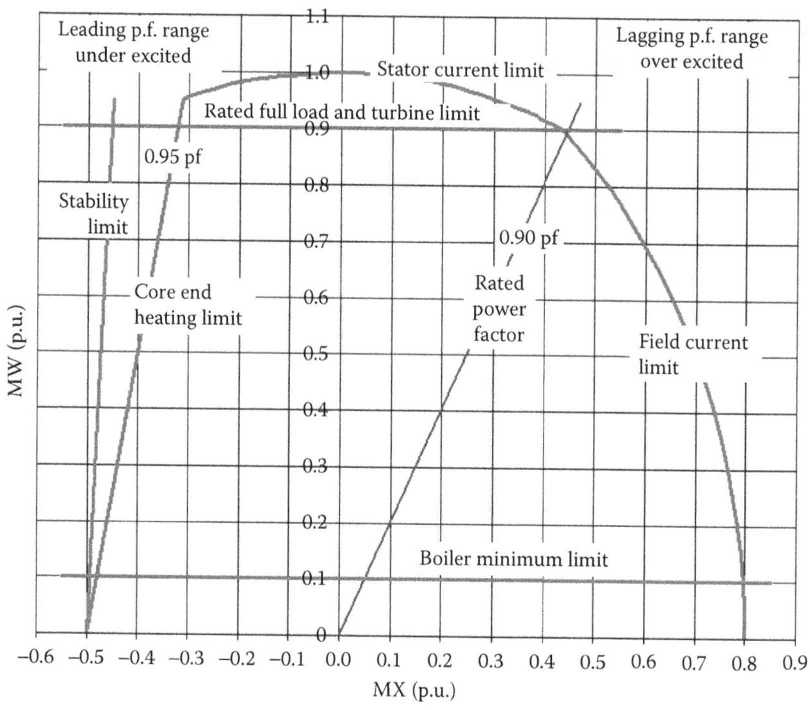

FIGURE 5.113
Typical capability curve (*English view*).

anything seriously affecting the operation of the solid-state excitation will immediately curtail the output of the generator.

The most common problem is a lack of sufficient cooling. This problem may arise from a number of sources, as shown below.

PROBABLE CAUSES:

- Failure of a cooling fan or fans in the power cabinets
- Failure of the heat-exchanger/air-conditioner unit:
    - Lack of flow of primary cooling water
    - Primary cooling water inlet temperature too hot
    - Blockage in the flow path of the primary cooling water
    - Failure of thermostat(s)
- Power cabinet door left open inadvertently
- Plugging of air filters
- Extended period of high rotor field overcurrent beyond the maximum rated capability of the excitation power section

POSSIBLE CONSEQUENCES:

- Failure of the electronic components driving the power solid-state devices

- Failure of the power solid-state devices (diodes/thyristors)
- Failure of insulation of cables in the cabinets
- Tripping the unit
- Causing a *loss-of-field* event

POSSIBLE SYMPTOMS:

- TEMPERATURE IN SOLID-STATE EXCITATION CUBICLE—HIGH **(S-E-15)**
- Temperature alarms from the excitation cabinets
- Alarms about *loss* of a fan
- Alarms about malfunction of cooling systems of the exciter cubicles

OPERATORS MAY CONSIDER:

1. Upon receiving indication that a loss of cooling situation is developing with the power section of the solid-state excitation, operators should consider immediately monitoring the *temperatures* therein, and making an allowance for reducing load, as necessary to keep them from reaching beyond maximum limits given by the OEM.
2. While monitoring the temperatures, troubleshooting the source of the problem.
3. Vendors provide (in the operation manuals of the equipment or upon request) maximum allowable operating times at rated load, with failed cooling. Operators ought to have this information handy in the control room, so that if a case like this one develops, they will have a time basis for troubleshooting the problem.

TEST AND VISUAL INDICATIONS:

Visual indications of fan damage, clogged filters, or any problems with the heat exchangers

REFERENCE

1. G. Klempner and I. Kerszenbaum. *Handbook of Large Turbo-Generator Operation and Maintenance*. Piscataway, NJ: IEEE Press, 2008, chapters 3 and 10.

M-E-09: MVAR TRANSIENTS

APPLICABILITY:

All generators when *synchronized* to the grid

MALFUNCTION DESCRIPTION:

This malfunction addresses the case when the main generator undergoes a reactive power output transient (*MVAR transient*). By *transient*, it means a fast swing of MVARs that is not a planned load change, but the result of a grid disturbance (such as a sudden voltage change, either high or low) or operator error or malfunction of the excitation system, or a fault on the field winding.

A large MVAR swing has the potential to induce a small MW transient in the output of the machine, as well as transients in the power angle of the generator.

Large MVAR swings have been known to result in new shorted turns in the rotor (due to high transient voltages induced in the rotor winding).

PROBABLE CAUSES:

- Grid disturbance
- AVR/excitation malfunctions
- Operator error
- Problem with the rotor winding (such as shorted turns, grounds, etc.)
- Under rare circumstances, the core may be momentarily over-excited due to voltage transients resulting from (or causing) the MVAR swing
- GROUNDED FIELD WINDING (M-R-13)
- OVEREXCITATION ONLINE (M-E-13)
- SHORTED TURNS IN FIELD WINDING (M-R-24)
- TERMINAL VOLTAGE EXCURSION—ONLINE (M-E-23)

POSSIBLE CONSEQUENCES:

Momentarily taken the unit beyond its capability curve (momentary overload):

- OPERATION OUTSIDE CAPABILITY CURVE—ONLINE (M-E-11)

Low probability for the development of shorted turns in the rotor:

- SHORTED TURNS IN FIELD WINDING (M-R-24)

Very low probability for the development of core faults:

- INTERLAMINAR INSULATION FAILURE (M-C&F-13)

POSSIBLE SYMPTOMS:

- MVAR TRANSIENT (S-E-08)
- Fast grid voltage changes
- Fast output current changes

OPERATORS MAY CONSIDER:

1. Operators should first identify the cause of the swings. Excitation malfunction tends to result in erratic MVAR changes that *don't go away*. Operator error can be readily recognized. Grid

disturbances can be identified from frequency and switchyard voltage records.

2. Once the cause of the MVAR swing is recognized, proper actions, if any needed, can be accomplished.
3. The most important action is bringing the machine back into its capability region, if it drifted out during the occurrence.
4. Notes should be taken of all important machine parameters, including temperatures of core and rotor vibration. If any of those shows an unusual change, adequate steps ought to be taken to eliminate the possibility of any further damage.

TEST AND VISUAL INDICATORS:

- EL-CID and full flux tests may uncover core faults, in the very low-probability scenario that the core suffered from a large MVAR/voltage transient (due to overfluxing).
- RSO, RFM, or other methods for uncovering new shorted turns in the rotor, in case this is suspected as being a consequence of the swing.

REFERENCE

1. G. Klempner and I. Kerszenbaum. *Handbook of Large Turbo-Generator Operation and Maintenance*. Piscataway, NJ: IEEE Press, 2008, chapters 4 and 12.

M-E-10: MW TRANSIENTS

APPLICABILITY:

All generators when *synchronized* to the grid

MALFUNCTION DESCRIPTION:

This malfunction addresses the case when the main generator undergoes an active (real) power output transient (*MW transient*). By *transient*, it means a fast swing of MW that is not the result of a planned operator control action.

Large MW transients are, invariable, accompanied by speed swings, MVAR swings, and often, frequency swings.

Large MW transients have the potential for damaging the generator (and/or turbines), in ways described below.

PROBABLE CAUSES:

- The most common cause of a MW transient is a grid disturbance. The resulting power swing can be significantly larger than the rated generator output.

- Governor malfunction.
- Slower MW swings can be the result of AVR malfunction or hunting with other generators—mostly within the same power plant.

POSSIBLE CONSEQUENCES:

- For large MW transients:
  - Stator endwindings damage
  - Damage to the stator endwinding support system
  - Damage to couplings
  - Increased vibrations
- SHORTED TURNS IN FIELD WINDING **(M-R-24)**
- VIBRATION OF ROTOR—HIGH **(M-R-30)**
- Bearing/hydrogen seal rubs:
  - BEARING MISALIGNMENT **(M-R-02)**
  - LOOSE GENERATOR FOOTING **(M-C&F-17)**
  - RUB BETWEEN BEARING AND JOURNAL **(M-R-20)**
  - RUB OF HYDROGEN SEAL **(M-R-21)**
  - RUB OF OIL WIPER **(M-R-22)**
- Unit trip:
  - LOSS OF STABILITY **(M-R-17)**

POSSIBLE SYMPTOMS:

Large MW swings observed in the control room instrumentation:

- MW TRANSIENTS **(S-E-09)**
- Loud noise transient in the vicinity of the generator
- Momentary shaking of the turbine deck

OPERATORS MAY CONSIDER:

1. There is basically nothing the operator can do to control transient MW swings as they happen.
2. Once a large swing has occurred, monitor vibration, temperatures, and other indicators to ascertain no internal or external damage has occurred to the generator.

TEST AND VISUAL INDICATIONS:

- If the machine is open, inspection of stator endwindings
- RSO test or alternatives for measuring rotor shorted turns

REFERENCE

1. G. Klempner and I. Kerszenbaum. *Handbook of Large Turbo-Generator Operation and Maintenance*. Piscataway, NJ: IEEE Press, 2008, chapter 4.

M-E-11: OPERATION OUTSIDE CAPABILITY CURVE—ONLINE

APPLICABILITY:

All generators when *synchronized* to the grid

MALFUNCTION DESCRIPTION:

This malfunction addresses the situation when a generator is operated outside its *capability curve*. The intent is not for a momentary excursion due to a transient, but to a period of time long enough to affect one or more components of the machine. Bear in mind that for *large* excursions outside the capability curve, a short duration may be too long, depending on the circumstances at the time. Even a short excursion in the leading power factor range, when the steady-state stability limit (SSSL) is reached can take the generator offline instantaneously.

Figure 5.112 shows a typical capability curve. Figure 112 represents the *English view* of the capability curve. In the *American view* or *European view*, the MW axis is shown in the horizontal position, and the MVAR (or MX) axis in the vertical, with lagging vars shown pointing toward the top of the page.

Regardless of the view, one can discern three main areas of the capability curve:

- First, the area between the MVAR (MX) axis and the rated power factor line. This area represents the thermal limits of the rotor winding.
- The next area, is that denoted by an arc with the center at the point where the MX and MW axis cross, and extending between the rated power factor line and the curve (shown as a straight line in Figure 5.112) roughly parallel to the MVAR (MX) axis, on the leading PF half of the graph. This area represents the stator winding thermal limit.
- Finally, the area between the stator thermal limit and the MVAR (MX) axis in the leading PF region, representing the core-end thermal limits.

From the above, it is obvious that any excursion beyond the capability curve will directly affect one or more critical components of the machine, keeping in mind that during large excursions it may take just a few seconds, or minutes at most, for winding and/or core insulation to be greatly damaged by the high temperatures developed during the event.

The References below provide a more in-depth description of this topic.

PROBABLE CAUSES:

- Operator error
- Governor and/or AVR malfunction
- Protection malfunction

POSSIBLE CONSEQUENCES:

Any number of components may be compromised, among those:

- ARCING BETWEEN CORE AND KEYBARS (M-C&F-01)
- BURNING/PYROLYSIS OF CORE INSULATION (M-C&F-05)
- BURNING/PYROLYSIS OF ROTOR INSULATION (M-R-04)
- BURNING/PYROLYSIS OF STATOR WINDING INSULATION (M-S-08)
- DAMAGE TO CONDUCTOR BAR INSULATION (M-S-15)
- DEGRADED/DAMAGED FIELD WINDING INSULATION (M-R-07)
- DYNAMIC ROTOR UNBALANCE (MECHANICAL) (M-R-08)
- GLOBAL STATOR TEMPERATURE ABOVE NORMAL (M-S-25)
- GLOBAL TEMPERATURE OF CORE ABOVE NORMAL (M-C&F-09)
- INTERLAMINAR INSULATION FAILURE (M-C&F-13)
- LOCAL CORE FAULT (M-C&F-15)
- LOCALIZED OVERHEATING OF STATOR CONDUCTOR (M-S-39)
- LOSS OF STABILITY (M-R-17)
- MVAR TRANSIENTS (M-E-09)
- MW TRANSIENTS (M-E-10)
- OVEREXCITATION ONLINE (M-E-13)
- OVERFLUXING ONLINE (M-E-15)
- OVERVOLTAGE (M-C&F-19)
- PHASE CURRENT HIGH (M-S-45)
- TEMPERATURE OF FIELD WINDING ABOVE NORMAL (M-R-26)
- TEMPERATURE OF SLIPRINGS/BRUSHGEAR ABOVE NORMAL (M-E-21)

POSSIBLE SYMPTOMS:

- Any number of indicators, depending on the type and extent of excursion beyond the capability curve.
- Protection devices alarms/trips.
- On modern VDUs, sometimes the capability curve is displayed in real time with the P–Q load point in it. Software can be written so that an alarm occurs when the load point moves on or beyond the envelope of the capability curve.

OPERATORS MAY CONSIDER:

1. First and foremost, the unit should be returned immediately to the normal operating region of the capability curve.
2. Monitoring all pertinent data (temperatures, vibrations, etc.) and responding accordingly.

3. As shown above, a wide range of components may be adversely affected by an excursion beyond the capability curve. If that happens, operators should ascertain, based on the duration and type of excursion, which components, if any, may have been impacted. Depending on the assessment, visual inspection and/or tests may be required.

TEST AND VISUAL INDICATIONS:

A wide range of possibilities depending on the type and duration of the excursion

REFERENCE

1. G. Klempner and I. Kerszenbaum. *Handbook of Large Turbo-Generator Operation and Maintenance*. Piscataway, NJ: IEEE Press, 2008, chapter 4.

M-E-12: OVEREXCITATION OFFLINE

APPLICABILITY:

All generators when *offline* (breaker open)

MALFUNCTION DESCRIPTION:

This malfunction addresses the situation where the unit is offline and the excitation output current is higher than the no load rotor field current. When this happens (at any speed) the volts per hertz limit will be reached and the stator core will be overexcited. This has the potential for serious consequences to the integrity of the machine, with the most likely result of stator core iron melting. This is also time related (very short time, as in seconds) and protected from happening by relay protection devices, and in some AVRs, with dedicated offline current limiters.

This is covered under the following heading:

- OVERFLUXING OFFLINE (M-E-14)

REFERENCE

1. G. Klempner and I. Kerszenbaum. *Handbook of Large Turbo-Generator Operation and Maintenance*. Piscataway, NJ: IEEE Press, 2008, chapters 1, 4, and 6.

M-E-13: OVEREXCITATION ONLINE

APPLICABILITY:

All generators when *synchronized* to the grid

MALFUNCTION DESCRIPTION:

This malfunction describes the situation where the generator is synchronized and the rotor field current exceeds its maximum value resulting in overheating of the rotor and stator windings (the generator is taken outside the region demarked by its capability characteristic). This problem does not describe a transient swing (covered under MVAR TRANSIENTS (M-E-09), but rather a slow or permanent excursion beyond maximum rotor field current value.

PROBABLE CAUSES:

- Operator error
- A combination of grid voltage drop and wrong setting of the *maximum rotor field current limit* in the AVR
- AVR/excitation malfunctions
- Loss of terminal-voltage feedback to the AVR

POSSIBLE CONSEQUENCES:

Overheating of the rotor and/or stator insulation:

- BURNING/PYROLYSIS OF ROTOR INSULATION **(M-R-04)**
- BURNING/PYROLYSIS OF STATOR WINDING INSULATION (M-S-08)
- GLOBAL STATOR TEMPERATURE ABOVE NORMAL **(M-S-25)**
- TEMPERATURE OF FIELD WINDING ABOVE NORMAL (M-R-26)
- DAMAGE/DEGRADATION OF THE BRUSHGEAR **(M-E-03)**

POSSIBLE SYMPTOMS:

- GENERATOR CORE CONDITION MONITOR—ALARM **(S-C&F-01)**
- Switchyard voltage too low (driving the AVR to supply large rotor field current)
- Switchyard voltage too high (driven by excessive rotor field current due to operator error or AVR-excitation problem)

OPERATORS MAY CONSIDER:

1. Controlling the machine's output with the purpose of restoring it to within the generator's capability curve
2. Monitoring all relevant machine parameters, in particular temperatures, currents, output voltages and verifying there are within limits and, taking immediate corrective actions if they are not

TEST AND VISUAL INDICATIONS:

If insulation degradation is suspected, electric testing and visual inspections may shed light on the extent of the problem, if any.

REFERENCES

1. G. Klempner and I. Kerszenbaum. *Handbook of Large Turbo-Generator Operation and Maintenance*. Piscataway, NJ: IEEE Press, 2008, chapters 1 and 4.
2. A. Murdoch, B. Gott, M.J. Antonio, G.E. Boukarim, and R.A Lawson. Generator Over Excitation Capability and Excitation System Limiters. In: *IEEE PES Winter Meeting*, Columbus, OH, 2001.

M-E-14: OVERFLUXING OFFLINE

APPLICABILITY:

All generators when *offline and energized* (field applied and main breaker open)

MALFUNCTION DESCRIPTION:

This malfunction addresses the situation when the stator core of a generator is overfluxed with the machine not synchronized to the grid; that is, the open circuit condition. Overfluxing is also known as an *excessive volts per hertz (V/Hz) event*.

The maximum value of the magnetic flux density in the core of large turbine generators (and for this purpose—any electric machine, including transformers), depends on the characteristics of the laminations that make the core and the specific design. The maximum flux density is chosen in such a way that the core will not saturate globally. Saturation of the core will negate the key factor in making a core from ferromagnetic materials, which is to maintain high permeability.

During the overfluxing of a core, a number of unwanted phenomena take place in addition to the lowering of the permeability of the laminations. For instance, the eddy currents induced in the core laminations grow exponentially and so does the loss they create. In extreme cases, this has a high probability of damaging the interlaminar insulation, inducing even higher losses and a runaway situation culminating with the destruction of the stator. Figure 5.114 shows such a case.

Excessive V/Hz can happen at any speed and circuit breaker condition, but it will almost always happen with the circuit breaker open (i.e., the unit not synchronized). The reason for this is that when the unit is operating synchronized to the grid, the voltage at the terminals of the generator is to a large extent controlled by the grid. Although the terminal's voltage can go somewhat above the maximum 105%, this excess will not be too large, thus affording the operator time to correct the situation. However, during run-ups before closing the main breaker(s), there is no such constraint from the grid, any malfunction of the AVR, protecting devices, or any operator error, can cause large V/Hz events.

One key characteristic of V/Hz events is that they have the capacity of render the stator core irreversibly damaged after just a few seconds.

FIGURE 5.114
Solid-state excitation system and the excitation transformer enclosure on the right.

In addition, the costs of fixing the problem can be extremely large in repairs and loss of production.

PROBABLE CAUSES:

- Operator error
- AVR malfunction
- Protection malfunction
- Loss of terminal voltage feedback to the AVR

POSSIBLE CONSEQUENCES:

- BURNING/PYROLYSIS OF CORE INSULATION **(M-C&F-05)**
- BYPRODUCTS FROM BURNING INSULATION **(M-C&F-06)**
- GROUND FAULT ON STATOR WINDING **(M-S-26)**
- INTERLAMINAR INSULATION FAILURE **(M-C&F-13)**
- LOCAL CORE FAULT **(M-C&F-15)**
- LOCALIZED OVERHEATING OF STATOR CONDUCTOR **(M-S-39)**
- TEMPERATURE OF A LOCAL CORE AREA ABOVE NORMAL **(M-C&F-20)**

POSSIBLE SYMPTOMS:

Alarm from the GCM:

- GENERATOR CORE CONDITION MONITOR—ALARM **(S-C&F-01)**
- LOCAL CORE TEMPERATURE ABOVE NORMAL **(S-C&F-05)**

- STATOR GROUND ALARM AND/OR TRIP **(S-S-17)**—(ground faults tend to occur sometime after the main V/Hz event; from hours to days and months, depending on the severity of the V/Hz occurrence [i.e., how long it lasted and how high its magnitude])
- Alarms and/or trips from V/Hz protection devices inside and outside the AVR
- RADIO FREQUENCY (RF) MONITOR HIGHT **(S-S-13)**
- ELECTROMAGNETIC INTERFERENCE (EMI) ACTIVITY HIGH **(S-S-03)**

OPERATORS MAY CONSIDER:

1. It is rare for an overfluxing event to present the operator with an opportunity to react in time to avoid serious consequences to the integrity of the stator. Thus, it is important that the V/Hz protection of the generator be set ahead of time to trip the unit immediately for such an occurrence. Similarly, the protection scheme should be *hardened* so that loss of a single PT will not render both the AVR and the V/Hz protection *blind* to a V/Hz event.
2. While ramping up the unit's excitation in *manual* mode, be very careful that at all speeds the generator terminal voltage divided by the speed, is no higher than rated terminal voltage divided by rated speed. For manual ramping, the safest practice is first to bring the unit to full speed and, and only then enable the excitation and built up the terminal voltage until the value required for synchronization is attained.
3. If a V/Hz event occurred, it is highly advisable to inspect the machine thoroughly and consult with the OEM and/or experts before attempting to restart the unit. Unfortunately, the immediate aftermath of an overfluxing event does not present many clues, and it is only after running for some time postevent, when the first signs show up. By then, most probable additional collateral damage has occurred.

TEST AND VISUAL INDICATIONS:

- Digital fault recorders are critical to assess the severity of the V/Hz event.
- EL-CID and/or *loop* flux tests may be able to show regions of the core where the interlaminar insulation has been damaged.
- Discoloration of laminations and insulation.

REFERENCES

1. G. Klempner and I. Kerszenbaum. *Handbook of Large Turbo-Generator Operation and Maintenance*. Piscataway, NJ: IEEE Press, 2008, chapters 1, 4, and 6.
2. CIGRE. Executive Summary: Preventing of Overfluxing of Generators. 2012.

M-E-15: OVERFLUXING ONLINE

APPLICABILITY:

All generators when *synchronized* to the grid

MALFUNCTION DESCRIPTION:

This malfunction addresses the case where an overfluxing event occurs while the unit is synchronized to the grid. This is also called a *volts per hertz* or simple *V/Hz* event.

The maximum value of the magnetic flux density in the core of large turbine generators (and for this purpose—any electric machine, including transformers), depends on the characteristics of the laminations that make the core and the specific design of the machine. The maximum flux density is chosen in such way that the core will not saturate globally. Saturation of the core will negate the key factor in making a core from ferromagnetic materials, which is to maintain high permeability.

A number of unwanted phenomena take place during the overfluxing of a core, in addition to the lowering of the permeability of the laminations. For instance, the eddy currents induced in the core laminations grow exponentially, and so do the losses they create, resulting, in extreme cases, in damage to the interlaminar insulation, more losses and a runaway situation culminating with the destruction of the stator.

Online overfluxing cannot reach the high magnitudes that offline overfluxing can, and therefore, the abnormal situation must be sustained for far longer periods of time (many minutes vs. seconds), for the core interlaminar insulation to become compromised. In actual fact, it is very difficult to overflux a stator core when the generator is connected to the system because the system voltage cannot easily be driven higher by a single machine connected to the system. Nevertheless, once the insulation between the laminations becomes damaged, the end results are the same, major destruction of the core and sometimes also stator winding and rotor.

PROBABLE CAUSES:

- Operator error
- AVR malfunction
- Loss of PT signal conveying terminal voltage levels to the AVR
- Protection malfunction
- Grid voltage increase beyond generator's maximum voltage
- Loss of excitation while the generator remains connected to the system

POSSIBLE CONSEQUENCES:

- BURNING/PYROLYSIS OF CORE INSULATION (M-C&F-05)
- BYPRODUCTS FROM BURNING INSULATION (M-C&F-06)
- GROUND FAULT ON STATOR WINDING (M-S-26)

- INTERLAMINAR INSULATION FAILURE **(M-C&F-13)**
- LOCAL CORE FAULT **(M-C&F-15)**
- LOCALIZED OVERHEATING OF STATOR CONDUCTOR **(M-S-39)**
- ROTOR SPEED BELOW NOMINAL—ONLINE **(M-E-17)**
- TEMPERATURE OF A LOCAL CORE AREA ABOVE NORMAL **(M-C&F-20)**
- On loss of excitation while the generator remains connected to the system, the generator will go into induction generator mode for a brief time, driven by the steam turbine. Then at some point the system will start to drive generator and turbine in motoring mode—the effect of this is that stator currents can double for a short time and cause core end overheating not to mention extensive damage to the rotor

POSSIBLE SYMPTOMS:

Alarm indicating terminal voltage too high:

- ANOMALOUS GENERATOR OUTPUT VOLTAGE **(S-E-01)**
- GENERATOR CORE CONDITION MONITOR—ALARM **(S-C&F-01)**
- LOCAL CORE TEMPERATURE ABOVE NORMAL **(S-C&F-05)**
- STATOR GROUND ALARM AND/OR TRIP **(S-S-17)** (ground faults tend to occur sometime after the main V/Hz event; from hours to days and months, depending on the severity of the V/Hz occurrence [i.e., how long it lasted and how high it went])
- Alarms and/or trips from V/Hz protection devices inside and outside the AVR
- RADIO FREQUENCY (RF) MONITOR HIGH **(S-S-13)**

OPERATORS MAY CONSIDER:

1. Responding immediately to overvoltage alarms greatly reduces the risk of an overfluxing event.
2. If a V/Hz event occurred, it is highly advisable to inspect the machine thoroughly and consult with the OEM and/or experts, before attempting to restart the unit. Unfortunately, the immediate aftermath of an overfluxing event does not present many clues, and it is only after running for some time postevent, when the first signs show up. By then, it is most probable that additional collateral damage has occurred.

TEST AND VISUAL INDICATIONS:

- EL-CID and/or flux tests may be able to show regions of the core where the interlaminar insulation has been damage.
- Discoloration of laminations and insulation.

REFERENCES

1. G. Klempner and I. Kerszenbaum. *Handbook of Large Turbo-Generator Operation and Maintenance.* Piscataway, NJ: IEEE Press, 2008, chapters 1, 4, and 6.
2. CIGRE. Executive Summary: Preventing of Overfluxing of Generators. 2012.

M-E-16: OVERHEATING OF SOLID-STATE EXCITATION POWER COMPONENT

APPLICABILITY:

All generators excited by solid-state exciters through brushes and sliprings

MALFUNCTION DESCRIPTION:

This malfunction describes the situation where a solid-state power component, such as rectifying diodes, thyristors, ***GTOs***[9] or high-current conducting busses and cables, reaches temperatures beyond design limits. These components are located inside the *excitation cabinets.*

Figure 5.115 shows a typical cabinet containing solid-state power devices (rectifying module) of an externally excited generator.

In some arrangements the source excitation transformer may be located adjacent to the rectifying cabinets (as in the right side of the above figure); in others it may be removed.

The rectification process releases significant amounts of heat. In large machines, the most efficient method for removing the heat generated by the rectifying elements (diodes or thyristors) is directly by water (fed

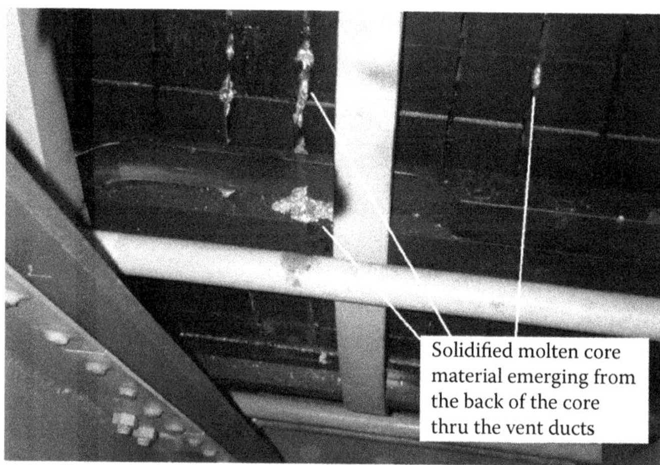

FIGURE 5.115
Melting of the core due to an overfluxing offline.

[9] GTO—Gate turn-off thyristor.

through the DC and AC busses). The rest of the cabinet is cooled by forced air. In other arrangements, the entire system is cooled by forced air.

Although the excitation cabinets may be somewhat removed from the generator, and thus may be seen as also removed in importance, the fact is that the entire operation of the generator depends on this static excitation being able to deliver the required field voltage and current. Thus, anything seriously affecting the operation of the solid-state excitation will immediately curtail the output of the generator.

There are a number of failure mechanisms that may result in a component or two to degrade from high temperature, and oftentimes, this situation will not be detected by global temperature pickups, but only pre-fault by infrared techniques.

PROBABLE CAUSES:

- Diodes/thyristors/GTOs age, and the aging may be accelerated by certain ambient and/or operating conditions. One consequence of this degradation is a reduction in the active area of the conducting junction, and subsequent elevated junction temperatures.
- Loose connections between solid-state device and heat sink, or between solid-state device and cables or busses, or between busses and cables, or between busses, and so forth.
- Localized lack of cooling:
- INSUFFICIENT COOLING OF SOLID-STATE EXCITATION CABINETS **(M-E-08)**

POSSIBLE CONSEQUENCES:

- Failure of power rectifier device (diode/thyristors/GTO)
- Failure of connection between busses/power cables
- Failure of insulation
- DC/AC ground faults
- Short-circuits
- Unit trip
- LOSS OF STABILITY **(M-R-17)**

POSSIBLE SYMPTOMS:

Unless an RTD/TC is located in the vicinity of the failing component, the likelihood the developing problem will be discover in time is slim, unless periodic infrared temperature measurements are carried out inside the cabinets.

OPERATORS MAY CONSIDER:

1. Upon discovering such a problem, operators should consider controlling rotor field current (and load) so that the temperatures of the affected devices remain within allowable limits, and then planning should be done to troubleshoot the situation ASAP.

2. Temperatures of the affected component(s) ought to be carefully and continuously monitored.
3. If the temperature of the affected devices cannot be controlled down to allowable limits or the situation appears to deteriorate fast, consideration should be given to taking the unit offline to troubleshoot the problem, avoiding further damage to the excitation power section.

TEST AND VISUAL INDICATIONS:

- Discoloration of power components/busses/cable insulation
- Insulation electric testing
- Electric testing of power rectifying devices
- Infrared mapping of the internal components (must be in operation)

REFERENCE

1. G. Klempner and I. Kerszenbaum. *Handbook of Large Turbo-Generator Operation and Maintenance*. Piscataway, NJ: IEEE Press, 2008, chapter 10.

M-E-17: ROTOR SPEED BELOW NOMINAL—ONLINE

APPLICABILITY:

All generators when *synchronized* to the grid

MALFUNCTION DESCRIPTION:

This malfunction addresses the situation when the speed of the rotor falls below nominal, while the unit is synchronized to the grid.

In large power systems, the frequency is always maintained very close to the nominal frequency, with exception, perhaps, during a major grid disruption; and even then, the deviation from the nominal frequency is measured in fractions of a Hertz. However, in smaller power systems—typically on islands, the frequency may drift to some extent from the nominal grid frequency.

Generators can operate slightly below their nominal speed, according to clearly defined criteria written in the IEEE and IEC standards. Any excursion below the allowable range can result in overfluxing of the stator core, with very detrimental consequences to the integrity of the generator. In particular there is such a risk, if the generator is operating at its maximum allowable voltage (typically: 5% above rated).

In general, the turbines are the most limiting component regarding operation below (or above) nominal speed. Different turbines have different withstand capabilities.

Manufacturers of turbines produce for each unit an under/overfrequency withstand capability chart that can be used to set the under/over frequency protection. When this protection is properly set, it will also cover the generator in almost every case.

Under/overfrequency protection was seldom implemented on the fleet of utilities connected to strong grids. However, this protection is quite commonly found on new generators, and also used by utilities connected to weak grids, and in particular in deregulated generating companies (e.g., independent power producers).

Those power plants operating at locations where significant deviations from nominal speed may occur, ought to have clear guidelines and protection settings that can mitigate any damage to the prime mover and generator during a low (or high) frequency event.

PROBABLE CAUSES:

Grid overload or other condition, resulting in a frequency drifting downward

POSSIBLE CONSEQUENCES:

- OVERFLUXING ONLINE **(M-E-15)**
- Damage to steam or combustion turbines

POSSIBLE SYMPTOMS:

- ROTOR SPEED BELOW NOMINAL—ONLINE **(S-E-13)**
- Protection activation:
    - Under frequency
    - V/Hz

OPERATORS SHOULD CONSIDER:

1. If the unit is tripped by underfrequency relays and/or V/Hz relays, station personnel should investigate the extent of the initiating condition and any possible negative effect on the equipment, before re-energizing the unit once the grid conditions return to normal.
2. In the event the unit has not tripped (underfrequency protection not implemented), the operator should assess the situation and take action as required.
3. It is possible that low frequency is the result of a degraded grid condition. Removing a generator from operation during such a time may aggravate the grid situation even further. Thus, in some cases it may be required to coordinate with dispatch or grid control before any drastic action (removing the unit from service) is carried out.
4. Keep in mind that turbines (blades) have a total accumulated time for allowable operation outside the permissive region. This total time is measured in minutes (10 min is a common number).

Thus, if in a particular station it is expected for machines to drift outside the permissive frequency region, the operators may strongly consider keeping track of the total accumulated time for each generator. This is carried out by dedicated *frequency-time* meters.

5. Implementing torsional vibration monitoring with the intent to track torsional and cumulative shaft fatigue.
6. If the utility is aware of severe operating conditions or long-line requirements when installing new units, they should consider purchasing a generator with a wider terminal voltage range. The additional capital cost may be found to be the best economical choice for such a machine.

TEST AND VISUAL INDICATIONS:

If damage is suspected to the stator core, EL-CID and full flux tests can be done to assess the condition of the core.

REFERENCES

1. G. Klempner and I. Kerszenbaum. *Handbook of Large Turbo-Generator Operation and Maintenance*. Piscataway, NJ: IEEE Press, 2008, chapters 2, 4, and 6.
2. I. Kerszenbaum. Guard against Over-Fluxing: Ensure Proper Generator Protection, Maintenance, CCJ Onsite (*Combined Cycle Journal*, August 6, 2015).

M-E-18: SUDDEN LARGE CHANGE IN GRID IMPEDANCE

APPLICABILITY:

All generators when *synchronized* to the grid

MALFUNCTION DESCRIPTION:

This malfunction addresses the situation when the grid, to which a turbogenerator is connected, undergoes a large change in impedance, as *seen* from the generator.

A system (grid) impedance change occurs anytime a breaker is opened or closed, or during a fault or transient. Most changes are small, and have hardly any impact on the generators. However, there are those few cases when the grid event is of a magnitude that is *seen* by those generators in the vicinity of the occurrence, with some practical implications. One of the manifestations of these unusual grid events is a significant sudden change in the impedance as *seen* from the generator. By *seen*, it means that the generator responds to the event by undergoing a speed, power output and/or field current and field voltage transient.

The generator is made—among other components—of stator bars, with their own impedance. An electron cruising along the stator does not *sense* the boundary line between the terminals of the generator and the grid. Thus, for all practical purposes, when a generator is connected to the grid, it becomes part of it. The grid impedance forms part of the equation used in calculating the maximum allowed leading power factor output of any given generator (SSSL[10]). Thus, it is clear that any change on the grid's impedance will affect the stability limit of the machine.

For a broader discussion, see the *References* section below.

PROBABLE CAUSES:

- Short-circuit in the electrical vicinity of the generator
- Large grid load rejection or generation rejection
- Trip of one or more of the power lines connecting the station to the grid

POSSIBLE CONSEQUENCES:

- A not-too-severe change will not cause more than a MW-MVAR swing of the generator, within its capability of riding it through.
- A large event—as in the case of a fault not too far (electrically speaking) from the machine, may cause the unit to trip (e.g., by the *out-of-step* relay or other protection devices).
- If the event is too large and there is no out-of-step protection, the unit can lose synchronism and slip poles, with significant potential damage to generator and turbines:
  - LOSS OF STABILITY **(M-R-17)**
  - MW TRANSIENTS **(M-E-10)**
  - MVAR TRANSIENTS **(M-E-09)**

POSSIBLE SYMPTOMS:

- ANOMALOUS GENERATOR OUTPUT VOLTAGE **(S-E-01)**
- MVAR TRANSIENTS **(S-E-08)**
- MW TRANSIENTS **(S-E-09)**
- Trip by the out-of-step protective function
- Trip by overcurrent, V/Hz or generator impedance protective functions

OPERATORS MAY CONSIDER:

1. In most cases, the unit will ride through these events without any one in the control room noticing them. In more severe cases, little more from the operator is required than making sure the equipment hasn't been permanently affected, by checking for
   a. Rotor-winding shorted turns
   b. Rotor vibration levels
   c. Water/hydrogen leaks

[10] SSSL—Steady-state stability limit.

2. In the event that the unit tripped by its protection, the operator can make use of DFR charts to evaluate the extent of the swings, and consult about the need for inspections.
3. In the event the unit did not trip by out-of-step protection, but tripped by other protection after slipping poles (loosing synchronization), a detailed inspection of areas prone to damage may be pertinent.

TEST AND VISUAL INDICATIONS:

- For visual inspections, the areas more prone for damage are the stator endwindings and their support systems, stator alignment, rotor alignment, couplings, and so on.
- Hydrogen and/or SCW leak tests.

REFERENCE

1. G. Klempner and I. Kerszenbaum. *Handbook of Large Turbo-Generator Operation and Maintenance.* Piscataway, NJ: IEEE Press, 2008, chapter 4.

M-E-19: SYNCHRONIZATION OUT OF PHASE

APPLICABILITY:

All generators

MALFUNCTION DESCRIPTION:

Synchronizing out of phase is a very serious malfunction due to the possibility of incurring major damage to the generator. It occurs at the moment of generator breaker closure and attempted connection to the power system. If the main generator breaker is closed too soon or too late, there will be a mismatch in the generator and system voltage angles and hence a mismatch in actual voltage levels. There are protections installed in virtually every machine to guard against such an occurrence, but if it occurs, the generator is likely to be rejected from connecting to the system, or take a very hard torsional *thump* trying to pull the rotor in line or synchronism with the system voltage.

PROBABLE CAUSES:

- Malfunction of the synchronizing equipment
- Operator error when manually synchronizing the generator to the system
- Holding down the synch-check relay button during attempts to synchronize to the system

POSSIBLE CONSEQUENCES:

Synchronizing out of phase can cause rotor components to be overly stressed mechanically and result in loosening or crack initiation in a rotor component. This in turn can result in inducing excessive rotor vibrations, bearing damage, H_2-seal damage, oil wiper damage, depending on the overall lateral movement of the rotor, and torsional effects on the rotor shaft.

- CRACKED SHAFT **(M-R-05)**
- DEGRADED/DAMAGED FIELD WINDING INSULATION **(M-R-07)**
- DYNAMIC ROTOR UNBALANCE (MECHANICAL) **(M-R-08)**
- FAILURE OF OIL-WIPER **(M-R-12)**
- GROUNDED SHAFT (MOMENTARILY) **(M-R-14)**
- MECHANICAL FAILURE OF ROTOR COMPONENT **(M-R-18)**
- RUB BETWEEN BEARING AND JOURNAL **(M-R-20)**
- RUB OF HYDROGEN SEAL **(M-R-21)**
- RUB OF OIL WIPER **(M-R-22)**
- VIBRATION OF ROTOR—HIGH **(M-R-30)**

The stator winding will see transient forces from the sudden voltage mismatch and likely result in high endwinding forces that can result in loosening of endwinding supports and also slot wedging. The amount of movement of the stator winding and damage incurred will depend on the degree of the voltage mismatch at the moment of breaker closure. Machines have been known to have a catastrophic failure due to these types of events.

- CRACKED CONDUCTOR STRANDS **(M-S-10)**
- DAMAGE TO CONDUCTOR BAR INSULATION **(M-S-15)**
- DAMAGE TO HV BUSHINGS **(M-S-17)**
- FAILURE OF RADIAL PHASE CONNECTOR **(M-S-19)**
- LOOSE GENERATOR FOOTING **(M-C&F-17)**
- LOOSENING OF STATOR ENDWINDING BLOCKING **(M-S-40)**
- RUPTURED WATER MANIFOLD (HEADER) **(M-S-47)**
- RUPTURED WATER PARALLEL MANIFOLD (HEADER) **(M-S-48)**
- STATOR COOLING WATER (SCW) LEAK INTO GENERATOR FROM MANIFOLDS **(M-S-50)**
- STATOR COOLING WATER (SCW) LEAK INTO GENERATOR FROM PHASE PARALLEL **(M-S-49)**
- VIBRATION OF STATOR ENDWINDING—HIGH **(M-S-56)**

POSSIBLE SYMPTOMS:

This will be very obvious due to the nature of the event, by reaction of the synchronizing protection. There may also be a noticeable shaking of the turbine generator deck if the synchronization is rejected or the synchronization is violent in terms of the lateral and torsional effect on the generator rotor and transient forces on the stator winding.

- VIBRATION OF ROTOR—HIGH **(S-R-10)**
- SHORTED TURNS IN FIELD WINDING **(S-R-07)**
- VIBRATION OF STATOR ENDWINDING HIGH **(S-S-21)**

OPERATORS MAY CONSIDER

If the out-of-phase synchronization is small in terms of voltage angles and magnitude difference between the system and the generator terminal voltage, there may be no concern in continuing operation. However, if there is a noticeable *thump* to the unit, operators may elect to take the unit offline at the earliest opportunity (if it actually did synchronize and was not rejected), and open the generator up for inspection of the stator endwinding and the rotor.

The following is an extract from an OEM's recommendation:

Generally, if the out-of-phase angle is 20° or less, and the unit is running OK (no high vibration or abnormal currents/temperatures), then we *might* not investigate. If out of phase is over 20°, we would generally open the unit to inspect for loose end windings on the stator.

TEST AND VISUAL INDICATIONS:

- Shutdown of the unit for IR and PI of the stator and rotor windings.
- Partial disassembly of the generator by removing the generator end doors for inspection of the stator endwinding area and the rotor.

REFERENCE

1. G. Klempner and I. Kerszenbaum. *Handbook of Large Turbo-Generator Operation and Maintenance*. Piscataway, NJ: IEEE Press, 2008, chapters 4, 8, and 9.

M-E-20: TEMPERATURE OF INLET COOLING AIR TO BRUSHGEAR—HIGH

APPLICABILITY:

All externally excited generators (via sliprings and brushes)

MALFUNCTION DESCRIPTION:

This malfunction addresses the situation where the temperature of the cooling air entering the brushgear compartment is too high, resulting in components of the brushgear overheating.

PROBABLE CAUSES:

- Ambient temperature too high
- Failure of heat exchangers—in those not-so-common cases where the air inlet to the brushgear is force-cooled

POSSIBLE CONSEQUENCES:

Overheating of the brushgear can lead, first, to excessive wear of the brushes and sliprings, and finally, to damage to insulation causing short-circuits between polarities and/or ground faults.

POSSIBLE SYMPTOMS:

- HIGH RATE OF BRUSH WEAR **(S-E-06)**
- PRESSURE DROP ACROSS BRUSHGEAR FILTER—HIGH **(S-E-12)**
- TEMPERATURE OF SLIPRINGS/BRUSHGEAR ABOVE NORMAL **(S-E-16)**

OPERATORS MAY CONSIDER:

1. Starting immediately troubleshooting the problem by first identifying the source of the problem.
2. If the troubleshooting is not expected to happen in short course, or the root cause is not readily identifiable, reducing load such that the field current is lowered.
3. Keeping track of temperature of brushes and sliprings with an infrared gun, and comparing with normal temperatures. If the temperatures are too high, mitigating the situation by lowering field current or shutting down the unit.

TEST AND VISUAL INDICATIONS:

- Ambient temperature too high
- Visual inspection of heat exchanger, if one exists

REFERENCE

1. G. Klempner and I. Kerszenbaum. *Handbook of Large Turbo-Generator Operation and Maintenance*. Piscataway, NJ: IEEE Press, 2008, chapters 3, 5, and 10.

M-E-21: TEMPERATURE OF SLIPRINGS/ BRUSHGEAR ABOVE NORMAL

APPLICABILITY:

All generators with external excitation

MALFUNCTION DESCRIPTION:

This malfunction addresses the situation where the brushgear (sliprings, brushes, busses, etc.) exhibit temperatures above *normal* during operation. By *normal*, it means temperatures that under same operating and ambient conditions have been measured lower in previous occasions on the same machine, or that they are above OEM-specified maximum temperatures, or above what is considered good industry practice.

Higher than normal temperatures can be the result of insufficient cooling of the brushgear compartment, or degraded commutation.

This issue is covered under the following malfunction descriptions:

- BLOCKAGE OF BRUSHGEAR COOLING AIR **(M-E-02)**
- DAMAGE/DEGRADATION OF THE BRUSHGEAR **(M-E-03)**

REFERENCE

1. G. Klempner and I. Kerszenbaum. *Handbook of Large Turbo-Generator Operation and Maintenance*. Piscataway, NJ: IEEE Press, 2008, chapters 1, 2, and 9.

M-E-22: TERMINAL VOLTAGE EXCURSION—OFFLINE

APPLICABILITY:

All generators at *rated speed but not synchronized* to the grid (breaker open)

MALFUNCTION DESCRIPTION:

This malfunction addresses the case where the machine is offline and running at its rated speed, and the voltage is below 95% or above 105% of its *nominal* value.

At other speeds (lower than rated, as the unit is brought up for synchronization), the limiting parameter is V/Hz.

At rated speed, allowing the open-circuit voltage to reach beyond *nominal + 5%* value will create a serious risk of overfluxing (unless the generator is designed for wider operational terminal voltage limits).

Note: The specific generator under scrutiny may have, by design, a higher maximum limit than *nominal + 5%*, in which case, *that* would be the limit not to be exceeded.

This topic is covered under

- OVERFLUXING OFFLINE **(M-E-14)**

POSSIBLE SYMPTOMS:

- ANOMALOUS GENERATOR OUTPUT VOLTAGE **(S-E-01)**

REFERENCE

1. G. Klempner and I. Kerszenbaum. *Handbook of Large Turbo-Generator Operation and Maintenance*. Piscataway, NJ: IEEE Press, 2008, chapters 4 and 6.

M-E-23: TERMINAL VOLTAGE EXCURSION—ONLINE

APPLICABILITY:

All generators when *synchronized* to the grid

MALFUNCTION DESCRIPTION:

This malfunction addresses the situation when the terminal voltage of the generator deviates beyond its maximum or minimum values.

Most generators follow the IEEE and IEC standards that stipulate that the generators must be capable of operating within ±5% of their rated terminal voltage. Some specially designed machines are manufactured for operation within a wider margin of voltages. In either case, operation beyond the maximum and minimum limits set by the OEM will most probably result in degradation of certain critical components.

Terminal voltage deviations may originate with the generator, or may be driven by the grid. In each case the response of the machine will be somewhat different, but the result from excursions beyond terminal voltage allowed limits, is the same. If the grid is driving the voltage, the generator AVR will try to respond by increasing or reducing the field current. If settings are not properly defined in the excitation protection, the field current may grow beyond its rated continuous maximum, or on the other hand, be reduced to the extent the unit loses stability.

PROBABLE CAUSES:

- Operator error
- grid voltage too low or too high
- AVR malfunction

POSSIBLE CONSEQUENCES:

- LOSS OF STABILITY **(M-R-17)**
- OVEREXCITATION ONLINE **(M-E-13)**
- OVERFLUXING ONLINE **(M-E-15)**

POSSIBLE SYMPTOMS:

- ANOMALOUS GENERATOR OUTPUT VOLTAGE **(S-E-01)**

Large MVAR changes:

- MVAR TRANSIENT **(S-E-08)**

OPERATORS MAY CONSIDER:

1. Bringing the voltage back into allowable operating limits immediately. This may require manual control of the field current, especially if the AVR is malfunctioning (unless there are specific station procedures requiring remaining operating in these conditions due to grid stability considerations).
2. Assessing any possible damage and if necessary, schedule appropriate tests and inspections.

TEST AND VISUAL INDICATIONS:

N/A

REFERENCE

1. G. Klempner and I. Kerszenbaum. *Handbook of Large Turbo-Generator Operation and Maintenance*. Piscataway, NJ: IEEE Press, 2008, chapter 6.

5.6 Hydrogen System Malfunctions

M-H-01: AIR OR HYDROGEN HEAT-EXCHANGER COOLING-WATER SUPPLY OFF

APPLICABILITY:

- All hydrogen-cooled generators
- Air-cooled generators with heat exchangers

MALFUNCTION DESCRIPTION:

This malfunction addresses the situation where the primary cooling water supply to an air or hydrogen cooler has been left closed.

In some instances, manually adjusting the supply of cooling water is not uncommon, in particular in cooler months, or in cooler areas, or both. In other instances, the valves are always fully open. It has happened more than once that operators or maintenance personnel wrongly closed a valve or number of valves isolating a cooler or all coolers in a generator. This is most commonly done during an outage or coming out of an outage.

PROBABLE CAUSES:

- Operator error
- Maintenance personnel error

POSSIBLE CONSEQUENCES:

- BLOCKAGE OF AIR COOLERS (M-H-02)
- BLOCKAGE OF HYDROGEN COOLERS (M-H-03)

POSSIBLE SYMPTOMS:

Refer to:

- BLOCKAGE OF AIR COOLERS (M-H-02)
- BLOCKAGE OF HYDROGEN COOLERS (M-H-03)

OPERATORS MAY CONSIDER:

Refer to:

- BLOCKAGE OF AIR COOLERS (M-H-02)
- BLOCKAGE OF HYDROGEN COOLERS (M-H-03)

TEST AND VISUAL INDICATIONS:

Closed air or hydrogen cooling water valves

REFERENCE

1. G. Klempner and I. Kerszenbaum. *Handbook of Large Turbo-Generator Operation and Maintenance*. Piscataway, NJ: IEEE Press, 2008, chapters 2 and 10.

M-H-02: BLOCKAGE OF AIR COOLERS

APPLICABILITY:

All air-cooled generators with heat exchangers

MALFUNCTION DESCRIPTION:

This malfunction addresses the situation where the primary cooling water to an air cooler(s) in an air-cooled generator is partially or totally blocked. This can occur due to a number of reasons, with similar result: inefficient removal of the heat generated in the windings, cores, leads, and so on.

Blockage may happen in only one cooler, or perhaps in all coolers inside the generator, depending on the location and type of problem. If the blockage is in only one out of two or four coolers, there is the distinct possibility of some thermal asymmetry introduced in the frame and stator, with somewhat different symptoms than if all coolers are blocked equally. The differences will show in the temperature measurements, and perhaps also in the magnitude of the vibrations and ultimately any damage incurred.

PROBABLE CAUSES:

- AIR or HYDROGEN HEAT-EXCHANGER COOLING-WATER SUPPLY OFF **(M-H-01)**
- GAS LOCKING OF COOLING WATER IN HYDROGEN OR AIR HEAT EXCHANGERS **(M-H-07)**
- Silting or fouling of the heat exchanger
- Leaks of cooling water external to the generator

POSSIBLE CONSEQUENCES:

- GLOBAL STATOR TEMPERATURE ABOVE NORMAL **(M-S-25)**
- GLOBAL TEMPERATURE OF CORE ABOVE NORMAL **(M-C&F-09)**
- TEMPERATURE OF FIELD WINDING ABOVE NORMAL **(M-R-26)**

POSSIBLE SYMPTOMS:

- When a cooler is ineffective, the differential temperature between the *hot gas* (entering the heat exchanger) and the *cold gas* (leaving the cooler) becomes very small.
- Alarms from the air cooling system.
- *Liquid-in-generator* alarm (if large amounts of primary cooling water leaks from the cooler).
- Increase in vibrations may happen when not all coolers are equally blocked.

OPERATORS MAY CONSIDER:

1. Operators may consider reducing load as required to maintain all component temperatures within normal limits. If this is possible, then the unit can be conveniently scheduled for troubleshooting.
2. If the temperatures keep increasing beyond allowable limits, consider shutting down the unit for repairs.

TEST AND VISUAL INDICATORS:

- Residuals from water leaks
- Pressure tests of heat exchangers and cooling water system
- Damage/stuck valves
- Indications of wrong alignment of the air cooling water system

REFERENCE

1. G. Klempner and I. Kerszenbaum. *Handbook of Large Turbo-Generator Operation and Maintenance*. Piscataway, NJ: IEEE Press, 2008, chapters 2 and 10.

M-H-03: BLOCKAGE OF HYDROGEN COOLERS

APPLICABILITY:

All hydrogen-cooled generators

MALFUNCTION DESCRIPTION:

This malfunction addresses the situation where the primary cooling water path of a hydrogen cooler(s) in a hydrogen-cooled generator is partially or totally blocked. This can occur due to a number of reasons, with similar results: inefficient removal of the heat generated in the windings, cores, leads, flux shields, and so on.

Blockage can happen in only one cooler, or perhaps in all coolers inside the generator, depending on the location and type of problem. If the blockage is in only one out of two or four coolers, there is the distinct possibility of some thermal asymmetry introduced in the frame and stator, with somewhat different indications than if all coolers are blocked equally. The differences will show in the temperature measurements, and perhaps also in the magnitude of the vibrations or damage incurred, if any.

Figure 5.116 shows a case of excessive fouling of H_2 coolers.

PROBABLE CAUSES:

- GAS LOCKING OF COOLING WATER IN HYDROGEN OR AIR HEAT-EXCHAGERS (M-H-07)
- Silting or fouling of the heat exchanger
- Leaks of cooling water external to the generator
- Operator error (not opening cooling water valves coming out an outage)
- Maintenance personnel error

FIGURE 5.116
Some of the internal components of a static excitation system.

POSSIBLE CONSEQUENCES:

- GLOBAL STATOR TEMPERATURE ABOVE NORMAL (M-S-25)
- GLOBAL TEMPERATURE OF CORE ABOVE NORMAL (M-C&F-09)
- TEMPERATURE OF FIELD WINDING ABOVE NORMAL (M-R-25)

POSSIBLE SYMPTOMS:

- Most hydrogen-cooled generators are designed with a *cold gas* temperature of close to 46°C, according to the pertinent ANSI/IEEE standard. An indication of a possible cooling problem would be cold gas temperature measuring significantly higher than that.
- When a cooler is ineffective, the differential temperature between the *hot gas* (entering the heat exchanger) and the *cold gas* (leaving the cooler) becomes very small.
- Alarms from the hydrogen cooling system.
- *Liquid-in-generator* alarm (if large amounts of primary cooling water leaks from the cooler).
- Increase in vibrations may happen when not all coolers are equally blocked.

OPERATORS MAY CONSIDER:

1. Operators may consider reducing load as required to maintain all component temperatures within normal limits. If this is possible, then the unit can be conveniently scheduled for troubleshooting.
2. If the temperatures keep increasing beyond allowable limits, consider shutting down the unit for repairs.

TEST AND VISUAL INDICATORS:

- Residuals from water leaks
- Pressure tests of heat exchangers and cooling water system
- Damage/stuck valves
- Indications of wrong alignment of the hydrogen cooling water system

REFERENCE

1. G. Klempner and I. Kerszenbaum. *Handbook of Large Turbo-Generator Operation and Maintenance*. Piscataway, NJ: IEEE Press, 2008, chapters 2 and 10.

M-H-04: DEGRADED HYDROGEN PURITY

APPLICABILITY:

All hydrogen-cooled generators

MALFUNCTION DESCRIPTION:

This malfunction addresses the situation where the purity of the hydrogen in the generator is degraded due to contamination, for instance, by moisture, CO_2, or air.

Purity in generators in maintained above 95%, preferably above 98%. Decreased purity has the immediate effect of increasing windage losses, resulting in lower efficiency and increased temperatures inside the generator. There is also the concern that, in the case of air, a mixture will be reached that is potentially explosive (between 4% and 74%).

As an example, a decrease in hydrogen purity affects the generator efficiency by increasing the windage losses. A typical 400 MW unit at 91% purity will have windage losses approximately three times higher than when operating at 99% purity. A typical 800 MW unit at 91% purity will have windage losses approximately seven times higher than when operating at 99% purity (Figure 5.117).

PROBABLE CAUSES:

- The purity may be degraded due to water leaks, air ingression through the bearings and seals, moisture, CO_2 (from purging gas), and so on.
- Degraded supply of hydrogen (*wet* hydrogen).

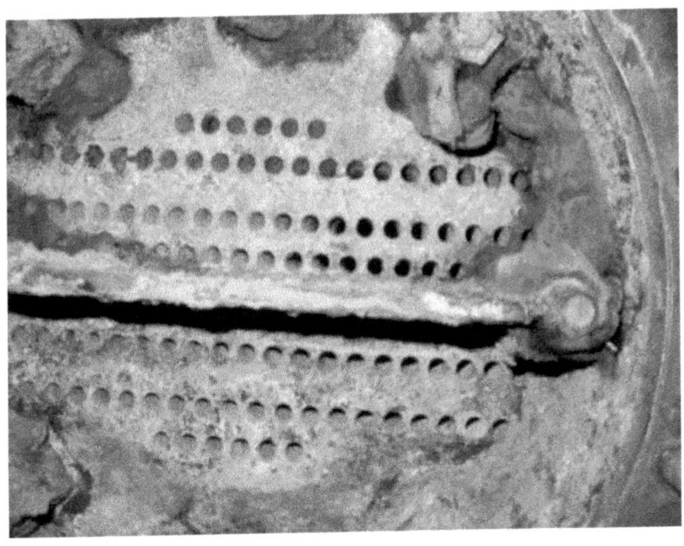

FIGURE 5.117
Heavy fouling of a hydrogen heat exchanger.

- INEFFECTIVE SEAL-OIL DETRAINING **(M-SOS-09)**
- LEAK OF SEAL OIL INTO GENERATOR **(M-SOS-12)**

POSSIBLE CONSEQUENCES:

- Very low purity due to high concentration of air could lead to an explosive atmosphere inside the casing
- Low efficiency
- Somewhat elevated temperatures due to extra windage losses

POSSIBLE SYMPTOMS:

- DEGRADED HYDROGEN PURITY **(S-H-02)**
- ELEVATED HYDROGEN DEW POINT **(S-H-03)**
- HYDROGEN-IN-STATOR COOLING WATER (SCW) HIGH **(S-S-04)**
- LIQUID-IN-GENERATOR ALARM **(S-C&F-04)**

Refer to:

- FAILURE OF HYDROGEN DRYER **(M-H-06)**
- LEAK OF HYDROGEN INTO H_2-COOLING WATER SYSTEM **(M-H-13)**
- MOISTURE IN HYDROGEN—HIGH **(M-H-14)**
- PHYSICAL BREACH OF HYDROGEN OR AIR HEAT EXCHANGERS **(M-H-15)**
- WATER LEAK INTO GENERATOR FROM AIR OR H_2 COOLER **(M-H-18)**

OPERATORS MAY CONSIDER:

1. Finding out the cause for the lower purity readings.
2. If the hydrogen supply is the source of the problem, may consider replacing the supply with high-purity dry hydrogen.
3. Purity can be increased by purging hydrogen with clean dry hydrogen (some machines do not have hydrogen driers and maintaining purity requires operators to purge and replace hydrogen periodically).

TEST AND VISUAL INDICATIONS:

Purity indicator if installed

REFERENCES

1. G. Klempner and I. Kerszenbaum. *Handbook of Large Turbo-Generator Operation and Maintenance*. Piscataway, NJ: IEEE Press, 2008, chapters 2, 4, and 5.
2. J. Speranza. The Importance of Pure, Dry Hydrogen Cooling Gas. Turbine Tech, ENERGY-TECH.com, June 2014.
3. M. Kolodziej, E. Borkey, and T. Reynolds. Water Contamination in Hydrogen-Cooled Generators Lurks as Serious Operational Threat. *Power Engineering magazine online*, August 2003.
4. J. Bothwell. Monitoring Moisture in Hydrogen Cooled Generators. In: *EPRI On-Line Monitoring and Condition Assessment Workshop*, Nashville, TN, August 26, 2003.

M-H-05: ELEVATED HYDROGEN DEW POINT

APPLICABILITY:

All hydrogen-cooled generators

MALFUNCTION DESCRIPTION:

This malfunction addresses the situation where the dew point of the hydrogen inside the generator casing is higher than desirable.

Hydrogen dew point temperature is an indicator of the moisture content in the hydrogen gas inside the generator. Moisture is undesirable for the stator and rotor insulation systems, since it can initiate insulation failure by electrical tracking, and for various steel components in the generator due to rusting and corrosive effects. The exposed rotor windings at the overhangs can be degraded by moisture condensation when the machine is offline. Hydrogen dew point may be monitored on a continuous basis by a dew point indicator and should be maintained at a value much lower than the expected cooling water temperature. It is recommended that the dew point be maintained at less than 0°C. An alarm sometimes is provided to alert operators if the dew point rises above this set point.

PROBABLE CAUSES:

- FAILURE OF HYDROGEN DRYER **(M-H-06)**
- DEGRADED CONDITION OF THE SEAL OIL **(M-SOS-03)**
- LEAK OF CONDUCTING-BAR COOLING WATER INTO GENERATOR **(M-S-36)**
- LEAK OF STATOR COOLING WATER (SCW) INTO GENERATOR **(M-S-37)**
- LEAK OF STATOR COOLING WATER (SCW) INTO GENERATOR FROM TERMINALS **(M-S-38)**
- WATER LEAK INTO GENERATOR FROM AIR- OR H_2 COOLER **(M-H-18)**
- Contaminated hydrogen supply
- Malfunction of the seal-oil system
- Malfunction of the vacuum treatment plant
- Leakage of RCW into the generator (generators with water-cooled rotors)
- Moisture collected during an open-casing maintenance activity or overhaul

POSSIBLE CONSEQUENCES:

- Corrosion of critical rotor components (e.g., 18Mn-5Cr retaining and zone rings)
- Electric tracking
- Some degradation of rotor windings

POSSIBLE SYMPTOMS:

- ELEVATED HYDROGEN DEW POINT **(S-H-03)**
- HYDROGEN INTO STATOR COOLING WATER (SCW) HIGH **(S-S-04)**
- LIQUID-IN-GENERATOR ALARM **(S-C&F-04)**
- Alarms from the hydrogen dryers

OPERATORS MAY CONSIDER:

1. The dew point can be reduced somewhat by purging periodically with dry hydrogen. This can keep the machine operating until the root cause is troubleshooted.
2. The cause for elevated moisture ought to be troubleshooted as soon as possible.

TEST AND VISUAL INDICATORS:

- Internal corrosion of ferrous metals
- Water deposits

REFERENCES

1. G. Klempner and I. Kerszenbaum. *Handbook of Large Turbo-Generator Operation and Maintenance*. Piscataway, NJ: IEEE Press, 2008, chapter 5.
2. J. Speranza. The Importance of Pure, Dry Hydrogen Cooling Gas. Turbine Tech, ENERGY-TECH.com, June 2014.
3. M. Kolodziej, E. Borkey, and T. Reynolds. Water Contamination in Hydrogen-Cooled Generators Lurks as Serious Operational Threat. *Power Engineering magazine online*, August 2003.
4. J. Bothwell. *Monitoring Moisture in Hydrogen Cooled Generators*, EPRI On-Line Monitoring and Condition Assessment Workshop, Nashville, TN, August 26, 2003.

M-H-06: FAILURE OF HYDROGEN DRYER

APPLICABILITY:

All hydrogen-cooled generators

MALFUNCTION DESCRIPTION:

This malfunction addresses the case where the dryer system (Figure 5.118) installed with the purpose of removing moisture from the hydrogen flowing into the casing of the generator fails to perform its designed function, partially or entirely.

Moisture can emanate from small water leaks from directly water-cooled stators (and/or water-cooled rotors), from seal oil or from a poor hydrogen supply. Moisture will elevate hydrogen dew point, increasing the corrosion risk of critical components, but also will reduce the electrical breakdown voltage of the hydrogen, which could under certain conditions lead to electrical tracking and short circuits.

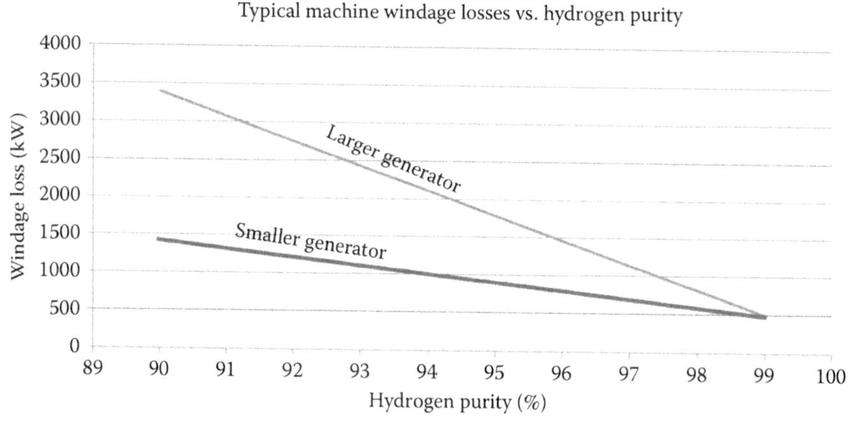

FIGURE 5.118
The impact on windage loss by the purity of the hydrogen for a relatively large and a smaller generator.

PROBABLE CAUSES:

- Degraded desiccant in the dryers
- Malfunction of other dryer components
- Degraded *wet* hydrogen supply

POSSIBLE CONSEQUENCES:

- ELEVATED HYDROGEN DEW POINT **(M-H-05)**

POSSIBLE SYMPTOMS:

- DEGRADED HYDROGEN PURITY **(S-H-02)**
- ELEVATED HYDROGEN DEW POINT **(S-H-03)**

OPERATORS MAY CONSIDER:

1. After confirming that it is a real event (not instrumentation), consider troubleshooting the dryer system as soon as possible
2. Dew point may be temporarily restored by periodically purging hydrogen and replacing it with dry hydrogen

TEST AND VISUAL INDICATIONS:

Test for degraded desiccant in the H_2 dryers

REFERENCES

1. G. Klempner and I. Kerszenbaum. *Handbook of Large Turbo-Generator Operation and Maintenance*. Piscataway, NJ: IEEE Press, 2008, chapters 9 and 10.
2. J. Speranza. The Importance of Pure, Dry Hydrogen Cooling Gas. Turbine Tech, ENERGY-TECH.com, June 2014.
3. M. Kolodziej, E. Borkey, and T. Reynolds. Water Contamination in Hydrogen-Cooled Generators Lurks as Serious Operational Threat. *Power Engineering magazine online*, August 2003.
4. J. Bothwell: *Monitoring Moisture in Hydrogen Cooled Generators*, EPRI On-Line Monitoring and Condition Assessment Workshop, Nashville, TN, August 26, 2003.

M-H-07: GAS LOCKING OF COOLING WATER IN HYDROGEN OR AIR HEAT EXCHANGERS

APPLICABILITY:

- All hydrogen-cooled generators
- All air-cooled generators with heat exchangers

MALFUNCTION DESCRIPTION:

This malfunction addresses the case where, due to a physical breach, enough hydrogen has entered the hydrogen cooling water system to basically create a bubble of the gas, blocking the flow of cooling water. Similar case applies for air locking the water cooling system of an air heat exchanger.

PROBABLE CAUSES:

- HYDROGEN INGRESSION INTO STATOR COOLING WATER (M-S-29)
- INADEQUATE FLOW OF COOLING WATER TO AIR- OR H_2 COOLERS (M-H-10)

POSSIBLE CONSEQUENCES:

- BLOCKAGE OF AIR COOLERS (M-H-02)
- BLOCKAGE OF HYDROGEN COOLERS (M-H-03)

POSSIBLE SYMPTOMS:

Refer to:

- BLOCKAGE OF AIR COOLERS (M-H-02)
- BLOCKAGE OF HYDROGEN COOLERS (M-H-03)

OPERATORS MAY CONSIDER:

Refer to:

- BLOCKAGE OF AIR COOLERS (M-H-02)
- BLOCKAGE OF HYDROGEN COOLERS (M-H-03)

TEST AND VISUAL INDICATIONS:

- Pressure tests on heat exchanger indicating a physical breach.
- Carbonate deposits of water from the heat exchanger leaking into the generator.

REFERENCE

1. G. Klempner and I. Kerszenbaum. *Handbook of Large Turbo-Generator Operation and Maintenance.* Piscataway, NJ: IEEE Press, 2008, chapters 2, 3, 10, and 11.

M-H-08: HYDROGEN ESCAPING THE GENERATOR

APPLICABILITY:

All hydrogen-cooled generators

MALFUNCTION DESCRIPTION:

This malfunction addresses the case where hydrogen escapes from the generator. The hydrogen inside the casing of hydrogen-cooled generators is at a higher-than-atmospheric pressure. Therefore, it must be kept inside the machine by effective sealing methods. A breach of any of those systems will cause the hydrogen to escape.

The hydrogen has two main functions: (1) cooling medium and (2) insulating medium. By escaping, pressure may not be maintained, and that may cause a loss of cooling and insulation capacity.

Another less common issue that occasionally happens, when there is a hydrogen gas leak from the generator casing, is self-ignition of the hydrogen. This is a very serious safety hazard to personnel and/ or equipment. As the hydrogen leaks from the casing under high pressure the temperature elevates to a level high enough to ignite it. The point of leakage is invisible except for a shimmering heat wave and dripping water at the leak point as the hydrogen mixes with air to form water.

PROBABLE CAUSES:

- Hydrogen seals not sealing (wrong clearances, cracks, misalignment, etc.)
- Casing joints leaking
- Through cracks in casing
- Porous welds in casing
- Joints to piping (hydrogen, water)
- Joints to electrical ports
- Ineffective gaskets in manholes, coolers, and so on
- Ineffective sealing at end-bracket joints
- Ineffective sealing of bushing joint
- Cracked bushings
- Elevated vibrations allowing H_2 to escape via H_2 seals
- Leaks into SCW
- Leaks into hydrogen heat exchangers cooling water
- External connections (e.g., GCM)

POSSIBLE CONSEQUENCES:

- HYDROGEN PRESSURE INSIDE GENERATOR BELOW NORMAL **(M-H-09)**
- Self-ignition of hydrogen when it mixes with air in certain proportions
- Accumulation in closed areas with risks of explosion (e.g., Isophase-bus, brushgear housing)

POSSIBLE SYMPTOMS:

- Hydrogen monitors in strategic locations (e.g., inside the IPB, or brushgear compartment).
- Water traces on the generator casing or hydrogen pipes that appear to come from *nowhere* may indicate a hydrogen leak. They are the result of self-ignition of the hydrogen as it mixes with air. The flame is invisible, so it must be approached with caution.

Loss of hydrogen pressure in casing:

- HYDROGEN PRESSURE INSIDE THE GENERATOR BELOW NORMAL **(S-H-06)**
- High rate of hydrogen make-up
- *Hydrogen-in-SCW* alarm
- *Hydrogen-in-hydrogen cooler cooling water* alarm

Elevated rotor vibrations:

- VIBRATION OF ROTOR—HIGH **(S-R-10)**
- Hot hydrogen seals

OPERATORS MAY CONSIDER:

1. Every hydrogen-cooled generator exhibits a certain amount of hydrogen loss. Normally, this quantity is monitored daily as a minimum, or continuously, and any unexplained increase should trigger an investigation.
2. The urgency of the operator response depends largely on the rate of loss of hydrogen, and its trending.
3. Keep in mind that hydrogen may be accumulating in areas that may become unsafe; for instance, inside the brushgear-excitation housing. Thus, any unexplained hydrogen leak should be treated seriously.
4. If the leak is very large, operators should consider shutting down the unit immediately, in particular when the source of the leak is unknown.
5. Monitor temperatures and hydrogen dew point inside the machine. A lowering dew point may indicate the hydrogen in-leaking into one of the cooling water systems, and water may be leaking into the machine.
6. A water trail starting at a casing joint may indicate a hydrogen flame (which is invisible) is there (great caution is advised!).

TEST AND VISUAL INDICATORS:

- Pressure/vacuum tests
- Other leak tests (such as snoop [Figure 5.119] or SF_6)
- Hydrogen seals damage/degradation

REFERENCE

1. G. Klempner and I. Kerszenbaum. *Handbook of Large Turbo-Generator Operation and Maintenance*. Piscataway, NJ: IEEE Press, 2008, chapters 6 and 11.

M-H-09: HYDROGEN PRESSURE INSIDE GENERATOR BELOW NORMAL

APPLICABILITY:

All hydrogen-cooled generators

MALFUNCTION DESCRIPTION:

This malfunction addresses the situation where the hydrogen in the casing of a hydrogen-cooled generator is below its *normal pressure*. By *normal*,

FIGURE 5.119
The figure shows a twin-dryer arrangement on a hydrogen skid.

it means any of the vendor-specified operating pressures as required for the operating output of the machine.

The hydrogen in the generator must meet a number of requisites. For instance, *purity*, *moisture content*, and *pressure* must be within limits specified by the manufacturer. Any deviation may have serious and onerous effects on the integrity of the machine. Hydrogen at pressures below that specified by the vendor will not provide the required heat-removal capacity the machine is designed for. Therefore, critical components will overheat. Most hydrogen-cooled generators are designed for operation in one or more given pressures. For instance, 45 psi and 60 psi, or 60 psi, 65 psi, and so on. The vendor provides for each of those pressures a corresponding capability curve. Figure 5.120 shows such an example.

FIGURE 5.120
Applying an innocuous soapy liquid for the detection of hydrogen leaks.

Figure 5.120 shows three different hydrogen pressures (30, 45, and 60 psi) with their corresponding capability curve. The load point of the generator must always fall inside the pertinent capability curve.

PROBABLE CAUSES:

- HYDROGEN ESCAPING THE GENERATOR (M-H-08)
- Malfunction of the hydrogen pressure regulator
- Not enough make-up hydrogen
- PRESSURE OF HYDROGEN SUPPLY—LOW (M-H-16)
- Maintenance/operator error

PROBABLE CONSEQUENCES:

- GLOBAL TEMPERATURE OF CORE ABOVE NORMAL (M-C&F-09)
- LEAK OF STATOR COOLING WATER (SCW) INTO GENERATOR (M-S-37)

- TEMPERATURE OF FIELD WINDING ABOVE NORMAL (M-R-26)
- TEMPERATURE OF A STATOR CONDUCTOR ABOVE NORMAL (M-S-53)

POSSIBLE SYMPTOMS:

- Temperature alarms from windings
- Temperature alarms from core
- Temperature alarms from hydrogen
- Pyrolysis byproducts (alarm from GCM)
- Alarm from hydrogen pressure sensors
- Alarms from hydrogen cooling skid

OPERATORS MAY CONSIDER:

1. Operators ought to consider immediately reducing load so that the machine does not operate beyond its capability curve. If only one curve is provided by the vendor, it is very doubtful the operators will have the knowledge and time to calculate and draw a second one based on the new degraded hydrogen pressure. Thus, in this situation, probably the most prudent action is to shutdown the unit unless the hydrogen pressure is restored to normal immediately.
2. Bear in mind that in directly water-cooled stators the SCW should be at a pressure of no less than 5 psi below that of the hydrogen in the casing. A drop in hydrogen pressure may cause existing small leaks of hydrogen-into-SCW to reverse and cause undesirable leaks of SCW into the generator. Some OEMs provide clear guidelines about how long it is allowed to operate with low differential pressures between the H_2 and the SCW in the casing, in particular when a significant leak exists.

TEST AND VISUAL INDICATORS:

No direct residual telling signs, unless the unit's temperature limits were exceeded for a significant length of time and by a wide margin. In this case, discoloration of components—in particular organics—could be seen.

REFERENCE

1. G. Klempner and I. Kerszenbaum. *Handbook of Large Turbo-Generator Operation and Maintenance*. Piscataway, NJ: IEEE Press, 2008, chapter 4.

M-H-10: INADEQUATE FLOW OF COOLING WATER TO AIR OR H_2 COOLERS

APPLICABILITY:

- Hydrogen-cooled generators
- Air-cooled generators with water-cooled heat exchangers (*TEWAC*[11] units)

MALFUNCTION DESCRIPTION:

This malfunction addresses the situation where the flow of cooling water to the air or hydrogen heat exchangers, is below the required to maintain effective cooling of the generator.

PROBABLE CAUSES:

- AIR OR HYDROGEN HEAT-EXCHANGER COOLING-WATER SUPPLY OFF **(M-H-01)**
- BLOCKAGE OF AIR COOLERS **(M-H-02)**
- BLOCKAGE OF HYDROGEN COOLERS **(M-H-03)**
- GAS LOCKING OF COOLING WATER IN HYDROGEN OR AIR HEAT EXCHANGERS **(M-H-07)**
- PHYSICAL BREACH OF HYDROGEN OR AIR HEAT EXCHANGER **(M-H-15)**
- Cooling-water pump failure
- Failure of the regulating valve on the air or hydrogen cooling water skid
- Silting/fouling of cooling water pipes
- Blockage of pipes due to maintenance error
- Blockage of pipes due to debris

POSSIBLE CONSEQUENCES:

Refer to:

- BLOCKAGE OF AIR COOLERS **(M-H-02)**
- BLOCKAGE OF HYDROGEN COOLERS **(M-H-03)**

POSSIBLE SYMPTOMS:

Refer to:

- BLOCKAGE OF AIR COOLERS **(M-H-02)**
- BLOCKAGE OF HYDROGEN COOLERS **(M-H-03)**

---

[11] TEWAC—Total enclosed water to air cooled.

OPERATORS MAY CONSIDER:

Refer to:

- BLOCKAGE OF AIR COOLERS **(M-H-02)**
- BLOCKAGE OF HYDROGEN COOLERS **(M-H-03)**

TEST AND VISUAL INDICATIONS:

Refer to:

- BLOCKAGE OF AIR COOLERS **(M-H-02)**
- BLOCKAGE OF HYDROGEN COOLERS **(M-H-03)**

REFERENCE

1. G. Klempner and I. Kerszenbaum. *Handbook of Large Turbo-Generator Operation and Maintenance*. Piscataway, NJ: IEEE Press, 2008, chapters 2 and 10.

M-H-11: INLET TEMPERATURE OF AIR OR HYDROGEN COOLING WATER—HIGH

APPLICABILITY:

- All hydrogen-cooled generators
- All air-cooled generators with water-cooled heat exchangers (TEWAC[12] units)

MALFUNCTION DESCRIPTION:

This malfunction describes the case where the temperature of the inlet cooling water to the air or hydrogen heat exchangers is high. This means that the effectiveness of the air or hydrogen coolers is compromised to the extent that in some cases, the temperature of critical components inside the machine cannot be maintained within allowable margins.

PROBABLE CAUSES:

The source of the primary (raw) cooling water to the air or hydrogen coolers is too warm.

POSSIBLE CONSEQUENCES:

- TEMPERATURE OF COLD GAS ABOVE NORMAL **(M-H-17)**

POSSIBLE SYMPTOMS:

- COOLING-GAS HEAT-EXCHANGER DIFFERENTIAL TEMPERATURE—HIGH **(S-H-01)**
- TEMPERATURE OF COLD GAS ABOVE NORMAL **(S-H-10)**

[12] TEWAC—Total enclosed water to air cooled.

OPERATORS MAY CONSIDER:

1. The most immediate action for consideration is reducing load so that all temperatures in the machine are maintained within their permissive range.
2. Depending on the source of the cooling water, the temperature may or may not be adjusted. In some stations, the source is greatly affected by the season (e.g., warm river or lake water), in others the water is cooled in large cooling towers, and the source of the problem may be a fault in one of the towers (for instance, loss of power to fan motors).

TEST AND VISUAL INDICATIONS:

- If natural source such as a river: measuring river/lake temperature.
- If cooling towers: there may be failed motors, fans, power supply fuses, and so on.

REFERENCE

1. G. Klempner and I. Kerszenbaum. *Handbook of Large Turbo-Generator Operation and Maintenance*. Piscataway, NJ: IEEE Press, 2008, chapters 4 and 5.

M-H-12: INSUFFICIENT PERFORMANCE OF AIR OR HYDROGEN HEAT EXCHANGERS

APPLICABILITY:

- All hydrogen-cooled generators
- Air-cooled generators with water-cooled heat exchangers (TEWAC[13] units)

MALFUNCTION DESCRIPTION:

This malfunction addresses the situation where the air or hydrogen heat exchanger is not performing its designed function, that is, the cold gas temperature is above its required range, despite the cooling gas going through the cooler(s).

PROBABLE CAUSES AND POSSIBLE CONSEQUENCES:

- AIR OR HYDROGEN HEAT-EXCHANGER COOLING-WATER SUPPLY OFF (M-H-01)
- BLOCKAGE OF AIR COOLERS (M-H-02)
- BLOCKAGE OF HYDROGEN COOLERS (M-H-03)
- FAILURE OF HYDROGEN DRYER (M-H-06)

[13] TEWAC—Total enclosed water to air cooled.

- GAS LOCKING OF COOLING WATER IN HYDROGEN OR AIR HEAT EXCHANGERS **(M-H-07)**
- INADEQUATE FLOW OF COOLING WATER TO AIR OR H_2 COOLERS **(M-H-10)**
- INLET TEMPERATURE OF AIR OR HYDROGEN COOLING WATER HIGH **(M-H-11)**
- PHYSICAL BREACH OF HYDROGEN OR AIR HEAT EXCHANGER **(M-H-15)**
- TEMPERATURE OF COLD GAS ABOVE NORMAL **(M-H-18)**

POSSIBLE SYMPTOMS:

- COOLING-GAS HEAT-EXCHANGER DIFFERENTIAL TEMPERATURE—HIGH **(S-H-01)**
- TEMPERATURE OF COLD GAS ABOVE NORMAL **(S-H-10)**

OPERATORS MAY CONSIDER:

1. Reducing load until all temperatures inside the machine are within their allowable range.
2. There are many possible causes for an underperforming air or hydrogen heat exchanger. The next activities should focus on identifying the source of the problem and troubleshooting it.

TEST AND VISUAL INDICATORS:

The spectra of possible causes for inadequate operation of air or hydrogen heat exchangers are numerous, and thus, the search may yield a number of symptoms, depending on each specific case. However, in most cases it can be reduced to warm raw water to the cooler, closed valves, damaged flow regulators, or leaks and/or blockages.

REFERENCE

1. G. Klempner and I. Kerszenbaum. *Handbook of Large Turbo-Generator Operation and Maintenance*. Piscataway, NJ: IEEE Press, 2008, chapters 2, 3, 4, 5, 9, 10, and 11.

M-H-13: LEAK OF HYDROGEN INTO H_2-COOLING WATER SYSTEM

APPLICABILITY:

All hydrogen-cooled generators

MALFUNCTION DESCRIPTION:

This malfunction addresses the situation where hydrogen is leaking into the raw (primary) water-cooling circuit of the hydrogen heat exchangers (coolers).

This topic is covered under the malfunction:

- PHYSICAL BREACH OF HYDROGEN OR AIR HEAT EXCHANGER (M-H-15)

REFERENCE

1. G. Klempner and I. Kerszenbaum. *Handbook of Large Turbo-Generator Operation and Maintenance*. Piscataway, NJ: IEEE Press, 2008, chapters 3, 5, and 8.

M-H-14: MOISTURE IN HYDROGEN HIGH

APPLICABILITY:

All hydrogen-cooled generators

MALFUNCTION DESCRIPTION:

This malfunction describes the situation where the moisture content of the hydrogen inside the generator is high, to the extent the dew point of the hydrogen is above normal, and its electrical withstand capability is reduced.

This malfunction is covered under:

- ELEVATED HYDROGEN DEW POINT (M-H-05)

REFERENCES

1. G. Klempner and I. Kerszenbaum. *Handbook of Large Turbo-Generator Operation and Maintenance*. Piscataway, NJ: IEEE Press, 2008, chapter 5.
2. EPRI Technical Report. Generator Retaining Ring Moisture Protection Guide, September 1993.
3. J.D. Albright and D.R. Albright. Generator Field Winding Shorted Turns: Moisture Effects. In: *EPRI—Steam Turbine-Generatortech Workshop and Vendor Exposition*, Nashville, TN, August 25–27, 2003.
4. M. Fenger and G.C. Stone. Investigations into the Effect of Humidity on Stator Winding Partial Discharges. *IEEE Transactions on Dielectrics and Electrical Insulation*, 12(2), April 2005.

M-H-15: PHYSICAL BREACH OF HYDROGEN OR AIR HEAT EXCHANGER

APPLICABILITY:

- Hydrogen-cooled generators
- Air-cooled generators with water-cooled heat exchangers (TEWAC[14] units)

[14] TEWAC—Total enclosed water to air cooled.

MALFUNCTION DESCRIPTION:

This malfunction addresses the situation where a physical failure of the hydrogen or air heat exchanger occurs, allowing primary cooling water to leak into the generator and in the case of hydrogen-cooled generators, hydrogen ingression into the primary water-cooling system.

In most air-cooled machines with water-cooled air, the location of the heat exchanger is such that a physical breach will not result in water directly falling over the stator windings and the rotor. However, the fast moving air inside the machine will carry moisture around, that may or may not condense depending on how much the unit is loaded (in a fully loaded machine with hot windings, condensation most probably will not happen), and what it the dew point of the hydrogen or air.

In hydrogen-cooled generators, the coolers may be located anywhere in the machine, depending on the cooler configuration. However, the hydrogen being at a higher pressure (usually 5 psi as a minimum) than the cooling water, hydrogen will tend to leak into the cooling water system when a physical breach occurs. Regardless, water may still leak into the machine due to *capillary action*:

- WATER LEAK INTO GENERATOR FROM AIR/H_2 COOLER **M-H-18**

PROBABLE CAUSES:

- A bad or corroded joint
- A bad or corroded tube
- Damaged tube or joint during removal or installation of the cooler
- Damaged tube or joint due to elevated frame vibrations

POSSIBLE CONSEQUENCES:

- GAS LOCKING OF COOLING WATER IN H_2 OR AIR HEAT EXCHANGERS **(M-H-07)**
- LEAK OF HYDROGEN INTO H_2 COOLING WATER SYSTEM **(M-H-13)**
- WATER LEAK INTO GENERATOR FROM AIR/H_2 COOLER **(M-H-18)**

POSSIBLE SYMPTOMS:

- HYDROGEN IS STATOR COOLING WATER (SCW) HIGH **(S-S-04)**
- LIQUID-IN-GENERATOR ALARM **(S-C&F-04)**

OPERATORS MAY CONSIDER:

1. Small leaks may be sustainable for some time, but the objective should be to troubleshoot them as soon as practically possible.
2. A large leak can be identified by the large amount of water removed from the generator drain valve and liquid leak detector, and/or a high rate of hydrogen-into-cooling water leaks. This may result in significant long-term damage and thus it may be prudent to shutdown the unit immediately for repairs.

TEST AND VISUAL INDICATORS:

- Visual inspection of damaged cooler
- Visual detection of lead carbonate or other water-produced deposits
- Pressure tests of heat exchanger

REFERENCE

1. G. Klempner and I. Kerszenbaum. *Handbook of Large Turbo-Generator Operation and Maintenance*. Piscataway, NJ: IEEE Press, 2008, chapters 2, 3, 5, 8, and 11.

M-H-16: PRESSURE OF HYDROGEN SUPPLY LOW

APPLICABILITY:

All hydrogen-cooled generators

MALFUNCTION DESCRIPTION:

This malfunction addresses the situation where the hydrogen make-up supply pressure is dropping, and cannot maintain the operational pressure of the hydrogen in the casing.

There are various types of supply systems. Some stations make their own hydrogen while others purchase it. The hydrogen supply consists of the pressurized reservoir, and the pressure regulators that maintain the hydrogen in the casing at its operational pressure.

PROBABLE CAUSES:

- The pressure regulators are faulty.
- The pressure regulators are set wrongly.
- There is an external leak in the piping between the source and the machine.
- The source of hydrogen is empty or almost empty (simple replenishing the supply is required).

This malfunction is described in:

- HYDROGEN PRESSURE INSIDE THE GENERATOR BELOW NORMAL **(M-H-09)**

REFERENCE

1. G. Klempner and I. Kerszenbaum. *Handbook of Large Turbo-Generator Operation and Maintenance*. Piscataway, NJ: IEEE Press, 2008, chapters 4 and 5.

M-H-17: TEMPERATURE OF COLD GAS ABOVE NORMAL

APPLICABILITY:

- All hydrogen-cooled generators
- All air-cooled generators with water-cooled heat exchangers (TEWAC units)

MALFUNCTION DESCRIPTION:

This malfunction addresses the case where the *cold gas* in hydrogen or air-cooled generators, respectively, is hotter than *normal*. By *normal*, it means temperatures that are substantially higher than in previous occasions under very similar conditions, or higher than recommended by the manufacturer and/or the pertinent standards.

For most generators, the typical cold gas temperature is between 30°C and 40°C. For hydrogen-cooled generators, the value provided by ANSI/IEEE standards is 46°C as maximum allowable temperature.

The causes, consequences, and operator-related activities covering this problem are contained in the following malfunction descriptions:

- AIR OR HYDROGEN HEAT-EXCHANGER COOLING WATER SUPPLY OFF **(M-H-01)**
- BLOCKAGE OF AIR COOLERS **(M-H-02)**
- BLOCKAGE OF HYDROGEN COOLERS **(M-H-03)**
- HYDROGEN PRESSURE INSIDE GENERATOR BELOW NORMAL **(M-H-09)**
- INADEQUATE FLOW OF COOLING WATER TO AIR- OR H_2 COOLERS **(M-H-10)**
- INLET TEMPERATURE OF AIR OR HYDROGEN COOLING WATER HIGH **(M-H-11)**

REFERENCE

1. G. Klempner and I. Kerszenbaum. *Handbook of Large Turbo-Generator Operation and Maintenance*. Piscataway, NJ: IEEE Press, 2008, chapter 4.

M-H-18: WATER LEAK INTO GENERATOR FROM AIR- OR H_2 COOLER

APPLICABILITY:

- All hydrogen-cooled generators
- All generators with water-cooled heat exchangers (TEWAC units)

MALFUNCTION DESCRIPTION:

This malfunction addresses the situation where cooling water is leaking into the machine from a hydrogen heat exchanger or a water-cooled air heat exchanger.

Figures 5.121 and 5.122 show typical hydrogen and air heat exchangers.

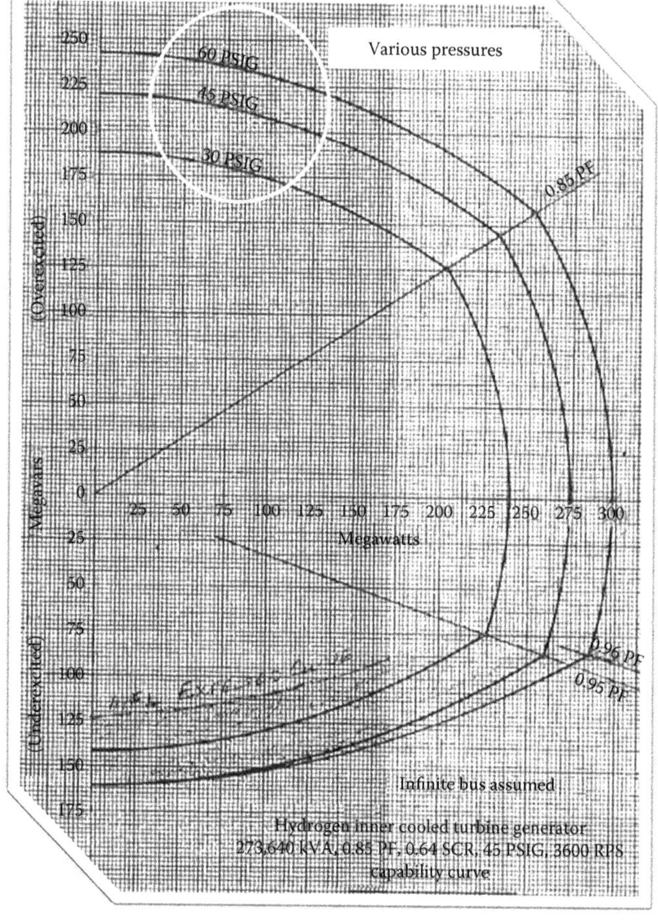

FIGURE 5.121
Capability curve shown permissive operating regions for three hydrogen pressures.

FIGURE 5.122
Vertical hydrogen cooler being removed from a large four-pole generator (there are four of those in the machine).

In the case of hydrogen-cooled generators, the pressure of the hydrogen inside the machine is by design no less than about 5 psi higher than the heat-exchanger cooling water. This is done with the purpose of avoiding a water leak into the machine in case a breach of the cooling path develops in the exchanger. In case of severe leaks, the higher hydrogen pressure will not eliminate water from leaking into the casing.

In the case of air-cooled machines, the air is at a lower pressure than the cooling water. However, in these machines it is not unusual to find the heat exchanger to be located in such a way that water will not leak directly onto the core and windings, though some of it is blown around the machine by the moving air (almost always the coolers are under the stator core).

PROBABLE CAUSES:

- Corrosion
- Weld failure
- Roll-joint failure of cooler tubes into the waterbox faceplate
- Bad *O-ring* or gasket
- Insufficient torque on bolts
- Inadequate design and/or manufacture of heat exchanger
- High frame vibration

POSSIBLE CONSEQUENCES:

- ELEVATED HYDROGEN DEW POINT **(M-H-05)**
- GAS LOCKING OF COOLING WATER IN H_2 OR AIR HEAT EXCHANGERS **(M-H-07)**
- GROUND FAULT OF STATOR WINDING **(M-S-26)**
- Corrosion of internal components

POSSIBLE SYMPTOMS:

- HYDROGEN MAKE-UP RATE ABOVE NORMAL **(S-H-05)**
- HYDROGEN PRESSURE INSIDE GENERATOR BELOW NORMAL **(S-H-06)**
- LIQUID-IN-GENERATOR ALARM **(S-C&F-04)**

OPERATORS MAY CONSIDER:

1. It is prudent to fix a water leak inside the generator as soon as possible. Water has the potential to causing a number of problems to critical components, in particular if the rotor has "18–5" retaining rings or "18–5" zone rings.
2. It is important to monitor hydrogen dew point and leak rate to evaluate the ongoing situation so that appropriate actions can be taken on time to avoid serious damage to the machine.

TEST AND VISUAL INDICATIONS:

- Water inside the generator
- Lead carbonate deposits on or around the heat exchanger
- Pressure tests of heat exchangers

REFERENCE

1. G. Klempner and I. Kerszenbaum. *Handbook of Large Turbo-Generator Operation and Maintenance*. Piscataway, NJ: IEEE Press, 2008, chapter 4.

5.7 Stator Cooling Water System Malfunctions

M-SCW-01: BLOCKAGE IN THE FLOW OF STATOR COOLING WATER (SCW)

APPLICABILITY:

All generators with direct water-cooled stator windings

MALFUNCTION DESCRIPTION:

This malfunction addresses the situation when the flow of SCW is partially or fully blocked. Figure 5.123 shows a typical SCW skid.

FIGURE 5.123
Air heat exchanger removed for maintenance during a major generator outage.

PROBABLE CAUSES:

- BLOCKAGE OF STATOR COOLING WATER (SCW) HEAT EXCHANGERS **(M-SCW-02)**
- BLOCKAGE OF STATOR COOLING WATER (SCW) STRAINERS/FILTERS **(M-SCW-03)**
- COPPER OXIDE FAULING OF WATER-CARRYING STRANDS **(M-SCW-06)**
- Gas locking of the SCW due to hydrogen leaking into the SCW (hydrogen is maintained at a higher pressure than the SCW, normally by no less than about 5 psi, so any physical breach of the SCW piping and/or water-carrying conductor will result in hydrogen ingression into the SCW system, rather than water leaking into the machine)
- GAS LOCKING OF STATOR COOLING WATER (SCW) SYSTEM **(M-SCW-08)**
- Fouling of the SCW piping/flow path
- Plugging due to lack of proper or diligent maintenance
- Plugging due to debris
- Valve(s) left closed or semiclosed due to operator error
- SCW pump(s) or pump-motor failure
- SCW pump supply failure

POSSIBLE CONSEQUENCES:

- BURNING/PYROLYSIS OF STATOR WINDING INSULATION **(M-S-08)**
- GLOBAL STATOR TEMPERATURE ABOVE NORMAL **(M-S-25)**
- GLOBAL TEMPERATURE OF CORE ABOVE NORMAL **(M-C&F-09)**

POSSIBLE SYMPTOMS:

- Stator conductor temperatures above normal
- GENERATOR CORE CONDITION MONITOR—ALARM (**S-C&F-01**)
- SCW system alarms (such as inlet pressure, cooler pressure differential, low flow, no electrical supply, etc.)
- HYDROGEN-IN-STATOR COOLING WATER (SCW) HIGH (**S-S-04**)
- HYDROGEN MAKE-UP RATE ABOVE NORMAL (**S-H-05**)

OPERATORS MAY CONSIDER:

1. Most important is to maintain all generator temperatures within their allowable operating range. Depending on how far the temperatures increase and the severity of the problem, this may require load reduction or shutdown.
2. If the temperatures of the winding and/or core exceeded their maximum limits, consider initiating an investigation to assess any possible damage to winding and core insulation, as well as flux-shield insulation, terminals, and any other component cooled by the SCW (such as exciter rectifier in some machines).
3. It is also important to note any significant deviations such as one or more stator winding temperatures being above the rest. All *slot* and/or *hose-outlet* temperature reading should lie within a certain temperature band. When such a deviation does occur, it is generally indicative of a slowly deteriorating situation and which allows more time to react.
4. The use of *dynamic temperature monitoring* (i.e., load-dependent/varying alarms rather than static Hi limits only) should be considered.

TEST AND VISUAL INDICATIONS:

- If temperatures largely exceeded maximum limits for some time, one of the results might be discoloration of the insulation, although the probability of this is not too high, and most probably will not be seen on the exterior of the coils, unless the event was very severe, or for a very long time.
- Hydrogen samples from the GCM may contain pyrolysis byproducts.
- Any blocking/plugging/fouling may be pressure tested.
- Any damage to the SCW system may be visually recognized.
- Blown fuses, cables on the SCW skid.

REFERENCE

1. G. Klempner and I. Kerszenbaum. *Handbook of Large Turbo-Generator Operation and Maintenance*. Piscataway, NJ: IEEE Press, 2008, chapter 5.

M-SCW-02: BLOCKAGE OF STATOR COOLING WATER (SCW) HEAT EXCHANGER

APPLICABILITY:

All generators with direct water-cooled stator windings

MALFUNCTION DESCRIPTION:

This malfunction addresses the situation where there is a blockage of the SCW inside the SCW heat exchanger(s). Figure 5.124 shows a typical SCW skid with the heat exchangers.

This problem is captured under:

- STATOR COOLING WATER FLOW BELOW NORMAL (M-SCW-14)

FIGURE 5.124
A typical stator cooling water skid. At the front the twin pump motors can be seen. At the left and back of the photo is shown the vertical degassing tank.

REFERENCE

1. G. Klempner and I. Kerszenbaum. *Handbook of Large Turbo-Generator Operation and Maintenance*. Piscataway, NJ: IEEE Press, 2008, chapters 3, 5, and 10.

M-SCW-03: BLOCKAGE OF STATOR COOLING WATER (SCW) STRAINERS/FILTERS

APPLICABILITY:

All generators with direct water-cooled stator windings

MALFUNCTION DESCRIPTION:

A generator SCW system will always have some type of filter or strainer, or both (Figures 5.125 and 5.126), to ensure debris is removed from the water. Strainers are generally associated with a mechanical type of device that is used for larger particles and can be back-flushed to discard the debris removed. Filters are associated with an organic type of filtering device in which smaller particles are removed and the filters periodically replaced.

Malfunction of either occurs when the strainers/filters become blocked (i.e., fouled or plugged) to the point where the amount of debris is excessive, and is now preventing them from working properly. This is an indication that the strainers require back-flushing or the filters may require replacement. This usually occurs as a gradual build-up of debris and could eventually cause a SCW flow blockage if the problem becomes severe enough.

FIGURE 5.125
A typical SCW skid. At the back, the twin horizontal heat exchangers are shown.

Stator cooling water strainer with cuprous-oxide (Cu_2O) deposits

FIGURE 5.126
A strainer contaminated with copper corrosion deposits.

PROBABLE CAUSES:

- Damaged gaskets and/or "O" rings in the SCW system, causing bits of them to come loose.
- Other foreign debris in the SCW.

POSSIBLE CONSEQUENCES:

- Partial SCW flow blockage.
- Total SCW flow blockage would be a low probability event.

Elevated stator winding temperatures:

- GLOBAL STATOR TEMPERATURE ABOVE NORMAL **(M-S-25)**

POSSIBLE SYMPTOMS:

- BLOCKAGE IN THE FLOW OF STATOR COOLING WATER (SCW) **(S-SCW-01)**
- SLOT TEMPERATURE OF STATOR CONDUCTOR BAR—ABOVE NORMAL **(CASE 1) (S-SCW-14)**
- SLOT TEMPERATURE OF STATOR CONDUCTOR BAR—ABOVE NORMAL **(CASE 2) (S-SCW-15)**
- STATOR COOLING WATER FLOW BELOW NORMAL **(S-SCW-12)**
- STATOR COOLING WATER PRESSURE DROP ACROSS SCW FILTER/STRAINER ABOVE NORMAL **(S-SCW-14)**

OPERATORS MAY CONSIDER

1. Check for fouled or plugged SCW filters or strainers
2. Back-flushing strainers
3. Replacing filters

TEST AND VISUAL INDICATIONS:

- The SCW flow check instrument will show a *low* flow condition.
- Inspection of the strainers/filters would show debris buildup that requires cleaning.

REFERENCE

1. G. Klempner and I. Kerszenbaum. *Handbook of Large Turbo-Generator Operation and Maintenance*. Piscataway, NJ: IEEE Press, 2008, chapters 3, 5, and 10.

M-SCW-04: CONDUCTIVITY OF STATOR COOLING WATER ABOVE NORMAL

APPLICABILITY:

All generators with stator windings directly cooled by water

MALFUNCTION DESCRIPTION:

This malfunction addresses the situation where the conductivity of the SCW is above *normal*. By *normal*, it means a value close to the widely accepted value of 0.1μS/cm, or any other value provided by the OEM.

A deionizing subsystem is required for maintaining low conductivity in the SCW. Failure of the SCW system deionizer will allow the conductivity of the SCW flowing through the stator winding to rise. The deionizer usually *polishes* a small percentage of the SCW on a continual basis (generally about 5% of the total SCW flow). If left uncorrected, the conductivity of the SCW will continue to rise slowly.

PROBABLE CAUSES:

- High conductivity in the SCW may be the result of a failure of the SCW deionizer.
- Internal erosion of the stator conductor bar strands from cavitation or chemical attack on the copper itself.
- Leak of the raw water into the demineralized water at the SCW coolers.

POSSIBLE CONSEQUENCES:

- High conductivity of the demineralized cooling water can cause electrical flashover to ground by tracking, particularly at the insulated hoses where an internal electric tracking path to ground is formed.

- Possible phase to ground leakage current through the increasingly conducting SCW.
- A ground fault is possible in the worst case.

POSSIBLE SYMPTOMS:

- CONDUCTIVITY OF STATOR COOLING WATER ABOVE NORMAL **(S-SCW-03)**

OPERATORS MAY CONSIDER:

1. Operation with conductivities above *normal* carries a risk for electrical tracking and eventually, ground faults. There is no *magic* limit. Any deviation above the OEM-recommended value puts the machine at risk, and obviously the risk increases as the conductivity becomes higher.
2. Consider troubleshooting the problem immediately by first finding the root cause and then fixing it.
3. Water analysis for copper content in the SCW can help find the source of the problem.

TEST AND VISUAL INDICATIONS:

Metal/mineral contents in the SCW

REFERENCES

1. G. Klempner and I. Kerszenbaum. *Handbook of Large Turbo-Generator Operation and Maintenance.* Piscataway, NJ: IEEE Press, 2008, chapter 5.
2. EPRI TR-1015669. Turbine Generator Auxiliary System Maintenance Guides—Volume 4: Generator Stator Cooling Water.
3. EPRI TR-107137s. Main Generator On-Line Monitoring and Diagnostics.

M-SCW-05: COOLER OUTLET TEMPERATURE OF SCW ABOVE NORMAL

APPLICABILITY:

All generators with direct water-cooled stator windings

MALFUNCTION DESCRIPTION:

This malfunction addresses the situation when the temperature of the SCW as it leaves the heat exchanger is above *normal*. By *normal*, it means that at that temperature or higher, the SCW will not remove all the heat it must remove to keep the stator temperature within normal range, with the unit at rated load.

PROBABLE CAUSES:

- PRIMARY WATER FLOW TO SCW COOLERS BELOW NORMAL **(M-SCW-12)**
- TEMPERATURE OF PRIMARY (RAW) WATER AT INLET OF SCW COOLERS ABOVE NORMAL **(M-SCW-16)**

POSSIBLE CONSEQUENCES:

- GLOBAL STATOR TEMPERATURE ABOVE NORMAL **(M-S-25)**

POSSIBLE SYMPTOMS:

- COOLER OUTLET TEMPERATURE OF SCW ABOVE NORMAL **(S-SCW-04)**
- GENERATOR CORE CONDITION MONITOR—ALARM **(S-C&F-01)**
- Stator temperatures above normal.
- Differential temperatures and/or pressures across the primary (raw) water to the SCW coolers above and below normal, respectively.

OPERATORS MAY CONSIDER:

1. First action for consideration is maintaining the stator temperatures within allowable operating range. This may require reducing load, or take the unit offline, depending on the severity of the problem.
2. Comparing the situation among the different coolers (if more than one exists). This exercise will help identifying the source of the problem.
3. Troubleshooting the problem.
4. Some machines have an *automatic runback scheme* for this situation.

TEST AND VISUAL INDICATORS:

If the situation is due to a blockage or fouled filter/strainer, visual inspection, and/or pressure/flow tests may uncover the problem.

REFERENCE

1. G. Klempner and I. Kerszenbaum. *Handbook of Large Turbo-Generator Operation and Maintenance*. Piscataway, NJ: IEEE Press, 2008, chapters 3 and 5.

M-SCW-06: COPPER-OXIDE BUILDUP IN WATER-CARRYING STRANDS

APPLICABILITY:

All generators with direct water-cooled stator windings

MALFUNCTION DESCRIPTION:

This malfunction concerns the corrosion rate of the hollow stator bar copper strands because of *cuprous* oxide buildup inside the stator bar strands due to the dissolved oxygen content getting out of control into the 200 to 300 μg/L range (Figure 5.127). If the corrosion rate is kept below 20 μg/L in a *low oxygen* system, corrosion will be minimized. Similarly, if the corrosion rate is kept above 2,000 μg/L in a *high oxygen* system, corrosion will be minimized (see Figure 5.128). Once fouling begins, it can get out of control very rapidly unless action is taken to arrest and correct the problem.

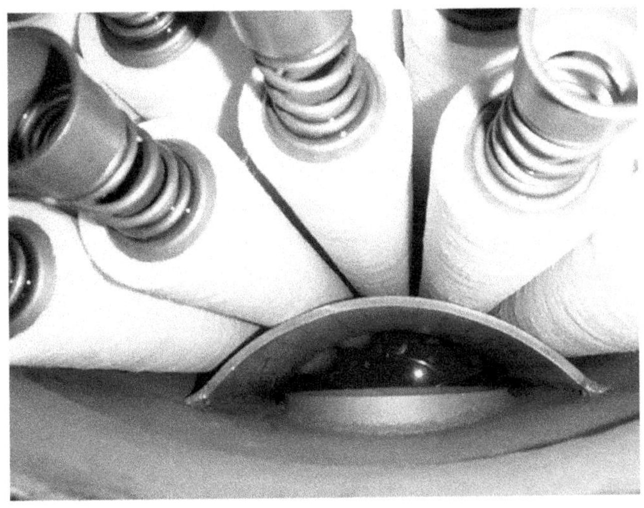

FIGURE 5.127
Typical organic (1–3 μm) filter components. Several dozens of these are located inside a tank, through which the stator cooling water is circulated and filtered.

FIGURE 5.128
A highly fouled stator bar due to copper oxide buildup.

Additionally, *cupric* oxide is not generally an issue with respect to fouling of the hollow strands. If the SCW system is kept in an alkaline state, even *cuprous* oxide becomes a nonissue (see Figure 5.128).

PROBABLE CAUSES:

SCW chemistry becoming out of control:

- Low oxygen systems having ingress of oxygen and rising into the 200 to 300 µg/L range.
- High oxygen systems losing oxygen and drifting into the 200 to 300 µg/L range.

POSSIBLE CONSEQUENCES:

- BURNING/PYROLYSIS OF STATOR WINDING INSULATION **(M-S-08)**
- DAMAGE TO CONDUCTOR BAR INSULATION **(M-S-15)**
- GLOBAL STATOR TEMPERATURE ABOVE NORMAL **(M-S-25)**
- GROUND FAULT OF STATOR WINDING **(M-S-26)**
- INSUFFICIENT FLOW OF COOLING MEDIUM IN A STATOR BAR **(M-S-32)**
- TEMPERATURE OF STATOR CONDUCTOR ABOVE NORMAL **(M-S-53)**

POSSIBLE SYMPTOMS:

- Stator temperatures above *normal*
- BLOCKAGE IN THE FLOW OF STATOR COOLING WATER (SCW) **(S-SCW-01)**
- BYPRODUCTS FROM STATOR INSULATION PYROLYSIS **(S-S-01)**
- GENERATOR CORE CONDITION MONITOR—ALARM **(S-C&F-01)**
- PARTIAL DISCHARGE ACTIVITY (PDA) HIGH **(S-S-11)**
- SLOT TEMPERATURE OF STATOR CONDUCTOR BAR—ABOVE NORMAL (CASE 1) **(S-S-14)**
- SLOT TEMPERATURE OF STATOR CONDUCTOR BAR—ABOVE NORMAL (CASE 2) **(S-S-15)**

OPERATORS MAY CONSIDER

1. The unit should have all temperatures within their allowable operating limits. If they are not, consider reducing load immediately until all temperatures are in their normal range.
2. If the temperatures exceed their maximum limits as given by the standards and/or the OEM, consider initiating an assessment of the conditions of the affected components.
3. Check the SCW chemistry for oxygen content.

TEST AND VISUAL INDICATIONS:

- Gas analysis from the GCM-captured gas samples
- Visual inspection of windings
- Megger tests

REFERENCES

1. G. Klempner and I. Kerszenbaum. *Handbook of Large Turbo-Generator Operation and Maintenance*. Piscataway, NJ: IEEE Press, 2008, chapters 3, 5, and 10.
2. K. Schleithoff and H.W.Emshoff. Optimization of Generator Cooling Water Conditioning (in German). In: *VGB conference on Chemie im Kraftwerk 1989, VGB Kraftwerkstechnik* 70(1990)9, pp. 794–798.
3. EPRI Report 1015669. December 2008, Turbine Generator Auxiliary System Maintenance Guides—Volume 4: Generator Stator Cooling System.
4. EPRI Report 1004704. Guidelines for Detecting and Removing Flow Restrictions of Water-Cooled Stator Windings, July 2002.
5. CIGRE. Guide on Stator Water Chemistry Management. Study Committee A1, WG A1.15, April 2010.
6. R. Svoboda and D. Palmer: Behaviour of Copper in Generator Stator Cooling-Water Systems. In: *ICPWS XV*, Berlin, Germany, September 8–11, 2008.

M-SCW-07: DEGRADED/RUPTURED "O" RINGS OR OTHER SCW PIPING CONNECTIONS

APPLICABILITY:

- All generators with direct water-cooled stator windings

MALFUNCTION DESCRIPTION:

This malfunction is concerned with the SCW "O" rings and/or gaskets installed on hoses carrying water from the manifolds to the stator bars, being ruptured or degraded to the extent they are leaking (Figure 5.129). This may allow hydrogen ingression into the SCW system.

This problem is captured under the following malfunction:

- HYDROGEN INGRESSION INTO STATOR COOLING WATER (M-S-29)

M-SCW-08: GAS-LOCKING OF STATOR COOLING WATER (SCW) SYSTEM

APPLICABILITY:

All generators with direct water-cooled stator windings

MALFUNCTION DESCRIPTION:

This malfunction addresses the case where a large leak develops inside the machine resulting in hydrogen ingression into the SCW piping system.

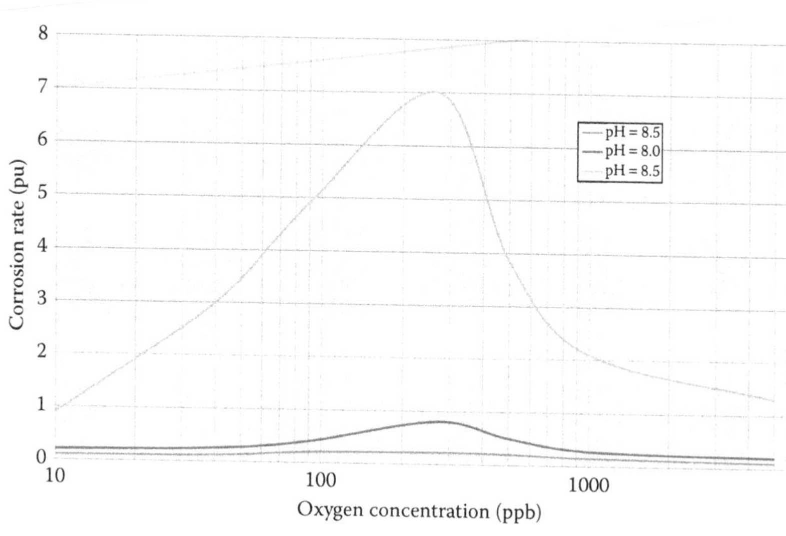

FIGURE 5.129
Copper corrosion rates versus oxygen concentration in the stator cooling water. (*From* Schleithoff, K. and Emshoff, H.W., *VGB Kraftwerkstechnik [VGB Conference on Chemie im Kraftwerk 1989]*, 70, 794–798, 1990).

Under certain conditions, gas locking of the system can occur, partially or totally blocking the flow of SCW.

PROBABLE CAUSES:

- HYDROGEN INGRESSION INTO STATOR COOLING WATER (M-S-29)

This problem is further captured by the following malfunction description:

- BLOCKAGE IN THE FLOW OF STATOR COOLING WATER (SCW) (M-SCW-01)

REFERENCE

1. G. Klempner and I. Kerszenbaum. *Handbook of Large Turbo-Generator Operation and Maintenance*. Piscataway, NJ: IEEE Press, 2008, chapter 5.

M-SCW-09: LEAK IN THE FLOW PATH OF THE STATOR COOLING WATER (SCW)

APPLICABILITY:

All generators with direct water-cooled stator windings

MALFUNCTION DESCRIPTION:

This malfunction addresses the situation where a leak has developed in the path of the SCW, inside the generator, before the generator, after the generator, and/or in the rectifier if the same SCW is used to cool the rectifying solid-state power devices.

PROBABLE CAUSES:

Any type of breach of a pipe and/or water-carrying conductor, by such phenomena as *magnetic termites*, short-circuit forces resulting in through cracks of conductors, cracked bushings, cracks due to high vibration, corrosion, physical unintentional damage of piping, and so on.

POSSIBLE CONSEQUENCES:

For leaks inside the generator:

- GAS LOCKING OF STATOR COOLING WATER (SCW) SYSTEM **(M-SCW-08)**

For leaks before the generator:

- GLOBAL STATOR TEMPERATURE ABOVE NORMAL **(M-S-25)**
- GLOBAL TEMPERATURE OF CORE ABOVE NORMAL **(M-C&F-09)**
- STATOR COOLING WATER FLOW BELOW NORMAL **(S-SCW-14)**

For large leaks only:

- Loss of control of oxygenation levels
- For leaks after the generator:
- Loss of SCW
- GLOBAL STATOR TEMPERATURE ABOVE NORMAL **(M-S-25)**
- GLOBAL TEMPERATURE OF CORE ABOVE NORMAL **(M-C&F-09)**

POSSIBLE SYMPTOMS:

- STATOR COOLING WATER OUTLET PRESSURE BELOW NORMAL **(S-SCW-13)**

High SCW makeup rate:

- LOW LEVEL OF SCW IN THE MAKEUP TANK **(S-SCW-10)**
- Increased stator temperatures. There are several symptoms under *stator symptoms*

Change in the oxygen and pH levels of the SCW:

- CONDUCTIVITY OF STATOR COLING WATER ABOVE NORMAL **(S-SCW-03)**

Change in the flowrate of the SCW:

- STATOR COOLING WATER FLOW BELOW NORMAL **(S-SCW-12)**

OPERATORS MAY CONSIDER:

1. It is most important to maintain stator temperatures within their allowable range.
2. Loss of pH and/or oxygen levels in generators where the SCW flows in copper strands has the potential to cause accelerated copper oxide deposits. In some documented cases, machines lost significant cooling capacity due to this phenomenon in a matter of days.
3. If the problem is significant and cannot be fixed quickly, it may be reasonable to shut down the unit to avoid copper oxide build up and/or overheating of the stator windings and stator core.

TEST AND VISUAL INDICATIONS:

- Visual/NDE inspection of cracks
- Pressure/vacuum tests
- Oxygen contents in the SCW
- SCW pH level

REFERENCE

1. G. Klempner and I. Kerszenbaum. *Handbook of Large Turbo-Generator Operation and Maintenance*. Piscataway, NJ: IEEE Press, 2008, chapters 2, 3, and 5.

M-SCW-10: LEAK OF SCW PRIMARY (RAW) COOLING WATER

APPLICABILITY:

All generators with direct water-cooled stator windings

MALFUNCTION DESCRIPTION:

This malfunction addresses the case where there is a leak in the flow path of the primary (raw) cooling water of the SCW system, before it reaches the heat exchanger.

This problem is addressed under the following malfunction:

- PRIMARY WATER FLOW TO SCW COOLERS BELOW NORMAL (M-SCW-12)

M-SCW-11: LEAKY SCW HOSES

APPLICABILITY:

All generators with direct water-cooled stator windings

MALFUNCTION DESCRIPTION:

This malfunction is concerned with SCW hoses being ruptured or degraded to the extent they are leaking, allowing hydrogen ingression into the SCW system.

This problem is captured under the following malfunction:

- HYDROGEN INGRESSION INTO STATOR COOLING WATER (M-S-29)

M-SCW-12: PRIMARY WATER FLOW TO SCW COOLERS—BELOW NORMAL

APPLICABILITY:

All generators with direct water-cooled stator windings

MALFUNCTION DESCRIPTION:

This malfunction addresses the case where the flowrate of primary (raw) cooling water to the SCW system is below *normal*. By *normal*, it means that there is not enough flow to provide sufficient cooling of the stator winding for rated output. This malfunction covers any unsatisfactory flow rate, all the way to zero flow.

The primary (raw) water to the SCW coolers can be either from a natural source, such as a river or lake, or from a cooling tower, or turbine plant cooling water.

PROBABLE CAUSES:

- Pump failure
- Valves blocked/closed due to maintenance/operator error
- Valves closed/semiclosed due to malfunction of the flow/temperature regulation system
- Partial or total blockage of supply pipes: Fouling/plugging/silting/debris/corrosion/and so on
- Serious leak of primary (raw) water

POSSIBLE CONSEQUENCES:

- BURNING/PYROLYSIS OF STATOR WINDING INSULATION (M-S-08)
- GLOBAL STATOR TEMPERATURE ABOVE NORMAL (M-S-25)
- GLOBAL TEMPERATURE OF CORE ABOVE NORMAL (M-C&F-09)

POSSIBLE SYMPTOMS:

Temperature indications from the stator (e.g., stator bar outlet SCW temperature) higher than normal:

- TEMPERATURE OF GLOBAL OUTLET STATOR COOLING WATER ABOVE NORMAL **(S-SCW-15)**
- DIFFERENTIAL PRESSURE ACROSS SCW HEAT EXCHANGER ABOVE NORMAL **(S-SCW-05)**

- GENERATOR CORE-CONDITON MONITOR—ALARM (S-C&F-01)
- Alarms from SCW cooling system

OPERATORS MAY CONSIDER:

1. The most pressing action is restoring the stator temperatures to normal range. This may require partially or fully unloading the machine.
2. Depending on the situation, a generator may be kept spinning and even synchronized with little or no load until the source of the problem is found, if the temperatures of the stator components remain within normal range (some machines have an *automatic runback scheme* for this purpose).
3. If the temperatures reach beyond the normal range, the case should be investigated for evaluation of any potential deterioration of the insulation and loss of life.
4. Troubleshooting the problem.

TEST AND VISUAL INDICATIONS:

- Indications of insulation abnormal degradation (discoloration due to thermal stress)
- Debris/leaks/closed valves/failed regulators/filters/pump fuses off/and so on clogging or blocking the flow of raw water to the coolers

REFERENCE

1. G. Klempner and I. Kerszenbaum. *Handbook of Large Turbo-Generator Operation and Maintenance*. Piscataway, NJ: IEEE Press, 2008, chapters 3, 5, and 10.

M-SCW-13: STATOR COOLING WATER (SCW) FILTER/STRAINER RESTRICTION

APPLICABILITY:

All generators with direct water-cooled stator windings

MALFUNCTION DESCRIPTION:

This malfunction addresses the situation where the flow of SCW is partially or fully blocked by restrictions in the SCW filters and/or strainers.

- Filters are mostly made of replaceable organic cartridges.
- Strainers are metallic and can be cleaned by back-flushing.
- Filters and strainers become fouled/plugged by debris, copper oxide residues, maintenance errors, and so on.

This problem is captured by the following malfunction:

- STATOR COOLING WATER FLOW BELOW NORMAL (M-SCW-14)

M-SCW-14: STATOR COOLING WATER FLOW BELOW NORMAL

APPLICABILITY:

All generators with direct water-cooled stator windings

MALFUNCTION DESCRIPTION:

This malfunction relates to the case where the flow of the SCWis below *normal*. By normal, it means that the flow is not enough to effectively remove all of the heat generated by the stator winding during operation at rated load. The flow reduction covered by this malfunction includes the case when there is no flow at all.

This malfunction will affect all those components cooled by the SCW, that is, the stator winding bars, circumferential connectors/busses, terminals, bushings, flux screens (flux shields), and water-cooled rectifiers (if it uses the same SCW than the generator itself).

The stator generates core losses and heat, plus conductor bar losses and heat. Cooling of the stator core is primarily done by the pressurized hydrogen gas in the generator. For stator bars that are not water-cooled, but have solid copper strands, the stator core is the heat sink for the winding, along with the hydrogen gas, indirectly. The SCW's main function is to remove the heat generated by water-cooled stator conductors. A fraction of it is also removed indirectly by the hydrogen, as it cools the core where the conducting bars lay. Each machine design has a different ratio between the heat removed by the SCW and the heat removed by the hydrogen. Thus, a lack of SCW will affect differently designed machines to dissimilar extents.

PROBABLE CAUSES:

- BLOCKAGE IN THE FLOW OF STATOR COOLING WATER (SCW) **(M-SCW-01)**
- GAS LOCKING OF COOLING WATER IN HYDROGEN OR AIR HEAT EXCHANGERS **(M-H-07)**
- LEAK IN THE FLOW PATH OF THE STATOR COOLING WATER (SCW) **(M-SCW-09)**
- STATOR COOLING WATER (SCW) FILTER/STRAINER RESTRICTION **(M-SCW-13)**
- Plugged pipes by CuOx, broken gaskets, and so on
- Plugged pipes by maintenance error
- Closed valves by maintenance error
- Malfunction of flow, pressure, and temperature regulators

POSSIBLE CONSEQUENCES:

- BURNING/PYROLYSIS OF STATOR WINDING INSULATION **(M-S-08)**
- DAMAGE TO CONDUCTOR BAR INSULATION **(M-S-15)**
- GLOBAL STATOR TEMPERATURE ABOVE NORMAL **(M-S-25)**
- GROUND FAULT OF STATOR WINDING **(M-S-26)**
- INSUFFICIENT FLOW OF COOLING MEDIUM IN A STATOR BAR **(M-S-32)**
- TEMPERATURE OF STATOR CONDUCTOR ABOVE NORMAL **(M-S-53)**
- Damage to the HV bushings

POSSIBLE SYMPTOMS:

- Stator temperatures above *normal*
- DIFFERENTIAL PRESSURE ACROSS SCW HEAT EXCHANGER ABOVE NORMAL **(S-SCW-05)**
- DIFFERENTIAL PRESSURE OF SCW ACROSS STATOR WINDING BELOW NORMAL **(S-SCW-07)**
- GENERATOR CORE CONDITION MONITOR—ALARM **(S-C&F-01)**
- HYDROGEN-IN-STATOR COOLING WATER (SCW) HIGH **(S-S-04)**
- Temperature differential of primary (raw) water across coolers too low

OPERATORS MAY CONSIDER:

1. The generator should have all temperatures within their allowable operating limits. If required, consider reducing load immediately until all temperatures are in their normal region.
2. If the temperatures exceeded their maximum limits as given by the standards and/or the OEM, consider initiating an assessment of the conditions of the affected components.
3. Troubleshooting the problem.
   - If the cause of the problem is a blockage of a SCW heat exchanger, operators can switch to a redundant cooler, if one is available.
   - If the cause is a plugging of the SCW filter or strainers, consider replacement or back-flush as needed. Some units require to take the machine offline for carrying out this maintenance activity.
4. It is also important to note the rate of rise of the stator winding temperature/s. All *slot* and *hose-outlet* temperature readings should lie within a certain temperature band. If there is a general slowly rising deviation of temperature, it is generally indicative of a slowly deteriorating situation and which allows more time to react. The use of *dynamic temperature monitoring* (i.e., load-dependent/varying alarms rather than static high limits only) should be considered.

TEST AND VISUAL INDICATIONS:

- Gas analysis from the GCM-captured gas samples
- Visual inspection of windings, flux shields, and any other affected component
- *Megger* tests

REFERENCE

1. G. Klempner and I. Kerszenbaum. *Handbook of Large Turbo-Generator Operation and Maintenance*. Piscataway, NJ: IEEE Press, 2008, chapters 5 and 10.

M-SCW-15: STATOR COOLING WATER HEAT EXCHANGER UNDERPERFORMING

APPLICABILITY:

All generators with direct water-cooled stator windings

MALFUNCTION DESCRIPTION:

This malfunction addresses the situation where the SCW coolers do not perform their designed duty, that is, the temperature of the stator winding is not kept within its allowable operating range under rated load conditions.

For *Probable Causes* and other details refer to:

- PRIMARY WATER FLOW TO SCW COOLERS BELOW NORMAL **(M-SCW-12)**
- TEMPERATURE OF PRIMARY (RAW) WATER AT INLET OF SCW COOLERS ABOVE NORMAL **(M-SCW-16)**

M-SCW-16: TEMPERATURE OF PRIMARY (RAW) WATER AT INLET OF SCW COOLERS ABOVE NORMAL

APPLICABILITY:

All generators with direct water-cooled stator windings

MALFUNCTION DESCRIPTION:

This malfunction describes the case where the inlet temperature of the primary (raw) water to the SCW heat exchanger is above *normal*. By *normal*, it means a temperature that renders the SCW system incapable of removing enough heat during operation at rated load, to maintain all stator temperatures within their allowable range (typically, the maximum allowable raw water temperature is also provided by the OEM).

The primary (raw) water to the SCW coolers can be either from a natural source, such as a river, sea or lake, or from a cooling tower, or turbine plant cooling water. Temperatures from rivers and lakes are seasonal, and in those cases, there is nothing the operator can do to lower the raw water temperature (in the summer) if that is the source.

PROBABLE CAUSES:

- River/lake temperature high
- Cooling tower malfunction (failed pumps, fans, make-up water, etc.)

POSSIBLE CONSEQUENCES:

- GLOBAL STATOR TEMPERATURE ABOVE NORMAL **(M-S-25)**
- STATOR COOLING WATER HEAT EXCHANGER UNDERPERFORMING **(M-SCW-15)**

POSSIBLE SYMPTOMS:

- Temperature readings from stator RTDs/TCs above normal
- TEMPERATURE OF PRIMARY (RAW) WATER AT INLET OF COOLERS ABOVE NORMAL **(S-SCW-16)**

OPERATORS MAY CONSIDER:

1. If the raw water is from a natural source (river, lake, sea), there is nothing that can be done other than reducing load until all temperatures are within normal range (derating). This is not an uncommon procedure in many plants that depend on this type of water source.
2. If the source is from a cooling tower, consider unloading the unit enough so that all temperatures are within their normal values and troubleshoot the problem.

TEST AND VISUAL INDICATORS:

None in the generator. All *causes* are external.

REFERENCE

1. G. Klempner and I. Kerszenbaum. *Handbook of Large Turbo-Generator Operation and Maintenance.* Piscataway, NJ: IEEE Press, 2008, chapter 5.

5.8 Seal-Oil System Malfunctions

M-SOS-01: BLOCKAGE OF SEAL-OIL FLOW

APPLICABILITY:

All hydrogen-cooled generators

MALFUNCTION DESCRIPTION:

This malfunction addresses the case where the flow of seal oil to the hydrogen seals is partially or totally blocked. The result is a seal-oil flow that is below that required to properly maintain effective sealing of the hydrogen, and the temperature of the seals within their design operating range.

Figure 5.130 shows seal-oil filter components. These may become clogged under certain conditions impeding the free flow of sealing oil.

FIGURE 5.130
Highly degraded gasket from a stator cooling water system.

PROBABLE CAUSES:

- Plugging or fouling of the seal-oil labyrinths and/or piping
- Plugging or fouling of the seal-oil coolers
- Plugging or fouling of the seal-oil filters

POSSIBLE CONSEQUENCES:

H_2-seal rub, causing

- TEMPERATURE OF HYDROGEN SEAL ABOVE NORMAL (M-R-27)
- Electrical arcing and pitting of the seals
- Rotor vibration may increase
- Possible H_2-seal failure
- Shaft ground

POSSIBLE SYMPTOMS:

- DIFFERENTIAL PRESSURE ACROSS SEAL-OIL FILTER ABOVE NORMAL **(S-SOS-02)**
- ELEVATED HYDROGEN DEW POINT **(S-H-03)**
- HYDROGEN MAKE-UP RATE ABOVE NORMAL **(S-H-05)**
- INLET PRESSURE OF SEAL OIL BELOW NORMAL **(S-SOS-04)**
- PRESSURE OF SEAL OIL ABOVE NORMAL **(S-SOS-06)**

- TEMPERATURE OF HYDROGEN SEAL ABOVE NORMAL (S-SOS-08)
- TEMPERATURE OF SEAL OIL ABOVE NORMAL (S-SOS-10)
- VIBRATION OF ROTOR—HIGH (S-R-10)

If a seal fails, vibrations may increase or change erratically, and/or loss of hydrogen from the generator may occur.

Shaft-ground alarm:

- SHAFT CURRENT ABOVE NORMAL (S-R-04)
- SHAFT VOLTAGE BELOW NORMAL (S-R-06)

OPERATORS MAY CONSIDER:

1. Degradation of the seal-oil system can have serious consequences to the integrity of the machine; therefore, immediate actions may be required to fix the situation or shutting the unit down.
2. The key parameters to monitor are the seal-face metal temperature and seal-oil pressure.

TEST AND VISUAL INDICATIONS:

- Blocked/clogged/fouled pipes, labyrinths, seal-oil filters
- Damaged hydrogen seal surfaces and/or shaft hydrogen sealing surfaces

REFERENCES

1. G. Klempner and I. Kerszenbaum. *Handbook of Large Turbo-Generator Operation and Maintenance*. Piscataway, NJ: IEEE Press, 2008, chapters 3 and 5.
2. EPRI Product 1026605. Generation Maintenance Applications Center: Combined-Cycle Combustion Turbine Lube Oil System Maintenance Guide, October 2010.

M-SOS-02: BLOCKAGE OF SEAL-OIL HEAT EXCHANGERS ON OIL SIDE

APPLICABILITY:

All hydrogen-cooled generators

MALFUNCTION DESCRIPTION:

This malfunction addresses the situation where the seal-oil flow is below normal due to a partial or total blockage.

This Malfunction is covered under:

- BLOCKAGE OF SEAL-OIL FLOW (M-SOS-01)

M-SOS-03: DEGRADED CONDITION OF THE SEAL OIL

APPLICABILITY:

All hydrogen-cooled generators

MALFUNCTION DESCRIPTION:

This malfunction relates to the situation where the seal-oil condition is degraded to the extent the seal-oil system cannot fulfill its designed function of effectively contain the hydrogen inside the generator casing.

PROBABLE CAUSES:

- INEFFECTIVE SEAL-OIL DETRAINING **(M-SOS-09)**
- Excessive moisture contamination
- Metallic contamination of the seal oil by a bad bearing in a pump
- Contamination from a loss of *FME*[15]
- Lack of attention to maintenance

POSSIBLE CONSEQUENCES:

- HYDROGEN ESCAPING THE GENERATOR **(M-H-08)**
- ELEVATED HYDROGEN DEW POINT **(M-H-05)**
- DEGRADED HYDROGEN PURITY **(M-H-04)**
- Abnormal erosion of seal and/or journal surfaces
- Contamination of the H_2 cooling gas

POSSIBLE SYMPTOMS:

- ELEVATED HYDROGEN DEW POINT **(S-H-03)**
- Contamination of cooling hydrogen with foreign material(s)
- Foaming in the main seal-oil tank—visible through the sight glass usually provided

OPERATORS MAY CONSIDER:

1. Depending on the severity of the problem, maintenance either can be postponed to the most convenient occasion, or must be performed in short notice.
2. Monitor the dew point and purity of the hydrogen.
3. Oil can be filtered, dried, or requiring replacement, depending on severity of the problem.
4. Seal-oil condition should always be checked after returning from a long period offline.

TEST AND VISUAL INDICATIONS:

- Several oil tests
- Foaming in the main seal-oil tank—visible through the sight glass usually provided

[15] FME—Foreign material exclusion.

REFERENCES

1. G. Klempner and I. Kerszenbaum. *Handbook of Large Turbo-Generator Operation and Maintenance*. Piscataway, NJ: IEEE Press, 2008, chapters 3, 5, and 10.
2. EPRI Product 1026605. Generation Maintenance Applications Center: Combined-Cycle Combustion Turbine Lube Oil System Maintenance Guide, October 2010.

M-SOS-04: EFFICIENCY OF SEAL-OIL COOLERS REDUCED

APPLICABILITY:

All hydrogen-cooled generators

MALFUNCTION DESCRIPTION:

This malfunction addresses the situation when the seal-oil cooling system is not effective in maintaining the temperature of the hydrogen seals within the allowable temperature range, as given by the OEM, or by accepted industry practice. Figure 5.131 shows a typical seal-oil system.

Several causes may result in reduced seal-oil cooling efficiency; these are covered under the following malfunction descriptions:

- BLOCKAGE OF SEAL-OIL FLOW (M-SOS-01)
- DEGRADED CONDITION OF THE SEAL OIL (M-SOS-03)

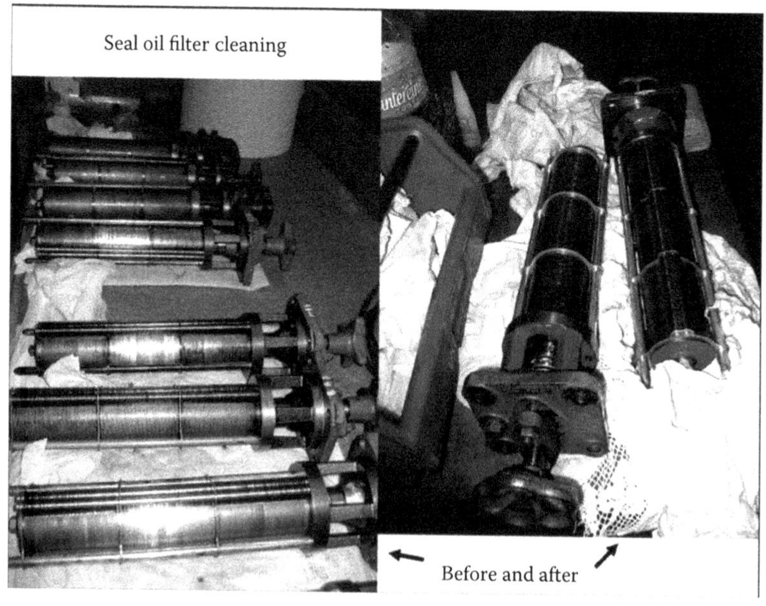

FIGURE 5.131
Seal-oil filters shown before and after a cleaning activity.

- FLOW OF COOLING WATER TO SEAL-OIL COOLERS BLOCKED **(M-SOS-07)**
- INEFFECTIVE SEAL-OIL DETRAINING **(M-SOS-09)**
- LEAK OF COOLING WATER BEFORE INLET TO SEAL-OIL COOLERS **(M-SOS-11)**
- SEAL-OIL COOLING-WATER INLET TEMPERATURE HIGH **(M-SOS-14)**

M-SOS-05: FAILURE OF SEAL-OIL PUMP

APPLICABILITY:

All hydrogen-cooled generators

MALFUNCTION DESCRIPTION:

This malfunction describes the instance where the main seal-oil pump or pumps fail suddenly.

PROBABLE CAUSES:

- Mechanical pump failure (such as bearing-seizure)
- Failure of the electric motor for the seal-oil pump
- Loss of electric power supply to pump motor

POSSIBLE CONSEQUENCES:

- In most if not all cases, a backup pump or pumps exist that will take over a failed pump. In the event of a loss of AC power (affecting all AC pump motors), a DC-powered pump motor will take over in most cases. The small seal-oil pressure transient does not affect the overall performance or operation of the generator. In some designs, a shaft-mounted oil pump will provide last-resort backup for most of the speed range from full speed downward.
- In those cases where all pumps fail to work (rare but not unheard of cases), the immediate loss of seal-oil flow will result in fast rise of the hydrogen seals temperature, and a quick loss of hydrogen from the generator's casing, including severe damage to the seals and rotor shaft. New carbon-seal technology is more forgiving to this type of incident.

POSSIBLE SYMPTOMS:

Rapid increase in hydrogen seal temperature:

- TEMPERATURE OF HYDROGEN SEAL ABOVE NORMAL **(S-SOS-08)**

Rapid decrease in seal-oil pressure:

- INLET PRESSURE OF SEAL OIL BELOW NORMAL **(S-SOS-04)**

Rapid loss of hydrogen from generator casing:

- HYDROGEN MAKE-UP RATE ABOVE NORMAL **(S-H-05)**
- HYDROGEN PRESSURE INSIDE GENERATOR BELOW NORMAL **(S-H-06)**

Increase of rotor vibration:

- VIBRATION OF ROTOR—HIGH **(S-R-10)**

OPERATORS MAY CONSIDER:

If an alternative seal-oil pump cannot be started immediately, the safest action is to trip the unit without delay. In most generators, pressure and flow gauges may have low-pressure or low-flow trip settings.

TEST AND VISUAL INDICATIONS:

- Damaged hydrogen seals and/or shaft seal running surfaces
- Faulty pump or pump motor
- Burned fuses or other electrical fault resulting in a loss of AC and/or DC power

REFERENCES

1. G. Klempner and I. Kerszenbaum. *Handbook of Large Turbo-Generator Operation and Maintenance*. Piscataway, NJ: IEEE Press, 2008, chapters 5 and 10.
2. EPRI Product 1026605. Generation Maintenance Applications Center: Combined-Cycle Combustion Turbine Lube Oil System Maintenance Guide, October 2010.

M-SOS-06: FAILURE OF SEAL-OIL VACUUM PUMP

APPLICABILITY:

All hydrogen-cooled generators

MALFUNCTION DESCRIPTION:

This malfunction addresses the situation where the seal-oil vacuum pump has failed.

Figure 5.132 shows a way to corroborate that there is a problem with gas removal from the seal oil.

PROBABLE CAUSES:

- Loss of electric power to the pump motor
- Failure of the pump motor
- And so on

POSSIBLE CONSEQUENCES:

- DEGRADED CONDITION OF THE SEAL OIL **(M-SOS-03)**
- ELEVATED HYDROGEN DEW POINT **(M-H-05)**
- Mechanical damage of seals and/or bearing damage if the seal oil is shared with the bearings

FIGURE 5.132
A typical seal-oil skid. The horizontal heat exchanger is located at the top of the skid, and the detraining tank is standing alone outside the skid on the right.

POSSIBLE SYMPTOMS:

- ELEVATED HYDROGEN DEW POINT **(S-H-03)**
- Alarm indicating loss of seal-oil vacuum pump operation
- Excessive foaming in the main seal-oil tank

OPERATORS MAY CONSIDER:

1. Some of the moist hydrogen coming into the generator from a faulty seal-oil detraining system is compensated by the hydrogen driers. However, if the dew point continues to deteriorate, operator must take into account the risk of operating under those conditions (Figure 5.133).
2. The situation ought to be fixed as soon as possible, and consideration ought to be given to removal of the unit from operation if the situation cannot be promptly remedied and the dew point remains high.

TEST AND VISUAL INDICATIONS:

- Damage to vacuum pumps, pump motor, burned fuse, and so on
- Excessive foaming in the main seal-oil tank

REFERENCES

1. G. Klempner and I. Kerszenbaum. *Handbook of Large Turbo-Generator Operation and Maintenance*. Piscataway, NJ: IEEE Press, 2008, chapters 3 and 5.
2. EPRI Product 1026605. Generation Maintenance Applications Center: Combined-Cycle Combustion Turbine Lube Oil System Maintenance Guide, October 2010.

FIGURE 5.133
A typical seal-oil de-foaming tank and a view port.

M-SOS-07: FLOW OF COOLING WATER TO SEAL-OIL COOLERS BLOCKED

APPLICABILITY:

All hydrogen-cooled generators

MALFUNCTION DESCRIPTION:

This malfunction addresses the situation where the raw cooling water to the seal-oil coolers is partially or fully blocked.

This malfunction is fully covered under the malfunction:

- FLOW OF COOLING WATER TO SEAL-OIL COOLERS BLOCKED (M-SOS-07)

M-SOS-08: FLOW OF COOLING WATER TO SEAL-OIL COOLERS BELOW NORMAL

APPLICABILITY:

All hydrogen-cooled generators

MALFUNCTION DESCRIPTION:

This malfunction addresses the situation where the flow of the primary (raw) cooling water to the seal-oil heat exchangers is below *normal*. By *normal,* it means that the temperature of the seal oil cannot be maintained below the maximum recommended by the OEM or equivalent industry practice.

PROBABLE CAUSES:

- Blockage due to error during a maintenance activity
- Blockage due to operator error
- Fouling/plugging of raw water piping
- Silting/fouling of the heat-exchanger tubing
- Leak upstream the heat exchanger

POSSIBLE CONSEQUENCES:

- DYNAMIC ROTOR UNBALANCE (MECHANICAL) **(M-R-08)**
- FAILURE OF HYDROGEN SEALS **(M-R-11)**
- RUB OF HYDROGEN SEAL **(M-R-21)**
- TEMPERATURE OF HYDROGEN SEAL ABOVE NORMAL **(M-R-27)**
- VIBRATION OF ROTOR—HIGH **(M-R-30)**

POSSIBLE SYMPTOMS:

- DIFFERENTIAL PRESSURE ACROSS SEAL-OIL FILTER ABOVE NORMAL **(S-SOS-02)**
- SEAL-OIL COOLING WATER INLET TEMPERATURE—HIGH **(S-SOS-07)**
- TEMPERATURE OF HYDROGEN SEAL ABOVE NORMAL **(S-SOS-08)**
- TEMPERATURE OF SEAL OIL ABOVE NORMAL **(S-SOS-10)**

OPERATORS MAY CONSIDER:

1. Depending on the severity of the situation (how low the flow of primary cooling water is), operators may have little time for troubleshooting the problem.
2. Seal-oil temperature is not significantly dependent on load, therefore unloading the unit is not going to be effective for this.
3. High temperature of the seal oil and hydrogen seals can have severe consequences to the integrity of the machine, thus it should be given immediate attention.

TEST AND VISUAL INDICATORS:

- Leaks, corroded pipes, fouled pipes, inoperable pump, and so on with possible problems with seal-oil system components
- Damaged hydrogen seals (in extreme cases)

REFERENCES

1. G. Klempner and I. Kerszenbaum. *Handbook of Large Turbo-Generator Operation and Maintenance*. Piscataway, NJ: IEEE Press, 2008, chapters 3, 5, and 10.
2. EPRI Product 1026605. Generation Maintenance Applications Center: Combined-Cycle Combustion Turbine Lube Oil System Maintenance Guide, October 2010.

M-SOS-09: INEFFECTIVE SEAL-OIL DETRAINING

APPLICABILITY:

All hydrogen-cooled generators

MALFUNCTION DESCRIPTION:

This malfunction relates to the situation where the hydrogen captured by the seal oil is not efficiently removed from the seal oil in the detraining tank or vented.

PROBABLE CAUSES:

Lack of sufficient vacuum in the seal-oil detraining tank.

POSSIBLE CONSEQUENCES:

- ELEVATED HYDROGEN DEW POINT **(M-H-05)**
- HYDROGEN ESCAPING THE GENERATOR **(M-H-08)**
- Accelerated leak of hydrogen from the casing
- Moisture content of hydrogen increasing
- Blockage in the venting path of the hydrogen removed from the seal oil in the detraining tank

POSSIBLE SYMPTOMS:

- DEGRADED HYDROGEN PURITY **(S-H-02)**
- ELEVATED HYDROGEN DEW POINT **(S-H-03)**
- HYDROGEN MAKE-UP RATE ABOVE NORMAL **(S-H-05)**
- PRESSURE IN SEAL-OIL DETRATINING TANK ABOVE NORMAL **(S-SOS-05)**
- Foaming in the main seal-oil tank—visible through the sight glass usually provided

OPERATORS MAY CONSIDER:

Increased moisture in the casing and an elevated dew point, depending on their levels, has the potential to resulting in medium- and long-term problems. Operators should attempt rectifying the problem ASAP.

TEST AND VISUAL INDICATIONS:

- Visual clues of seal-oil vacuum pump malfunction or venting blockage or similar problem.
- Elevated moisture in the casing—causes foaming in the main seal-oil tank—visible through the sight glass usually provided.

REFERENCES

1. G. Klempner and I. Kerszenbaum. *Handbook of Large Turbo-Generator Operation and Maintenance*. Piscataway, NJ: IEEE Press, 2008, chapters 3 and 5.
2. EPRI Product 1026605. Generation Maintenance Applications Center: Combined-Cycle Combustion Turbine Lube Oil System Maintenance Guide, October 2010.

M-SOS-10: LEAK FROM SEAL-OIL SYSTEM

APPLICABILITY:

All hydrogen-cooled generators

MALFUNCTION DESCRIPTION:

This malfunction refers to the situation where a leak has occurred from the seal-oil system, such that the flow of seal oil is adversely affected.

PROBABLE CAUSES:

- Leak through a gasket
- Leak at the seal-oil heat exchanger
- Leak from the seal-oil tank
- And so on

POSSIBLE CONSEQUENCES:

- BLOCKAGE OF SEAL-OIL FLOW **(M-SOS-01)**
- FAILURE OF HYDROGEN SEALS **(M-R-11)**

POSSIBLE SYMPTOMS:

- SHAFT VOLTAGE BELOW NORMAL **(S-R-06)**
- TEMPERATURE OF HYDROGEN SEAL ABOVE NORMAL **(S-SOS-08)**
- VIBRATION OF ROTOR—HIGH **(S-R-10)**
- Shaft-ground alarm if a low seal-oil condition causes the hydrogen seals to rub

OPERATORS MAY CONSIDER:

1. Loss of seal oil has an immediate deleterious effect on the generator (see FAILURE OF HYDROGEN SEALS [M-R-11]). Therefore, operator response ought to be also immediate.

2. In some cases, the leak can be isolated. When this is not possible, and if the leak is serious enough to compromise the operation of the hydrogen seals, consider taking the unit off load.
3. Monitoring the leak and the temperature of the seals face metal and seals outlet oil can provide an idea of the severity of the problem and how it is trending.

TEST AND VISUAL INDICATIONS:

- Oil deposits
- Oil traces
- Damaged running seal and journal surfaces

REFERENCES

1. G. Klempner and I. Kerszenbaum. *Handbook of Large Turbo-Generator Operation and Maintenance*. Piscataway, NJ: IEEE Press, 2008, chapters 5 and 10.
2. EPRI Product 1026605. Generation Maintenance Applications Center: Combined-Cycle Combustion Turbine Lube Oil System Maintenance Guide, October 2010.

M-SOS-11: LEAK OF COOLING WATER BEFORE INLET TO SEAL-OIL COOLERS

APPLICABILITY:

All hydrogen-cooled generators

MALFUNCTION DESCRIPTION:

This malfunction addresses the situation where a leak has developed from the seal-oil cooling system, upstream the inlet port to the seal-oil heat exchanger.

PROBABLE CAUSES:

- Leaking valve
- Corroded-leaking pipe
- Leaking gasket, "O" ring, and so on

POSSIBLE CONSEQUENCES:

- FLOW OF COOLING WATER TO SEAL-OIL COOLERS BLOCKED **(M-SOS-07)**

POSSIBLE SYMPTOMS:

- COOLING-WATER PRESSURE AT INLET TO SEAL-OIL HEAT EXCHANGER—LOW **(S-SOS-01)**
- TEMPERATURE OF SEAL OIL ABOVE NORMAL **(S-SOS-10)**

OPERATORS MAY CONSIDER:

Refer to:

- FLOW OF COOLING WATER TO SEAL-OIL COOLERS BLOCKED (M-SOS-07)

If more than one seal-oil cooler exists (one fully redundant), switching to the *good* cooler may allow continuing operation without reducing output.

TEST AND VISUAL INDICATIONS:

- FLOW OF COOLING WATER TO SEAL-OIL COOLERS BLOCKED (M-SOS-07)

REFERENCES

1. G. Klempner and I. Kerszenbaum. *Handbook of Large Turbo-Generator Operation and Maintenance*. Piscataway, NJ: IEEE Press, 2008, chapters 3, 5, and 10.
2. EPRI Product 1026605. Generation Maintenance Applications Center: Combined-Cycle Combustion Turbine Lube Oil System Maintenance Guide, October 2010.

M-SOS-12: LEAK OF SEAL OIL INTO GENERATOR

APPLICABILITY:

All hydrogen-cooled generators

MALFUNCTION DESCRIPTION:

This malfunction relates to the situation where oil from the hydrogen seals leaks into the generator. In general, it is not uncommon that a very small amount of seal oil leaks past the oil wipers and into the casing. This problem describes that situation where seal-oil ingression to the generator is larger than normal.

PROBABLE CAUSES:

- PRESSURE OF SEAL OIL ABOVE NORMAL (M-SOS-13)
- FAILURE OF OIL WIPER (M-R-12)

For a full description of the problem of oil ingression into the generator casing, refer to:

- INGRESSION OF OIL INTO GENERATOR (M-R-16)

M-SOS-13: PRESSURE OF SEAL OIL ABOVE NORMAL

APPLICABILITY:

All hydrogen-cooled generators

MALFUNCTION DESCRIPTION:

This malfunction addresses the situation where the pressure of the seal oil is above the maximum recommended by the manufacturing, with the potential of resulting in excessive seal-oil leakage into the generator.

PROBABLE CAUSES:

Malfunction of the seal-oil regulating pressure valve

POSSIBLE CONSEQUENCES:

- INGRESSION OF OIL INTO GENERATOR **(M-R-16)**

POSSIBLE SYMPTOMS:

- LIQUID-IN-GENERATOR ALARM **(S-C&F-04)**
- DEGRADED HYDROGEN PURITY **(S-H-02)**

OPERATORS MAY CONSIDER:

1. Oil ingression into the casing can be a long-term problem, as described in
    - INGRESSION OF OIL INTO GENERATOR **(M-R-16)**
2. Small amounts of oil leaking into the casing pose no major or short-term problem, and this fact provides the operators with some flexibility to keep the unit running and troubleshoot the problem online.
3. Continuous monitoring of the hydrogen purity.

TEST AND VISUAL INDICATIONS:

- Oil inside the generator
- Faulty seal-oil pressure regulator

REFERENCES

1. G. Klempner and I. Kerszenbaum. *Handbook of Large Turbo-Generator Operation and Maintenance*. Piscataway, NJ: IEEE Press, 2008, chapters 3, 5, and 10.
2. EPRI Product 1026605. Generation Maintenance Applications Center: Combined-Cycle Combustion Turbine Lube Oil System Maintenance Guide, October 2010.

M-SOS-14: SEAL-OIL COOLING-WATER INLET TEMPERATURE HIGH

APPLICABILITY:

All hydrogen-cooled generators

MALFUNCTION DESCRIPTION:

This malfunction addresses the situation where the seal-oil cooling water enters the seal-oil heat exchanger with a temperature above normal operating range.

PROBABLE CAUSES:

- FLOW OF COOLING WATER TO SEAL-OIL COOLERS BLOCKED **(M-SOS-07)**
- LEAK OF COOLING WATER BEFORE INLET TO SEAL-OIL COOLERS **(M-SOS-11)**
- Seal-oil cooling water source could be seasonal, with temperatures too high at certain times
- Failure of cooling towers

POSSIBLE CONSEQUENCES:

- TEMPERATURE OF HYDROGEN SEAL ABOVE NORMAL **(M-R-27)**

POSSIBLE SYMPTOMS:

- TEMPERATURE OF HYDROGEN SEAL ABOVE NORMAL **(S-SOS-08)**
- TEMPERATURE OF SEAL OIL ABOVE NORMAL **(S-SOS-10)**
- VIBRATION OF ROTOR—HIGH **(S-R-10)**

OPERATORS MAY CONSIDER:

1. The temperature of the raw water cooling the seal oil is in many cases not fully controllable. It may indirectly come from a natural source of water, such as a river, lake, or ocean. Thus, it may be seasonal. On the other hand, the temperature of the seals is in general not related to the loading of the generator. Therefore, there is little or no control by the operator of this situation. Reducing the hydrogen pressure inside the generator concurrent with a reduction of load may allow the operator to find an operating point where the seal-oil temperature returns to the normal range. In any case, operating the unit with hydrogen seal temperatures above the limits established by the OEM can be very onerous to the machine.
2. Most units are designed in such a way that seasonal temperature differences do not require reducing the load on the generator due

to the operation of the hydrogen seals (or bearings). However, if such a situation is encountered, the plant should look into long-term remedies that will eliminate the need for load curtailment.

TEST AND VISUAL INDICATIONS:

None, unless seal running surfaces are damaged.

REFERENCES

1. G. Klempner and I. Kerszenbaum. *Handbook of Large Turbo-Generator Operation and Maintenance*. Piscataway, NJ: IEEE Press, 2008, chapter 5.
2. EPRI Product 1026605. Generation Maintenance Applications Center: Combined-Cycle Combustion Turbine Lube Oil System Maintenance Guide, October 2010.

M-SOS-15: TEMPERATURE OF SEAL-OIL COOLER-OUTLET—ABOVE NORMAL

APPLICABILITY:

All hydrogen-cooled generators

MALFUNCTION DESCRIPTION:

This malfunction addresses the case when seal oil emerging from the seal-oil heat exchanger is above *normal* temperature. By *above normal,* it means that the seal oil is not being cooled sufficiently and is not capable of maintaining the temperature of the seal face metal of the hydrogen seals below the maximum temperature recommended by the manufacturer.

PROBABLE CAUSES:

- BLOCKAGE OF SEAL-OIL FLOW **(M-SOS-01)**
- EFFICIENCY OF SEAL-OIL COOLERS REDUCED **(M-SOS-04)**
- FLOW OF COOLING WATER TO SEAL-OIL COOLERS BLOCKED **(M-SOS-07)**
- LEAK FROM SEAL-OIL SYSTEM **(M-SOS-10)**
- LEAK OF COOLING WATER BEFORE INLET TO SEAL-OIL COOLERS **(M-SOS-11)**
- SEAL-OIL COOLING-WATER TEMPERATURE HIGH **(M-SOS-14)**

POSSIBLE CONSEQUENCES:

- FAILURE OF HYDROGEN SEALS **(M-R-11)**

POSSIBLE SYMPTOMS:

- TEMPERATURE OF HYDROGEN SEAL ABOVE NORMAL **(S-SOS-08)**
- VIBRATION OF ROTOR—HIGH **(S-R-10)**
- Any of a number of additional indications, depending on what the cause of the malfunction is

OPERATORS MAY CONSIDER:

1. Hydrogen seals are one of the critical components of the generator. Failure of a seal can result in major damage to the machine, risk to personnel, and long forced outages.
2. Operators should try rapidly to restore the temperature of the seal oil to its acceptable range. Depending on the cause of the problem, this may or may not be possible.
3. If restoration of the seal-oil temperature to its allowable range is not possible in a short time, consider shutting down the unit for repairs.

TEST AND VISUAL INDICATIONS:

Depending on the source of the problem, there may be a number of physical clues that can be found during an inspection of the machine and the seal-oil skid, such as inadvertently closed valves, leaks, and so on.

REFERENCES

1. G. Klempner and I. Kerszenbaum. *Handbook of Large Turbo-Generator Operation and Maintenance*. Piscataway, NJ: IEEE Press, 2008, chapter 5.
2. EPRI Product 1026605. Generation Maintenance Applications Center: Combined-Cycle Combustion Turbine Lube Oil System Maintenance Guide, October 2010.

5.9 Lubrication (Lube-Oil) System Malfunctions

M-LOS-01: BLOCKAGE OF LUBE-OIL FLOW

APPLICABILITY:

All generators

MALFUNCTION DESCRIPTION:

This alfunction addresses the case where the flow of lube oil to the bearings is partially or totally blocked. The result is lube-oil flow which is below the required flow rate to maintain bearing temperatures within their designed operating range.

PROBABLE CAUSES:

- Plugging or fouling of the bearing labyrinths and/or piping
- Plugging or fouling of the lube-oil coolers
- Plugging or fouling of the lube-oil filters
- Maintenance error
- Operator error

POSSIBLE CONSEQUENCES:

- TEMPERATURE OF BEARING ABOVE NORMAL **(M-R-25)**

POSSIBLE SYMPTOMS:

- DIFFERENTIAL PRESSURE ACROSS LUBE-OIL COOLER ABOVE NORMAL **(S-LOS-01)**
- PRESSURE OF LUBE OIL BELOW NORMAL **(S-LOS-05)**
- TEMPERATURE OF BEARING ABOVE NORMAL **(S-R-08)**
- TEMPERATURE OF LUBE-OIL COOLER OUTLET ABOVE NORMAL **(S-LOS-07)**
- VIBRATION OF ROTOR—HIGH **(S-LOS-10)**

OPERATORS MAY CONSIDER:

1. Degradation of the lube-oil system can have serious consequences to the integrity of the machine, therefore immediate actions may be required to fix the situation or shut the unit down.
2. The key parameters to monitor are the bearing temperature and lube-oil flow rate.

TEST AND VISUAL INDICATIONS:

- Blocked–clogged–fouled pipes, labyrinths, lube-oil filters
- Damaged bearing and/or journal surfaces

REFERENCES:

1. G. Klempner and I. Kerszenbaum. *Handbook of Large Turbo-Generator Operation and Maintenance*. Piscataway, NJ: IEEE Press, 2008, chapters 3, 4, 5, and 10.
2. EPRI Product 1026605. Generation Maintenance Applications Center: Combined-Cycle Combustion Turbine Lube Oil System Maintenance Guide, October 2010.
3. G.E. Totten, S.R. Westbrook, and R.J. Shah (editors). *Fuels and Lubricants Handbook: Technology, Properties, Performance and Testing*. Manual Series: MNL37WCD. West Conshohocken, PA: ASTM International.
4. R.H. McCloskey. Troubleshooting Bearing and Lube Oil System Problems. In: *Proceedings of the Twenty-Fourth Turbomachinery Symposium*, pp. 147–165.

M-LOS-02: EFFICIENCY OF LUBE-OIL COOLERS—DEGRADED

APPLICABILITY:

All generators

MALFUNCTION DESCRIPTION:

This malfunction addresses the situation where the lube-oil cooling system is not effective in maintaining the temperature of the bearing(s) below maximum allowable temperature.

There are a number of causes that may result in reduced lube-oil cooling efficiency; those are covered under the following malfunction descriptions:

- BLOCKAGE OF LUBE-OIL FLOW **(M-LOS-01)**
- FAILURE OF LUBE-OIL PUMP **(M-LOS-03)**
- FLOW OF COOLING WATER TO LUBE-OIL COOLERS BELOW NORMAL **(M-LOS-04)**
- FLOW OF COOLING WATER TO LUBE-OIL COOLERS BLOCKED **(M-LOS-05)**
- FLOW OF LUBE-OIL TO BEARINGS BELOW NORMAL **(M-LOS-06)**
- INLET TEMPERATURE OF LUBE-OIL COOLING WATER—HIGH **(M-LOS-07)**
- LEAK FROM LUBE-OIL SYSTEM **(M-LOS-08)**
- LEAK OF COOLING WATER BEFORE INLET TO LUBE-OIL COOLERS **(M-LOS-09)**

M-LOS-03: FAILURE OF LUBE-OIL PUMP

APPLICABILITY:

All generators

MALFUNCTION DESCRIPTION:

This malfunction describes the instance where a lube-oil pump or pumps fail suddenly.

PROBABLE CAUSES:

- Mechanical pump failure (such as bearing seizure)
- Electric pump-motor failure
- Loss of electric power supply to pump motor

POSSIBLE CONSEQUENCES:

- In most if not all cases, a backup pump or pumps exist that will take over a failed pump. In the event of a loss of AC power (affecting all AC pump motors), a DC-powered pump motor will take over. The small lube-oil pressure transient does not affect the overall performance or operation of the generator. In some units, shaft-mounted pumps can provide lubrication with certain limitations (they depend on the speed of the unit).
- In those cases where all pumps fail (rare but not unheard), the immediate loss of lube-oil flow will result in significant damage to the affected bearings and other components, such as shaft-mounted exciters and/or PMGs.
- FLOW OF LUBE OIL TO BEARINGS BELOW NORMAL (M-LOS-06)

POSSIBLE SYMPTOMS:

- *Loss-of-pump* alarm
- If all pumps fail:
  - Rapid decrease in lube-oil pressure
  - Increase of rotor vibrations followed by major shaking of the turbine deck
- PRESSURE OF LUBE OIL BELOW NORMAL **(S-LOS-05)**

OPERATORS MAY CONSIDER:

If an alternate lube-oil pump cannot be started (or does not start) immediately, consider shutting down the unit without delay, if this hasn't been already done by the pertinent protection systems.

TEST AND VISUAL INDICATORS:

- Faulty pump or pump motor
- Burned fuses or other electrical fault resulting in a loss or AC and/or DC power
- Damaged bearing and journal running surfaces

REFERENCES

1. G. Klempner and I. Kerszenbaum. *Handbook of Large Turbo-Generator Operation and Maintenance*. Piscataway, NJ: IEEE Press, 2008, chapters 3 and 4.
2. EPRI Product 1026605. Generation Maintenance Applications Center: Combined-Cycle Combustion Turbine Lube Oil System Maintenance Guide, October 2010.
3. G.E. Totten, S.R. Westbrook, and R.J. Shah (editors). Fuels and Lubricants Handbook: Technology, Properties, Performance and Testing. Manual Series: MNL37WCD. West Conshohocken, PA: ASTM International.
4. R.H. McCloskey. Troubleshooting Bearing and Lube Oil System Problems. In: *Proceedings of the Twenty-Fourth Turbomachinery Symposium*, pp. 147–165.
5. F.B. Wilcox. Lubrication Systems for Turbomachinery. In: *Proceedings of the Seventh Turbomachinery Symposium*, pp. 189–192.

M-LOS-04: FLOW OF COOLING WATER TO LUBE-OIL COOLERS BELOW NORMAL

APPLICABILITY:

All generators

MALFUNCTION DESCRIPTION:

This malfunction addresses the situation where the flow of cooling (raw) water to the lube-oil heat exchanger(s) is below *normal*. By *normal*, it means that the rate of flow required for maintaining the lube-oil temperature within allowable operating range.

PROBABLE CAUSES:

Major raw cooling water leak:

- LEAK OF COOLING WATER BEFORE INLET TO LUBE-OIL COOLERS **(M-LOS-09)**
- Operator error (valve[s] closed)
- Maintenance error

POSSIBLE CONSEQUENCES:

- TEMPERATURE OF BEARING ABOVE NORMAL **(M-R-25)**

POSSIBLE SYMPTOMS:

- Lube-oil pump *stopped* alarm
- FLOW OF COOLING WATER TO LUBE-OIL COOLERS BELOW NORMAL **(S-LOS-02)**
- TEMPERATURE OF BEARING ABOVE NORMAL **(S-R-08)**
- VIBRATION OF ROTOR—HIGH **(S-R-10)**

OPERATORS MAY CONSIDER:

1. The temperature of the lube oil should be brought back to its allowable operating range as soon as possible, to avoid damage to bearing and journals.
2. Identifying as soon as possible the cause for the drop of cooling water flow.

TEST AND VISUAL INDICATIONS:

- Depending on the root cause, possible indications can be damaged such as motors, pumps, pressure regulators, and so on.
- Blown fuses.

REFERENCES

1. G. Klempner and I. Kerszenbaum. *Handbook of Large Turbo-Generator Operation and Maintenance*. Piscataway, NJ: IEEE Press, 2008, chapters 3, 4, 5, and 10.
2. EPRI Product 1026605. Generation Maintenance Applications Center: Combined-Cycle Combustion Turbine Lube Oil System Maintenance Guide, October 2010.
3. G.E. Totten, S.R. Westbrook, and R.J. Shah (editors). Fuels and Lubricants Handbook: Technology, Properties, Performance and Testing. Manual Series: MNL37WCD. West Conshohocken, PA: ASTM International.
4. R.H. McCloskey. Troubleshooting Bearing and Lube Oil System Problems. In: *Proceedings of the Twenty-Fourth Turbomachinery Symposium*, pp. 147–165.
5. F.B. Wilcox. Lubrication Systems for Turbomachinery. In: *Proceedings of the Seventh Turbomachinery Symposium*, pp. 189–192.
6. EPRI Product CS-4555. Guidelines for Maintaining Steam Turbine Lubrication, July 1986.

M-LOS-05: FLOW OF COOLING WATER TO LUBE-OIL COOLERS BLOCKED

APPLICABILITY:

All generators

MALFUNCTION DESCRIPTION:

This malfunction addresses the situation where the flow of cooling (raw) water to the lube-oil coolers is partially or fully blocked.

This malfunction is fully covered under:

- FLOW OF COOLING WATER TO LUBE-OIL COOLERS BELOW NORMAL (M-LOS-04)

M-LOS-06: FLOW OF LUBE OIL TO BEARINGS BELOW NORMAL

APPLICABILITY:

All generators

MALFUNCTION DESCRIPTION:

This malfunction describes the situation where the flow of lube oil to the bearings is below *normal*. By *normal*, it means flowrate required for maintaining the operating temperature of the bearings and journals within allowable range.

PROBABLE CAUSES:

Failure of lube-oil pump or pump motor:

- FAILURE OF LUBE-OIL PUMP (M-LOS-03)

Loss-of-pump-motor electrical supply:

- FAILURE OF LUBE-OIL PUMP (M-LOS-03)

Major leak of lube oil:

- LEAK FROM LUBE-OIL SYSTEM (M-LOS-08)

Flow blockage:

- BLOCKAGE OF LUBE-OIL FLOW (M-LOS-01)
    - Debris
    - Loss of FME[16]
    - Maintenance error
    - Operator error

PROBABLE CONSEQUENCES:

- TEMPERATURE OF BEARING ABOVE NORMAL (M-R-25)

[16] FME—Foreign material exclusion.

POSSIBLE SYMPTOMS:

- FLOW OF LUBE OIL BELOW NORMAL **(S-LOS-03)**
- PRESSURE OF LUBE OIL BELOW NORMAL **(S-LOS-05)**
- TEMPERATURE OF BEARING ABOVE NORMAL **(S-R-08)**

OPERATORS MAY CONSIDER:

1. Bearing temperatures above *normal* require immediate operator attention. If the temperature is below maximum allowable:
   a. Monitoring bearings temperatures.
   b. Monitoring and comparing oil inlet and outlet temperatures.
   c. Troubleshooting the lube-oil system.
   d. Looking for shaft current issues.
2. If the bearings temperatures approach or breach the maximum allowable temperature, consider bringing the unit offline for troubleshooting.

TEST AND VISUAL INDICATIONS:

- Damage to bearing babbitt and rotor shaft journals
- Blockage of lube-oil passages
- Metal impurities in oil
- Problems with lube-oil system pumps, pump motors, and loss of electrical supply

REFERENCES

1. G. Klempner and I. Kerszenbaum. *Handbook of Large Turbo-Generator Operation and Maintenance.* Piscataway, NJ: IEEE Press, 2008, chapters 4 and 5.
2. EPRI Product 1026605. Generation Maintenance Applications Center: Combined-Cycle Combustion Turbine Lube Oil System Maintenance Guide, October 2010.
3. G.E. Totten, S.R. Westbrook, and R.J. Shah (editors). Fuels and Lubricants Handbook: Technology, Properties, Performance and Testing. Manual Series: MNL37WCD. West Conshohocken, PA: ASTM International.
4. R.H. McCloskey. Troubleshooting Bearing and Lube Oil System Problems. In: *Proceedings of the Twenty-Fourth Turbomachinery Symposium*, pp. 147–165.
5. F.B. Wilcox. Lubrication Systems for Turbomachinery. In: *Proceedings of the Seventh Turbomachinery Symposium*, pp. 189–192.
6. EPRI Product CS-4555. Guidelines for Maintaining Steam Turbine Lubrication, July 1986.
7. F.Y. Simma, R.J. Chetwynd, and S.A. Rowe. Turbine Generator Lubrication Oil Supply Reliability Improvements at Southern California Edison's San Onofre Nuclear Generating Station In: *ASME 2006 Power Conference*, Chicago, IL, April 2005, PWR2005-50149, pp. 445–452.

M-LOS-07: INLET TEMPERATURE OF LUBE-OIL COOLING WATER—HIGH

APPLICABILITY:

All generators

MALFUNCTION DESCRIPTION:

This malfunction addresses the situation where the lube-oil cooling water enters the lube-oil heat exchanger(s) with a temperature above normal operating range.

PROBABLE CAUSES:

- Lube-oil cooling water source could be seasonal, with temperatures too high during summer months
- Failure of cooling towers

POSSIBLE CONSEQUENCES:

- TEMPERATURE OF BEARING ABOVE NORMAL **(M-R-25)**

POSSIBLE SYMPTOMS:

- INLET TEMPERATURE OF LUBE-OIL COOLING WATER HIGH **(S-LOS-04)**
- TEMPERATURE OF BEARING ABOVE NORMAL **(S-R-08)**
- TEMPERATURE OF LUBE-OIL COOLER OUTLET ABOVE NORMAL **(S-LOS-07)**

OPERATORS MAY CONSIDER:

1. The temperature of the water cooling the lube oil is, in many cases, not fully controllable. It may indirectly come from a natural source of water, such as a river, lake, or ocean. Thus, it may be seasonal. On the other hand, the temperature of the bearings is, in general, almost not dependent on the loading of the generator. Therefore, there is little or no control by the operator of this situation.
2. Other than monitoring the lube oil and bearing temperatures, is little the operator can do, the problem is to prevention.
3. Most units are designed in such a way that seasonal temperature differences do not require de-loading the generator. If such a situation is encountered, the plant should consider looking at long-term remedies that will eliminate the need for load curtailment.

TEST AND VISUAL INDICATIONS:

None, unless bearing and/or journal running surfaces are damaged during an overheating event.

REFERENCES

1. G. Klempner and I. Kerszenbaum. *Handbook of Large Turbo-Generator Operation and Maintenance*. Piscataway, NJ: IEEE Press, 2008, chapters 3 and 5.
2. EPRI Product 1026605. Generation Maintenance Applications Center: Combined-Cycle Combustion Turbine Lube Oil System Maintenance Guide, October 2010.
3. G.E. Totten, S.R. Westbrook, and R.J. Shah (editors). Fuels and Lubricants Handbook: Technology, Properties, Performance and Testing. Manual Series: MNL37WCD. West Conshohocken, PA: ASTM International.
4. R.H. McCloskey. Troubleshooting Bearing and Lube Oil System Problems. In: *Proceedings of the Twenty-Fourth Turbomachinery Symposium*, pp. 147–165.
5. F.B. Wilcox. Lubrication Systems for Turbomachinery. In: *Proceedings of the Seventh Turbomachinery Symposium*, pp. 189–192.
6. EPRI Product CS-4555. Guidelines for Maintaining Steam Turbine Lubrication, July 1986.

M-LOS-08: LEAK FROM LUBE-OIL SYSTEM

APPLICABILITY:

All generators

MALFUNCTION DESCRIPTION:

This malfunction refers to the situation where a leak has sprung on the lube-oil system, such that the flow of seal oil is adversely affected.

PROBABLE CAUSES:

- Leak through a gasket
- Leak at the lube-oil heat exchanger
- Leak from the lube-oil tank
- And so on

POSSIBLE CONSEQUENCES:

- BLOCKAGE OF LUBE-OIL FLOW **(M-LOS-01)**
- RUB BETWEEN BEARING AND JOURNAL **(M-R-20)**
- RUB OF OIL WIPER **(M-R-22)**
- TEMPERATURE OF BEARING ABOVE NORMAL **(M-R-25)**

POSSIBLE SYMPTOMS:

- TEMPERATURE OF BEARING ABOVE NORMAL **(S-R-08)**

OPERATORS MAY CONSIDER:

1. Loss of lube oil through a leak may have immediate deleterious effect on the generator. Therefore, operator response also ought to be quick. Refer to
   - TEMPERATURE OF BEARING ABOVE NORMAL (M-R-25)
2. In some cases, the leak can be isolated. When this is not possible, and if the leak is serious enough to compromise the operation of the bearings, consider shutting down the unit without further delay.
3. Monitoring the leak and the temperature of the bearings can give an idea of the severity of the problem and how it is trending.

TEST AND VISUAL INDICATIONS:

- Oil deposits inside the generator
- Oil traces on windings and other internal components
- Damaged running surfaces of bearings and journals

REFERENCE

1. G. Klempner and I. Kerszenbaum. *Handbook of Large Turbo-Generator Operation and Maintenance*. Piscataway, NJ: IEEE Press, 2008, chapter 4.

M-LOS-09: LEAK OF COOLING WATER BEFORE INLET TO LUBE-OIL COOLERS

APPLICABILITY:

All generators

MALFUNCTION DESCRIPTION:

This malfunction addresses the situation where a leak has developed on the lube-oil cooling system, upstream the inlet port to the lube-oil heat exchanger.

PROBABLE CAUSES:

- Leaking valve
- Corroded or leaking pipe
- Leaking gasket, "O"-ring, and so on

POSSIBLE CONSEQUENCES:

- FLOW OF COOLING WATER TO LUBE-OIL COOLERS BELOW NORMAL (M-LOS-04)
- TEMPERATURE OF BEARING ABOVE NORMAL (M-R-25)

POSSIBLE SYMPTOMS:

- FLOW OF COOLING WATER TO LUBE-OIL COOLERS BELOW NORMAL **(S-LOS-02)**

OPERATORS MAY CONSIDER:

1. If more than one lube-oil cooler exists (one fully redundant), switching to the *good* cooler may allow for continuing operation without reducing output.
2. Inspecting all valves and piping to look for leaks.

TEST AND VISUAL INDICATIONS:

Leaks at some point in the system piping and valving.

REFERENCE

1. G. Klempner and I. Kerszenbaum. *Handbook of Large Turbo-Generator Operation and Maintenance*. Piscataway, NJ: IEEE Press, 2008, chapter 5.

M-LOS-10: LUBE-OIL IMPURITIES

APPLICABILITY:

All generators

MALFUNCTION DESCRIPTION:

This malfunction addresses the case where impurities in the lubricating oil are significant to the extent that the function of the lube-oil system is adversely affected or may become adversely affected.

PROBABLE CAUSES:

- Gasket material decomposition or debris
- Metallic particles left over from maintenance activities
- Metallic particles from a bearing rub
- Moisture
- Salts or other chemicals ingression via leaks in the heat exchangers
- Bacterial growth
- Other causes

POSSIBLE CONSEQUENCES:

- Hard particles can score the babbitt material.
- Moisture and other contaminants can oxidize/degrade the oil resulting in long-term degradation of the bearings surface.
- Contaminants can cause partial or total blockage of bearing labyrinths and/or other oil passages.
- The insulation capability of the oil decreases increasing the risk of bearing currents.

POSSIBLE SYMPTOMS:

- TEMPERATURE OF BEARING ABOVE NORMAL **(S-R-08)**
- TEMPERATURE OF LUBE OIL AT BEARING OUTLET—HIGH **(S-R-06)**
- VIBRATION OF ROTOR—HIGH **(S-R-10)**

OPERATORS MAY CONSIDER:

1. Bearings are one of the most critical components of the generator. Failure of a bearing can result in major damage to the machine and long forced outages.
2. If the bearings are operating at higher-than-normal temperatures, operators ought to try to restore normal operating conditions.
3. Depending on the severity of the problem, a bearing inspection may be required, or just treating or replacing the oil.

TEST AND VISUAL INDICATORS:

- Debris in filters and/or strainers
- Scored bearing babbitt
- Chemical–physical analysis of the lube oil

REFERENCES

1. G. Klempner and I. Kerszenbaum. *Handbook of Large Turbo-Generator Operation and Maintenance*. Piscataway, NJ: IEEE Press, 2008, chapter 3.
2. H.P. Bloch. *Practical Lubrication for Industrial Facilities*. Chapter 17.
3. H.P. Bloch. *Improving Machinery Reliability*. Chapter 12.

M-LOS-11: TEMPERATURE OF LUBE-OIL COOLER-OUTLET—ABOVE NORMAL

APPLICABILITY:

All generators

MALFUNCTION DESCRIPTION:

This malfunction addresses the case when lube oil at the lube-oil heat-exchanger outlet is *above normal* temperature. By *above normal*, it means that the hot lube oil is not capable of maintaining the temperature of the bearings below the maximum recommended by the manufacturer.

PROBABLE CAUSES:

- BLOCKAGE OF LUBE-OIL FLOW **(M-LOS-01)**
- EFFICIENCY OF LUBE-OIL COOLERS REDUCED **(M-LOS-02)**
- FLOW OF COOLING WATER TO LUBE-OIL COOLERS BELOW NORMAL **(M-LOS-04)**
- INLET TEMPERATURE OF LUBE-OIL COOLING WATER—HIGH **(M-LOS-07)**

- LEAK FROM LUBE-OIL SYSTEM **(M-LOS-08)**
- LEAK OF COOLING WATER BEFORE INLET TO LUBE-OIL COOLERS **(M-LOS-09)**

PROBABLE CONSEQUENCES:

- TEMPERATURE OF BEARING ABOVE NORMAL **(M-R-25)**

POSSIBLE SYMPTOMS:

- TEMPERATURE OF BEARING ABOVE NORMAL **(S-R-08)**
- TEMPERATURE OF LUBE OIL AT BEARING OUTLET—HIGH **(S-LOS-06)**
- Any of a number of additional indications, depending on what is the cause of the problem.

OPERATORS MAY CONSIDER:

1. Bearings are one of the most critical components of the generator. Failure of a bearing can result in major damage to the machine and long forced outages.
2. Operators should rapidly try to restore the temperature of the lube-oil to its acceptable range. Depending on the cause of the problem, this may or may not be possible.
3. If restoration of the lube-oil temperature to its allowable range is not possible in a short time, consider shutting the unit down for repairs.

TEST AND VISUAL INDICATORS:

Depending on the source of the problem, there may be a number of physical clues that can be found during an inspection of the machine and the lube-oil skid.

REFERENCES

1. G. Klempner and I. Kerszenbaum. *Handbook of Large Turbo-Generator Operation and Maintenance*. Piscataway, NJ: IEEE Press, 2008, chapters 3 and 5.
2. EPRI Product 1026605. Generation Maintenance Applications Center: Combined-Cycle Combustion Turbine Lube Oil System Maintenance Guide, October 2010.
3. G.E. Totten, S.R. Westbrook, and R.J. Shah (editors). Fuels and Lubricants Handbook: Technology, Properties, Performance and Testing. Manual Series: MNL37WCD. West Conshohocken, PA: ASTM International.
4. R.H. McCloskey. Troubleshooting Bearing and Lube Oil System Problems. In: *Proceedings of the Twenty-Fourth Turbomachinery Symposium*, pp. 147–165.
5. F.B. Wilcox. Lubrication Systems for Turbomachinery. In: *Proceedings of the Seventh Turbomachinery Symposium*, pp. 189–192.
6. EPRI Product CS-4555. Guidelines for Maintaining Steam Turbine Lubrication, July 1986.

6

Relations between Symptoms and Malfunctions

6.1 Definite versus Bayesian Approaches

Within the context of the issues covered in this book, relationships between symptoms and malfunctions are deterministic in nature, but stochastic in the way they are presented to us. As an example, if the generator (core) condition monitor—GCM (also known as the core condition monitor)—produces an alarm proven to be real, there is a definite and unique association with a specific faulty or degraded component, for instance, an overheated stator conductor. However, when the operator is presented with such an alarm, and without any other direct information, he/she can only guess the possible component(s) causing the alarm, from a number of components that may develop pyrolysis byproducts, such as core insulation, flux-shield insulation, the rotor winding, or the stator.

The actual search for the degraded or faulty component is stochastic in nature. The most basic approach would be to rely on historic probabilistic data. For example, in the case of the GCM alarm, previous experience may indicate that the probability that the offending components are the rotor or stator windings is four times more likely than the core being the cause, which translates into an 80% probability. In reality, such probabilistic indices are not to be found in a reliable form for the vast majority of components in the generator. And if they were, they had to be strongly associated with a given make, age, rating, and design.

The next stochastic approach is that based on Bayesian statistics.[1] The concept of Bayesian probability is that it represents the *belief* that a certain event may occur, given the conditions in which it may occur. For example, in simple probabilistic mode, one would say the probability it will rain at this location is x%. The Bayesian variant would be if it is cloudy, the belief that it will rain is y%. In the present example, a Bayesian approach would be: If the

[1] Bayesian statistics is a subset of the field of statistics in which the evidence about the true state of the world is expressed in terms of degrees of belief, instead of a probability as frequency or likelihood for an event.

GCM is alarming and a stator RTD[2] or TC[3] is showing elevated temperatures, then the likelihood the degraded component is in the stator is much higher than the other alternatives. One could give that scenario or *belief*, a particular probability; say 90%.

The Bayesian approach has been adopted by a number of organizations trying to develop *intelligent* computer-based algorithms to help troubleshooting large generators (and other large power plant components). None has been developed to the stage that was introduced into the industry as commercially available programs (as of 2015). The troubleshooting procedure suggested to the readers of this book is a sort of Bayesian approach, but without the actual probabilities, given that this would require numerous details that only a computer can manage. On the other hand, Bayesian likelihoods are something humans are naturally predisposed to, and actually use all the time in their daily lives.

6.2 Establishing Links between Symptoms and Malfunctions

The following example shows how the reader can streamline the troubleshooting process, gaining focus on what might be the most likely source (component) of a problem, in a shorter time, by avoiding too many unproductive searches. To continue with the previous motive, an alarm originated by the GCM is assumed. Let us also assume the machine is a typical air-cooled generator. A quick perusal of the list of symptoms can be found under Section 4.2.1, there is a symptom (S-C&F-01) called *generator core-condition monitor—alarm* (see Figure 6.1).

The next step is finding those malfunctions that may cause a GCM alarm. In Chapter 4, under the "Generator Core-Condition Monitor—Alarm" symptom, the following list of possible malfunctions can be found (Table 6.1).

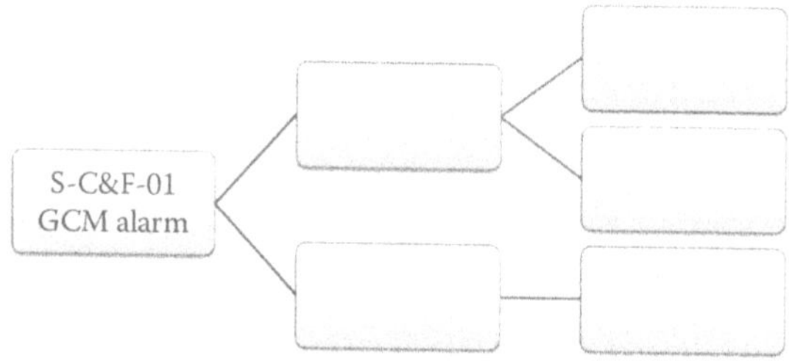

FIGURE 6.1
Step 1: Identifying the first symptom to a component fault or degradation.

[2] RTD—Resistance temperature detector.
[3] TC—Thermocouple.

TABLE 6.1

Step 2: Listing All Malfunctions Related to the Original Symptom

RELATED MALFUNCTIONS	NUMBER
ARCING BETWEEN CORE AND KEYBARS	M-C&F-01
BLOCKAGE OF PARALLEL-PHASE CONNECTOR COOLING PATH	M-S-03
BLOCKAGE OF PHASE-CONNECTOR COOLING-PATH	M-S-05
BLOCKAGE OF PHASE-COOLANT FLOW	M-S-04
BLOCKED COOLING VENT IN STATOR CORE	M-C&F-02
BURNING/PYROLYSIS OF CORE INSULATION	M-C&F-05
BURNING/PYROLYSIS OF FLUX-SHIELD INSULATION	M-C&F-04
BURNING/PYROLYSIS OF ROTOR INSULATION	M-R-04
BURNING/PYROLYSIS OF STATOR WINDING INSULATION	M-S-08
BYPRODUCTS FROM BURNING INSULATION	M-C&F-06
GAS LOCKING OF PHASE-PARALLEL COOLANT	M-S-23
GAS LOCKING OF PHASE-COOLANT	M-S-22
INSUFFICIENT FLOW OF PHASE-PARALLEL COOLANT	M-S-34
INSUFFICIENT FLOW OF TERMINAL COOLANT	M-S-35
LOOSE CORE LAMINATIONS	M-C&F-16
TEMPERATURE OF A PARALLEL CIRCUIT ABOVE NORMAL	M-S-51

This long list may be discouraging at first glance. However, it can be readily reduced for the case of indirectly cooled windings, yielding Table 6.2. Also, this air-cooled machine is too small for having flux shields installed, and thus, this malfunction is not included in Table 6.2.

If no additional information is available, and the alarm is proven to be real, then the troubleshooter would have to consider what appears to be a long list of possible causes (Figure 6.2). However, in most instances some additional information is available, either from recent history, or coetaneous to the original symptom. Let us look at a few different examples,

TABLE 6.2

Step 3: Removing All Non-Germane Malfunctions from the Original List of Step 2

RELATED MALFUNCTIONS	NUMBER
ARCING BETWEEN CORE AND KEYBARS	M-C&F-01
BLOCKED COOLING VENT IN STATOR CORE	M-C&F-02
BURNING/PYROLYSIS OF CORE INSULATION	M-C&F-05
BURNING/PYROLYSIS OF ROTOR INSULATION	M-R-04
BURNING/PYROLYSIS OF STATOR WINDING INSULATION	M-S-08
BYPRODUCTS FROM BURNING INSULATION	M-C&F-06
LOOSE CORE LAMINATIONS	M-C&F-16
TEMPERATURE OF A PARALLEL CIRCUIT ABOVE NORMAL	M-S-51

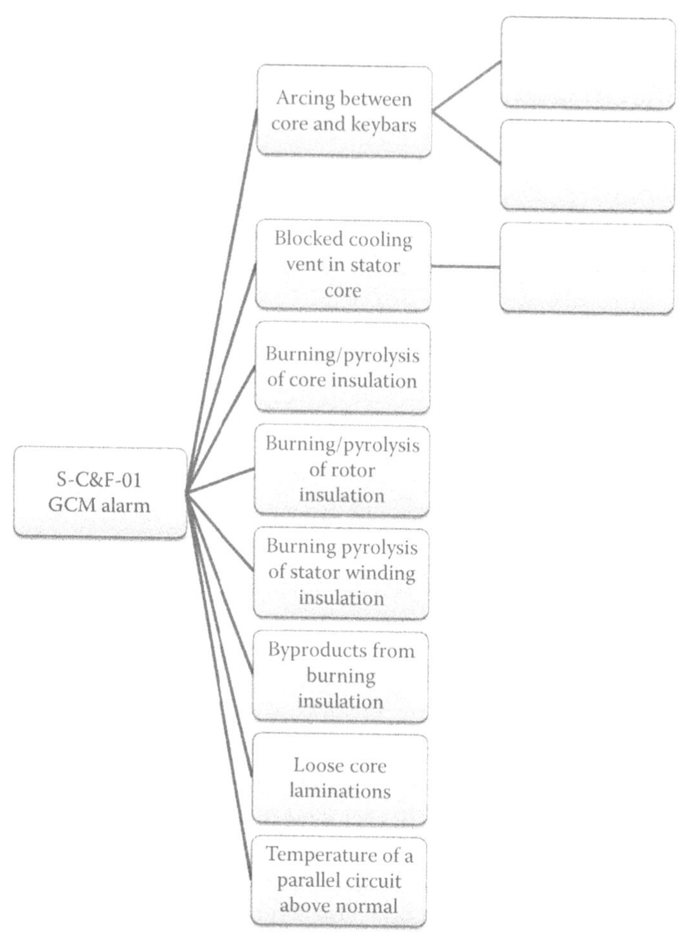

FIGURE 6.2
Possible malfunctions related to the GCM alarm.

Example 6.1

Over a number of weeks, days, or hours before the GCM alarm, one or more RTDs in the stator winding showed increased temperatures with the machine under load. Following a similar approach as with the GCM alarm, a search of the symptoms list uncovers Symptom (S-S-15): "Slot Temperature of Stator Conductor Bar—Above Normal (Case 2)."

Table 6.3 shows the germane list of malfunctions related to a hot stator conductor, after all references to inner- and hydrogen-cooled machines are removed. Also, some malfunctions can be readily checked, such as "Air or

TABLE 6.3

Malfunctions Related to a Higher-than-Normal Reading of an RTD Embedded in a Stator Slot

RELATED MALFUNCTIONS	NUMBER
BLOCKED COOLING VENT IN STATOR CORE	M-C&F-02
FOREIGN BODY INTERACTING WITH STATOR CORE	M-C&F-07
INTERLAMINAR INSULATION FAILURE	M-C&F-13
LOCAL CORE FAULT	M-C&F-15
LOCALIZED OVERHEATING OF STATOR CONDUCTOR	M-S-39
TEMPERATURE OF CORE END ABOVE NORMAL	M-C&F-21
TEMPERATURE OF LOCAL CORE AREA ABOVE NORMAL	M-C&F-20
TEMPERATURE OF SINGLE PHASE ABOVE NORMAL	M-S-52
TEMPERATURE OF STATOR CONDUCTOR ABOVE NORMAL	M-S-53

Hydrogen Heat-Exchanger Cooling Water Supply Off," "Operation Outside Capability Curve—Online," and "Phase Current High." These can also be readily removed from the list after a quick operator check. In addition, global temperatures can be immediately discounted given other RTDs and/or TCs do not show that to be the case. Hence, these malfunctions can also be discounted. Same thing can be said for the inlet temperature of the water to the air coolers and temperature of the cold gas. These are readily measured and can be also discounted.

Comparing the malfunctions listed on Table 6.3 with those listed on Table 6.2, a number of malfunctions common to both symptoms become apparent. These may be identical (such as "Blocked Cooling Vent in Stator Core"), or closely related, such as "Burning Pyrolysis of Stator Winding Insulation" with "Temperature of Stator Conductor above Normal." This is shown in Figure 6.3. A number of malfunctions *disconnected* from both symptoms can be readily removed. For example, the arcing between cores and keybars will not result in a slot RTD to read higher than normal temperature, hence, it can be removed. A similar analysis will apply to a rotor problem. Figure 6.4 shows the reduced relationships.

Inspection of Figure 6.4 clearly points to a core issue, mainly a blocked vent or other type of localized core fault, or rather a stator winding problem. Separating core from winding problems may be easier than thought. For example, reducing load will reduce the temperature measured by the RTD embedded in the slot, because the losses go as the square of the current, while the core losses do not significantly depend on stator current. On the other hand, a local core problem will result in the temperature of the RTD to go down in a smaller measure with a reduction in current. Comparing the RTD's behavior with others embedded in other sections of the winding provides a way to discriminating between a core and a

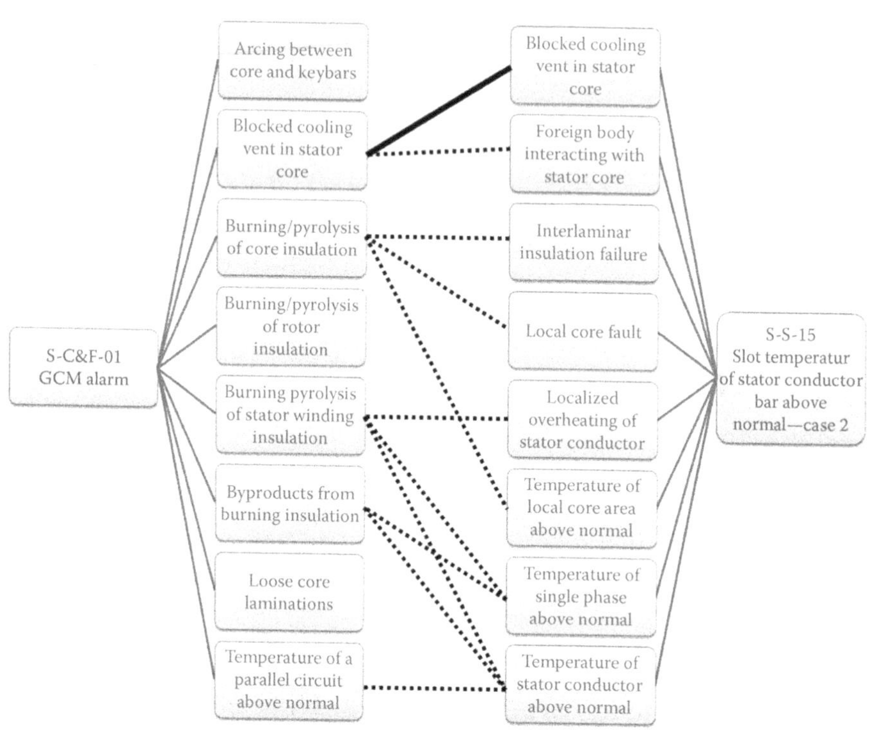

FIGURE 6.3
Relationships between the malfunctions related to the GCM alarm and the stator RTD reading.

winding problems. Eventually, the machine will have to be opened, given both cores and/or winding problems will require a visual inspection and specific testing.

Example 6.2

A few hours, days, weeks, months, or even years before the GCM alarm, the unit experienced a significant overfluxing event. A search on the list of symptoms comes up with Symptom S-E-11: "Overfluxing (Volts per Hertz) Too High Offline" with the list of related malfunctions shown in Table 6.4.

Figure 6.5 shows the links that can be established between the GCM alarm and a previous overfluxing event. It is clear from inspection of the relationships between the various malfunctions related to both symptoms, that the possibility of having a core hotspot related to an overfluxing event that happened in past time cannot be ignored.

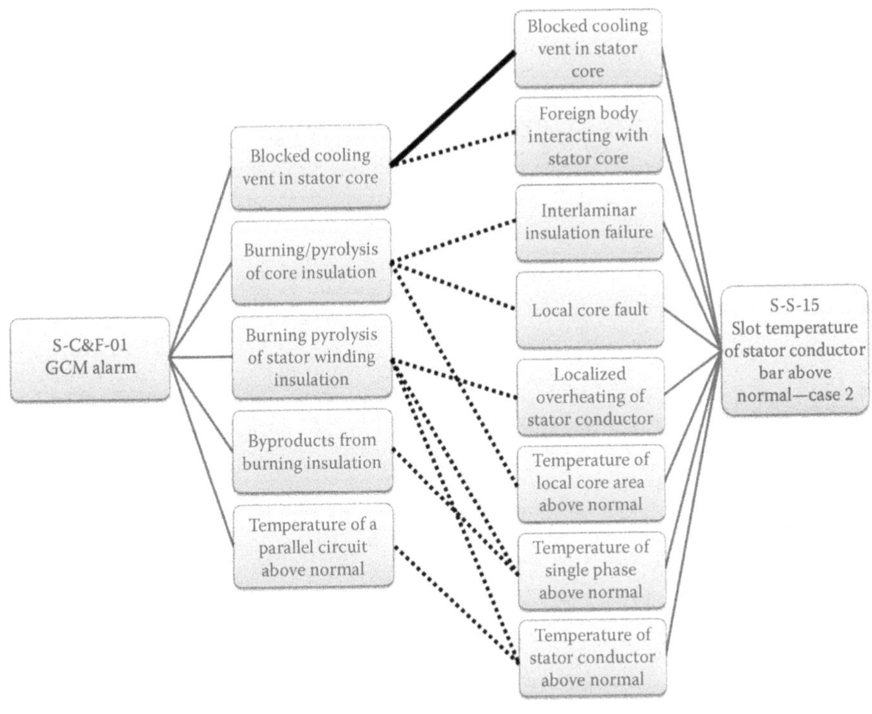

FIGURE 6.4
The figure shows the reduced list of relationships for the case studied.

TABLE 6.4
Malfunctions that May Be Related to an Overfluxing Event

RELATED MALFUNCTIONS	WEIGHT
BURNING/PYROLYSIS OF CORE INSULATION	M-C&F-05
LOCAL CORE FAULT	M-C&F-15
OVERFLUXING OFFLINE	M-E-14
TERMINAL VOLTAGE EXCURSION—OFFLINE	M-E-22

The examples shown above illustrate the first layer of links or connections between different symptoms. In trying to troubleshoot a real case, one may have to follow deeper layers of relationships. For instance, a particular symptom will relate a number of malfunctions, and some or all of these may direct the reader to another group of malfunctions. Eventually one may hope some strong relationships become obvious, pointing to a particular degraded or faulty component.

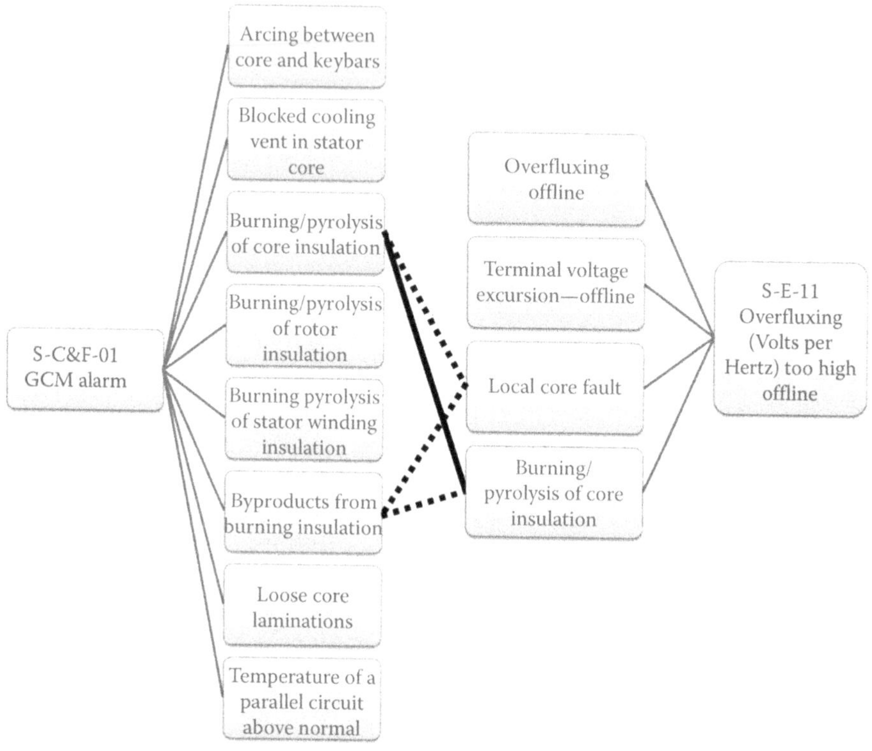

FIGURE 6.5
The figure shows the links between a recent GCM alarm and a past overfluxing event.

6.3 Multistage Events versus Single-Stage Simplification

In presenting the examples above the approach followed in some cases is to lead from a *cause* to a *symptom* directly, while in reality there may be a number of stages in between. The following example (see Section 6.3.1) will shed light on this matter.

6.3.1 Multistage Event

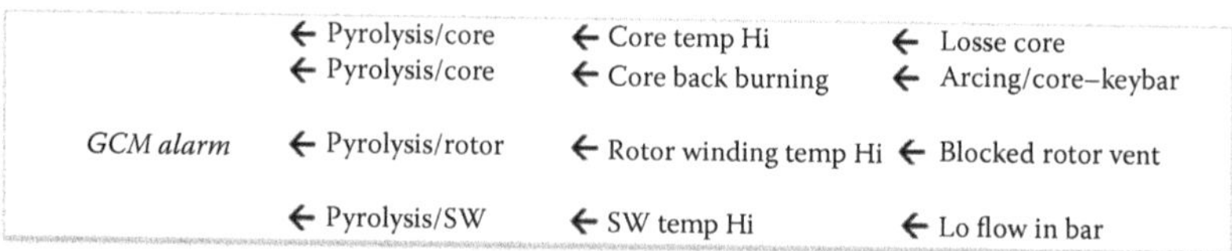

In the flowchart above, it can be seen that between the original *root* cause (malfunction, [e.g., a loose core], and the symptom (GCM[4] alarm), there are at least two intermediate stages. The first intermediate stage is a high temperature (localized) core due to shorted laminations caused by continuous relative movement and wear of the interlaminar insulation, which leads to pyrolysis of the interlaminar insulation and eventually a GCM alarm due to the intervening pyrolysis byproducts. However, the way it is shown in our analysis is as follows:

	← Core temp Hi
	← Loose core
	← Pyrolysis/core
	← Core back burning
	← Arcing/core–keybar
GCM alarm	← Pyrolysis/rotor
	← Rotor winding temp Hi
	← Blocked rotor vent
	← Pyrolysis/SW
	← SW temp Hi
	← Lo flow in bar

Clearly, the intervening stages are not captured entirely. Nevertheless, when engaged in a real troubleshooting effort, these intervening stages become obvious from the write-up under each malfunction.

Most often than not, generator-related problems do not present themselves with *a smoking gun* malfunction, and the search for the source(s) of the existing symptom(s) is rather complex, requiring technical personnel with expertise in these type of machines to become involved in the process. Operators, technicians, and engineers with direct responsibility for the operation and maintenance of large turbogenerators should maintain and update their knowledge base and expertise with periodic training and continuous education. During the multitude of visits to power plants over many years, the authors of this book have recognized a need for enhancing this type of training and education, and they hope that this book will go some way in assisting with that goal.

[4] GCM—Generator core (condition) monitor.

7

Monitored Parameters

7.1 Tables of Useful Variables

This chapter includes a number of ready-to-use tables to be filled with typical machine parameters, which the operator/engineer can use to type in all the relevant operating ranges, including alarm and trip values, so that they are handy during a troubleshooting activity. In their many years of experience, the authors have encountered numerous occasions when values of critical parameters that were required during a troubleshooting activity were not readily available, and required spending valuable time poring over many documents to find them, if at all. In later years, with the increasing installation of computerized acquisition and control units at the control room, many of these variables can be found "buried" in the system. Nevertheless, often the operator on duty is not conversant with the fine details of mining data in these computer files. So, it is handy to have all the important variables tabulated where they are readily accessible when needed in a hurry. In a station with several generators, each unit has a set of tables prepared and kept in an accessible place, preferably in the control room, either accessible on a screen, or as a hardcopy, or both.

The tables shown below are for a generic machine. Each station must change/add/delete to adapt it to their specific equipment.

Following is a list of the tables included in this chapter:

1. Nameplate
2. Main generator checks
3. Stator checks
4. Rotor and excitation checks
5. Hydrogen system checks
6. Lube-oil system checks
7. Seal-oil system checks
8. Stator cooling water system checks

NAMEPLATE INFORMATION

NAME OF POWER STATION [ ]

UNIT DENOMINATION [ ]

MANUFACTURER [ ]

TYPE [ ]

SERIAL NUMBER [ ]

DATE OF MANUFACTURE [ ]

COMMISSIONING DATE [ ]

3-PHASE GENERATOR; Y-CONNECTED

RATED POWER OUTPUT [ ] [MVA]

TYPE OF COOLING (Mark with "x" those applicable):

(__) Air Cooled

(__) Hydrogen Cooled

(__) Stator Inner-Cooled by Hydrogen

(__) Stator Inner-Cooled by Water

(__) Stator Inner-Cooled by Oil

(__) Rotor Inner-Cooled by Water

HYDROGEN PRESSURE/UNIT OF PRESSURE [ ] / [ ]

RATED VOLTAGE [ ] [VOLTS]

RATED CURRENT [ ] [AMPERES]

FREQUENCY [ ] [HZ]

NUMBER OF POLES / RPM [ ] / [ ]

RATED POWER FACTOR [ ]

EXCITATION VOLTAGE [ ] [VOLTS]

EXCITATION CURRENT [ ] [AMPERES]

STATOR WINDING INSULATION CLASS / TRADE NAME [ ] / [ ]

ROTOR WINDING INSULATION CLASS / TRADE NAME [ ] / [ ]

STATOR WINDING TEMPERATURE RISE [ ] [KELVIN]

ROTOR WINDING TEMPERATURE RISE [ ] [KELVIN]

ROTOR WEIGHT [METRIC TONS]

STATOR WEIGHT [METRIC TONS]

ROTATIONAL INERTIA / UNITS /

INERTIA CONSTANT [MW.s / MVA]

THE FOLLOWING PER-UNIT VALUES ARE BASED ON RATED MVA AND RATED KV

Xd DIRECT SYNCHRONOUS REACTANCE

Xq QUADRATURE SYNCHRONOUS REACTANCE

X'd DIRECT SATURATED TRANSIENT REACTANCE

X'q QUAD. SATURATED TRANSIENT REACTANCE

X"d DIRECT SAT. SUBTRANSIENT REACTANCE

X"q QUAD. SAT. SUBTRANSIENT REACTANCE

X2 NEGATIVE SEQ. SATURATED REACTANCE

X2 NEGATIVE SEQ. UNSATURATED REACTANCE

R2 NEGATIVE SEQUENCE RESISTANCE

Ra POSITIVE SEQUENCE ARMATURE RESISTANCE

T'do DIRECT TRANSIENT OPEN CIRCUIT TIME CONST.

T'd DIRECT TRANSIENT SHORT-CIRCUIT TIME CONST.

T"do DIRECT SUBTRANSIENT OPEN CIRC. TIME CONST.

Ta SHORT CIRCUIT TIME CONSTANT

X'o ZERO SEQUENCE REACTANCE

(**SCR**) SHORT CIRCUIT RATIO

MAIN GENERATOR CHECKS

UNIT DENOMINATION

DESCRIPTION	EXPECTED RANGE	MINIMUM VALUE / (TRIP: Y/N?)	MINIMUM ALARM VALUE	MAXIMUM VALUE	MAXIMUM ALARM VALUE / (TRIP: Y/N?)	UNITS	COMMENTS	REFERENCES
TERMINAL VOLTAGE						VOLTS		
LINE CURRENT						AMPERES		
GROSS REACTIVE POWER OUTPUT						MVAR		
GROSS ACTIVE POWER OUTPUT						MW		
STATOR NEUTRAL (% GROUND)						%		
MAIN TRANSFORMER NEUTRAL CURRENT						AMPERES		
OPERATING HYDROGEN PRESSURE								
SPEED						RPM		

STATOR CHECKS

UNIT DENOMINATION | |

DESCRIPTION	EXPECTED RANGE	MINIMUM VALUE / (TRIP: Y/N?)	MINIMUM ALARM VALUE	MAXIMUM VALUE	MAXIMUM ALARM VALUE / (TRIP: Y/N?)	UNITS
CORE-TEETH TEMPERATURE						
BACK OF CORE TEMPERATURE						
CORE MID-SECTION TEMPERATURE						
END-PLATE BODY TEMPERATURE						
END-PLATE FINGERS TEMPERATURE						
FLUX SHIELD OR SHUNT TEMPERATURE						
HOT GAS TEMPERATURE						
COLD GAS TEMPERATURE						
STATOR INLET COOLING WATER TEMPERATURE						
STATOR OUTLET COOLING WATER TEMPERATURE						
TEMP. OF RTDs EMBEDDED BETWEEN COILS IN THE SLOT						
TEMPERATURE OF WINDING TERMINALS						
OPERATIONAL HYDROGEN PRESSURE						
END-WINDING VIBRATION						
PARTIAL DISCHARGE ACTIVITY (PDA)						
RADIO FREQUENCY (RF) ACTIVITY						

ROTOR & EXCITATION CHECKS

UNIT DENOMINATION

DESCRIPTION	*EXPECTED RANGE*	*MINIMUM VALUE / (TRIP: Y/N?)*	*MINIMUM ALARM VALUE*	*MAXIMUM VALUE*	*MAXIMUM ALARM VALUE / (TRIP: Y/N?)*	*UNITS*
EXCITATION VOLTAGE						VOLTS
EXCITATION CURRENT						AMPERES
NO-LOAD FIELD CURRENT AT RATED RPM AND VOLTAGE						AMPERES
ROTOR SPEED						RPM
ROTOR VIBRATION						
FIELD WINDING TEMPERATURE						
BRUSHGEAR TEMPERATURE						
TEMPERATURE DIFFERENTIAL ACROSS BRUSHGEAR						
BRUSH GRADE						
SLIPRINGS ROUGHNESS NUMBER						
PRESSURE DIFF. ACROSS BRUSHGEAR FILTER						
SHAFT VOLTAGE (AC)						VOLTS
SHAFT VOLTAGE (DC)						VOLTS
PILOT EXCITER (PMG, HFG, Etc.) VOLTAGE						VOLTS
MAIN EXCITER FIELD VOLTAGE						VOLTS
MAIN EXCITER FIELD CURRENT						AMPERES
MAIN EXCITER INLET AIR TEMPERATURE						
MAIN EXCITER RAW WATER TO AIR-COOLER INLET TEMP.						
AVR CHANNEL A & B BALANCE INDICATOR						VOLTS
CHANNEL A & B THYRISTOR VOLTAGE						VOLTS
PSS (POWER SYSTEM STABILIZER) OUTPUT VOLTAGE						VOLTS
RECTIFIER INLET AIR-TEMPERATURE						
MAIN ROTOR INLET COOLING WATER TEMPERATURE (rotors directly cooled by water)						
MAIN ROTOR OUTLET COOLING WATER TEMPERATURE (rotors directly cooled by water)						
MAIN ROTOR COOLING WATER INLET PRESSURE (rotors directly cooled by water)						

HYDROGEN SYSTEM CHECKS

UNIT DENOMINATION

DESCRIPTION	*EXPECTED RANGE*	*MINIMUM VALUE / (TRIP: Y/N?)*	*MINIMUM ALARM VALUE*	*MAXIMUM VALUE*	*MAXIMUM ALARM VALUE / (TRIP: Y/N?)*	*UNITS*
HYDROGEN VOLUME IN GENERATOR CASING						
HYDROGEN PRESSURE INSIDE THE GENERATOR						
BULK HYDROGEN SUPPLY PRESSURE						
HYDROGEN-FAN DIFFERENTIAL PRESSURE						
COLD HYDROGEN GAS TEMPERATURE						
HOT HYDROGEN GAS TEMPERATURE						
COLD-HOT GAS TEMPERATURE DIFFERENTIAL						
HYDROGEN DEW-POINT TEMPERATURE						
HYDROGEN PURITY						%
HYDROGEN MAKE-UP OR LEAKAGE RATE						
PRESSURE OF RAW WATER AT AIR/H_2-COOLERS INLET						
PRESSURE OF RAW WATER AT AIR/H_2-COOLERS OUTLET						
TEMP. OF RAW WATER AT AIR/H_2-COOLERS INLET						
TEMP. OF RAW WATER AT AIR/H_2-COOLERS OUTLET						

LUBE-OIL SYSTEM CHECKS

UNIT DENOMINATION

DESCRIPTION	*EXPECTED RANGE*	*MINIMUM VALUE / (TRIP: Y/N?)*	*MINIMUM ALARM VALUE*	*MAXIMUM VALUE*	*MAXIMUM ALARM VALUE / (TRIP: Y/N?)*	*UNITS*
LUBE-OIL COOLER INLET TEMP. (BEARING OUTLETS)						
LUBE-OIL COOLER OUTLET TEMP. (BEARING INLETS)						
BEARING METAL TEMPERATURE						
LUBE-OIL PUMP, OIL OUTLET PRESSURE						
LUBE-OIL PRESSURE AT BEARING INLETS						
LUBE-OIL FILTER DIFFERENTIAL OIL PRESSURE						
LUBE-OIL TANK LEVEL						
LUBE-OIL FLOW						
RAW WATER INLET PRESSURE TO LUBE-OIL COOLERS						
RAW WATER FLOWRATE INTO LUBE-OIL COOLERS						
RAW WATER INLET TEMP. TO LUBE-OIL COOLERS						
RAW WATER OUTLET TEMP. FROM LUBE-OIL COOLERS						

SEAL OIL SYSTEM CHECKS

UNIT DENOMINATION

DESCRIPTION	EXPECTED RANGE	MINIMUM VALUE / (TRIP: Y/N?)	MINIMUM ALARM VALUE	MAXIMUM VALUE	MAXIMUM ALARM VALUE / (TRIP: Y/N?)	UNITS
SEAL-OIL COOLER INLET TEMP. (H_2 SEAL OUTLETS)						
SEAL-OIL COOLER OUTLET TEMP. (H_2 SEAL INLETS)						
SEAL METAL TEMPERATURE						
OIL PRESSURE AT OUTLET OF SEAL-OIL PUMP						
SEAL-OIL PRESSURE AT INLET OF H_2-SEALS						
DIFFERENTIAL PRESSURE ACROSS SEAL-OIL FILTER						
SEAL-OIL TANK LEVEL						
SEAL-OIL FLOW						
SEAL-OIL VACUUM TANK PRESSURE						
RAW WATER INLET PRESS. TO SEAL-OIL COOLERS						
RAW WATER OUT. PRESS. FROM SEAL-OIL COOLERS						
RAW WATER INLET TEMP. TO SEAL-OIL COOLERS						
RAW WATER OUTLET TEMP. FROM SEAL-OIL COOLERS						

STATOR COOLING WATER (SCW) CHECKS

UNIT DENOMINATION

DESCRIPTION	EXPECTED RANGE	MINIMUM VALUE / (TRIP: Y/N?)	MINIMUM ALARM VALUE	MAXIMUM VALUE	MAXIMUM ALARM VALUE / (TRIP: Y/N?)	UNITS
CONDUCTIVITY OF BULK STATOR COOLING WATER						μS/cm
CONDUCTIVITY OF SCW AT DEMINERALIZER OUTLET						μS/cm
DISOLVED OXYGEN						ππβ
ELECTROCHEMICAL POTENTIAL						μς
FLOW RATE OF SCW ACROSS STATOR						
FLOW RATE OF SCW ACROSS THE DEIONIZER						
FLOW RATE OF PRIMARY WATER THRU SCW COOLERS						
PRESSURE DROP ACROSS SCW FILTER						
PRESSURE DROP ACROSS SCW STRAINER						
PRESSURE DROP ACROSS HEAT EXCHANGER						
HYDROGEN-SCW PRESSURE DIFFERENTIAL						
HYDROGEN-INTO-WATER LEAK RATE						
COPPER CONCENTRATION						
COPPER IN ION EXCHANGER						
STATOR INLET COOLING WATER TEMPERATURE						
STATOR INLET PRESSURE						
STATOR OUTLET COOLING WATER TEMPERATURE						
STATOR OUTLET PRESSURE						
PRESSURE DROP ACROSS FULL STATOR						
STATOR BAR OUTLET COOLING WATER TEMPERATURE						
HEAT-EXCHANGER SCW INLET TEMPERATURE						
HEAT-EXCHANGER SCW OUTLET TEMPERATURE						
HEAT-EXCHANGER SCW DIFFERENTIAL TEMPERATURE						
HEAT-EXCHANGER PRIMARY-WATER INLET TEMP.						

HEAT-EXCHANGER PRIMARY-WATER OUTLET TEMP.						
HEAT-EXCHANGER PRIMARY-WATER DIFF. TEMP.						
RECTIFIER COOLING WATER FLOW RATE						
TERMINAL BUSHING COOLING WATER FLOW RATE						
WATER TANK LEVEL						
MAKE UP OF STATOR COOLING WATER						

Index

Note: Page numbers followed by f and t refer to figures and tables, respectively.

L

R

S

T

U

V

W

Y

Z

Milton Keynes UK
Ingram Content Group UK Ltd.
UKHW051013071024
449327UK00012B/222

9 780367 655907